Hatfield Polytechnic

Hatfield Campus
College Lane, Hatfield
Herts AL10 9AD

This book is in heavy demand and is due back strictly by the last date stamped below.

ONE WEEK LOAN

**Fundamentals of
Optical, Spectroscopic
and
X-Ray Mineralogy**

FUNDAMENTALS OF OPTICAL, SPECTROSCOPIC AND X-RAY MINERALOGY

Sachinath Mitra
Professor and Former Head
Department of Geological Sciences
Jadavpur University, Calcutta
India

JOHN WILEY & SONS
New York Chichester Brisbane Toronto Singapore

First Published in 1989 by
WILEY EASTERN LIMITED
4835/24 Ansari Road, Daryaganj
New Delhi 110 002, India

Distributors:

Australia and New Zealand
JACARANDA WILEY LTD.
GPO Box 859, Brisbane, Queensland 4001, Australia

Canada
JOHN WILEY & SONS CANADA LIMITED
Baffins Lane, Chichester, West Sussex, England

South East Asia
JOHN WILEY & SONS, INC.
05-04, Block B, Union Industrial Building
37 Jalan Pemimpin, Singapore 2057

Africa and South Asia
WILEY EXPORTS LIMITED
4835/24 Ansari Road, Daryaganj
New Delhi 110 002, India

North and South America and rest of the world
JOHN WILEY & SONS, INC.
605 Third Avenue, New York, NY 10158, USA

Copyright © 1989, WILEY EASTERN LIMITED
New Delhi, India

Library of Congress Cataloging-in-Publication Data

Mitra, Sachinath.
　　Fundamentals of optical, spectroscopic and x-ray mineralogy.

　　Bibliography: p.
　　　1. Optical mineralogy.　2. X-rays—Diffraction.
I. Title.
QE369.06M58　　1989　　549'.125　　87-37140

ISBN 0-470-21043-5　　John Wiley & Sons, Inc.
ISBN 81-224-0060-4　　Wiley Eastern Limited

Printed in India at Composers, New Delhi.

To
   my mother
     Smt. Uma Rani Mitra

....have faith in life.
—U.R. MITRA

Preface

In a dazzling scenario of high technology today a teacher of mineralogy is bound to feel the over-bearing urge of updating his class lectures, as the cross-over lines between mineralogy and major disciplines of science fade out.

In this spectrum of convergence, the Mineral Physics Committee of the American Geophysical Union has pulled together the mineralogists, crystallographers, geophysicists, geochemists and petrologists for collaborative studies on minerals to explain their properties and behaviour and their relevance to the better understanding of the earth's physiology. Modern techniques of X-ray, spectroscopy, electron and neutron diffraction are being used to explore complicated mineral systems.

The classical books on optical mineralogy written till fifties still offer the most important guides to mineral studies. In those, the phenomena like interference colours etc. shown by minerals under crossed-micols are well explained, but satisfactory answers to 'queer' questions from a sophomore relating the cause of colour of mineral etc., were not readily available.

While introducing the optics of minerals one would naturally feel that the students be taught the fundamental principles of the subject, treating it in transmitted as well as in reflected light (Chapters I, II and IV) and extending beyond both ends of the visible range of the electromagnetic spectrum; x-ray at one end and infrared on the other (Chapters VI and VII).

Optical studies using wavelengths of x-ray—visible—infrared have become almost a routine procedure in advanced countries and is increasingly being introduced of late in less advanced countries. But there is no comprehensive book covering the transmission and reflection optics of minerals in this range of the electromagnetic spectrum. This book introduces the subject of studying minerals using electromagnetic radiation of $>4000\text{Å}$ to 2Å (infrared-visible-x-ray) wavelengths. The students have also been made aware of the 'secondary' radiation phenomena like fluorescence, phosphorescence etc.

This book has been written for the graduate students who had been introduced to light optics but were not introduced to the Crystal Field Theory (CFT) at pre-University level. With a brief statement of the principles of CFT, its importance in explaining the mineral optics has been outlined (Chapter III). The theoretical approach, admittedly, is a straight-jacketed canonical one, aimed to reach the practical arena avoiding to a large extent the frightfully admired group theoretic energy band notations.

Optical spectroscopy provides information on the electronic states, coordination, site occupancies, defects, and thermodynamic variables in minerals. Because of crystal field effects, absorption spectral colours can indicate the nature of defects or ions present even in traces, and can relate certain physical properties and site occupancies. In determining the site-occupancies the optical spectroscopy stands fairly well as a method with respect to the standard procedure of x-ray diffraction (section 7.25.1).

From petrologists' point of view the study on cation disorder in a mineral phase is of great importance as it provides the information on the P-T conditions of the mineral formation. Besides the x-ray diffraction technique, this disordering phenomenon can be studied by optical, Mössbauer and NMR spectroscopy. Measurement of Al/Si disorder in aluminosilicates by XRD becomes difficult because of similar scattering of Al and Si. However, thermochemical studies and NMR can help evaluating the XRD signatures for the order-disorder phenomena in minerals†. Thermodynamic data can also be obtained from spectroscopic measurements.††

Techniques of electron-diffraction, X-ray and neutron diffraction (Chapter VII) are used to explore complicated mineral systems. However, some minerals, typically felspars, show modulated structures having different periodicity from the basic unit cell and generate satellite reflections in diffraction patterns. These have recently been theoretically modelled by using group theory.†††

Mineralogists are increasingly using Raman and Mössbauer spectroscopy, electron spin resonance (ESR) and nuclear magnetic resonance (NMR) methods for characterisation of minerals. But for the reason stated earlier these fall outside the scope of this book.

Chapter VII is intended to serve as an introduction to x-ray mineralogy and stands far from being a comprehensive treatise on x-ray crystallography, for which excellent books by Buerger, Phillips, Nuffield and others are available. But this book can claim an edge over those primarily in citing more mineralogical examples and particularly in covering the x-ray characterisation of mineral groups. Also, the x-ray intensities of common minerals have been added as a ready reference to those students who may not reach at JCPDS or ASTM X-Ray Data Cards easily.

In conclusion, I hope the students would find this book useful not solely for their examination purposes but as well for equipping themselves with a background for understanding such literature they are gradually being innundated by the kindred of American Mineralogist, Physics and Chemistry of Minerals and such others in the train (or offing), filled with jargons of solid-state physics.

---

† Carpenter, M.A. McConnell, J.D.C. and Navrotsky, A. (1984) Thermochemistry of ordering and mixing in plagioclase felspars. *Progress in Experimental Petrology* (N.E.R.C.) *6*, 190-194.

†† Burns, R.G. (1985) Thermodynamic data from crystal field spectra, *in* Microscopic to Macroscopic, edited by S.W. Kieffer and A. Navrotsky. Reviews in Mineralogy, Vol *14* (Min. Soc. America), 277-316.

††† McConnell, J.D.C. and Heine, V (1984) An aid to the structural analysis of incommensurate phases. *Acta Crystallographica, A40,* 473-482.

Acknowledgements

In preparing the manuscript the textbooks that were consulted are acknowledged below citing the authors' names and the corresponding chapters: Bettey (1972, I-II), Bloss (1961, I-II), Cameron (1961, IV), Galopin and Henry (1972, IV), Kerr (1959, I-II), Wahlstrom (1969, II), Winchell and Winchell (1951, I-II). The details are stated in the References and Selected Readings. The diagrams and tables used have been acknowledged with due references at appropriate places. JCPDS is thanked for allowing to copy the card of aluminous chromite in chapter VII. Thanks are due to NASA-Johnson Space Centre, Houston, for allowing to use the lunar rock slides.

Thanks are acknowledged to my research students, Sarvashree Tarak Nath Pal, Tapan Pal, Samsuddin Ahmed, Prabal Maity, Santanu Biswas, to my son Punyaslok and above all to my wife, Ashima, for their untiring help in proof reading and indexing. For valuable suggestions at the proof stage the author is indebted to Drs. R.K. Sen, G.B. Mitra and N. Roychoudhury of the Indian Association for the Cultivation of Science, Calcutta and to Mr. Sailen Pal, an ebullient colleague of the author.

SACHINATH MITRA

Contents

Preface   *vii*

Acknowledgements   *ix*

1. **General Optics: Interactions of Light with Matter**   1
   - 1.1 Properties of Light   *1*
   - 1.2 Refractive Index   *2*
     - Polarisability of ions and refractive index   *5*
     - Lustre   *7*
     - Lustre and R.I.   *8*
   - 1.3 Determination of R.I.   *10*
   - 1.4 Vision and Colour Sensation   *14*
   - 1.5 Colour of Minerals   *15*
     - Chromophores in minerals   *16*

2. **Study in Polarised Light**   19
   - 2.1 Polarisation of Light   *19*
   - 2.2 Nicol Polariser   *20*
     - Polarisation by absorption   *20*
     - Commercial polaroids   *20*
   - 2.3 Other Uses of Polarised Light   *22*
   - 2.4 The Polarising Microscope   *23*
   - 2.5 Wave Properties   *24*
   - 2.6 Pleochroism   *26*
     - Polarisation colour and pleochroism for mineral identification   *26*
   - 2.7 Anisotropism   *27*
     - Optics and structure   *28*
     - Interference colours   *29*
     - Orders of interference colours   *29*
   - 2.8 Birefringence Chart   *30*
   - 2.9 Optical Indicatrix   *30*
     - Uniaxial indicatrix   *30*
     - Biaxial indicatrix   *32*
   - 2.10 Extinction Angle   *33*
   - 2.11 Accessory Plates   *33*
     - Calculation of phase difference   *35*
     - Use of accessory plates   *35*
     - Compensators   *35*
   - 2.12 Determination of Fast/Slow Vibration Directions   *36*
   - 2.13 Interference Figures   *36*
     - Uniaxial interference figure   *37*

xii Contents

- 2.14 Biaxial Interference Figures *40*
  - Acute bisectrix (Bx$_a$) figure *40*
  - Obtuse bisectrix (Bx$_o$) figure *41*
  - Optic Axis figure *41*
  - Flash figures *41*
- 2.15 2V Estimation *41*
- 2.16 Dispersion *44*
- 2.17 Universal Stage *45*
  - The U-stage components *46*
  - Measurement of 2V *47*
  - Petrofabric analysis *48*
- 2.18 Petrographic Studies *48*
  - Microtextures *49*
  - Some sample studies examples *50*

3. Optical (Absorption) Spectroscopic Studies of Minerals 52

- 3.1 Introduction *52*
- 3.2 Electronic Building of Elements *52*
- 3.3 Transition Elements *53*
  - Transition metal ions (Ti, V, Cr, Mn, Fe, Co, Ni)
- 3.4 Crystal Field Theory (CFT) *55*
- 3.5 Weak Field and Strong Field *56*
- 3.6 CFT, MOT and VBT *58*
- 3.7 CFT in Mineralogy: A Resume *59*
- 3.8 Factors Controlling the Crystal Field Spectra *60*
- 3.9 CFSE in Fe Distribution in Silicates *61*
- 3.10 Optical Absorption Spectroscopy *61*
  - Selection rules *62*
  - Positions of bands *63*
  - Intensities *63*
  - Widths *63*
  - Instrument used *64*
- 3.11 Study in Polarised Spectra *64*
  - Polarised spectra of olivines *64*
  - Polarised spectra of orthopyroxene (hypersthene) *66*
  - Importance in characterisation of 'dubious' ions in minerals *67*
- 3.12 Colours of Minerals: Fundamentals Principles *68*
  - Electronic processes causing colours in minerals *69*
  - Colour centres *73*
  - Molecular orbital transitions *74*
  - Bond length in CT colour *75*
  - Other molecular transitions causing colour *76*
  - Combination of transitions *76*
  - Metals, semiconductors and insulators *77*
- 3.13 Characteristic Absorption Lines of Some Gem Minerals *79*

　　　　3.14　Optical Properties of Minerals due to Excitation of
　　　　　　　Crystal Energy　*79*
　　　　3.15　UV Fluorescence of Minerals　*83*

4. **Reflection Optics**　　　　　　　　　　　　　　　　　　　　　　　85
　　　　4.1　Dielectrics and Opaques　*85*
　　　　4.2　Dielectrical Properties of Matter　*85*
　　　　4.3　Reflectance　*85*
　　　　4.4　Colours of Opaques　*86*
　　　　　　　Band structures　*87*
　　　　　　　Colour related to interband transitions　*88*
　　　　　　　Absorption edge effect　*88*
　　　　　　　Internal reflection　*88*
　　　　4.5　Colour Perception　*89*
　　　　4.6　Chromacity Diagram　*89*
　　　　4.7　Colours of Strongly Reflecting Surfaces　*89*
　　　　4.8　Microscopy for Opaque's Study　*90*
　　　　　　　Ore microscope　*91*
　　　　　　　Reflecting and interference colour　*92*
　　　　4.9　Bireflectance and Reflection Pleochroism　*93*
　　　　　　　Internal reflection colours　*93*
　　　　　　　Anisotropism　*94*
　　　　4.10　The Hardness　*94*
　　　　　　　Set-up　*95*
　　　　　　　Shapes of indentation　*96*
　　　　　　　Microhardness determination　*96*
　　　　　　　Scale of equivalence　*96*
　　　　　　　Microhardness anisotropy　*97*
　　　　　　　Microhardness variations　*97*
　　　　4.11　Crystal Form and Habit　*97*
　　　　4.12　Cleavage and Parting　*98*
　　　　4.13　Twinning　*98*
　　　　4.14　Measurement of Reflectivity　*98*
　　　　4.15　Reflection Spectral Dispersion　*101*
　　　　4.16　Studies in Reflected Light　*102*
　　　　4.17　Some Illustrations　*103*

5. **Reflection Spectroscopy**　　　　　　　　　　　　　　　　　　　　107
　　　　5.1　Introduction　*107*
　　　　5.2　Laboratory Study　*108*
　　　　5.3　Mixtures and Modal Analysis　*112*

6. **Infrared Spectroscopy: An Outline**　　　　　　　　　　　　　　　114
　　　　6.1　Introduction　*114*
　　　　6.2　Theory　*114*
　　　　6.3　IR Spectrometers　*115*
　　　　6.4　Finger Printing　*118*

	6.5	Characterisation *118*	
		H$_2$O bands *118*	
		Hydroxyl of OH$^-$ Groups *119*	
		H$_2$O & OH$^-$ Spectra in Silicate Minerals: examples *119*	
	6.6	C-N-O-H Bands *122*	
	6.7	CO$_3^-$ Bands *123*	
7.	**X-Ray Optics**		**126**
	7.1	Nature and Production of X-rays *126*	
	7.2	X-ray Spectra *126*	
	7.3	X-ray Generation *130*	
		Filters *130*	
		Mass absorption *131*	
	7.4	Optical vs. X-ray Absorption *131*	
	7.5	Choice of a Target *132*	
	7.6	X-ray Diffraction *133*	
	7.7	Bragg Law *134*	
	7.8	Laue Pattern *134*	
	7.9	Powder Method *135*	
		The Debye-Scherrer camera *135*	
		Film distortion correction *137*	
		Film measurement *137*	
		Calculation for d-spacing & Cell edge *138*	
	7.10	Line Intersities *139*	
	7.11	Use of Standards *139*	
	7.12	X-ray Diffractometry *140*	
		X-ray Diffractometer *142*	
		Powder photographs vs. diffractometer record *143*	
		Line intensities and spectrometer charts *143*	
	7.13	Determination of the Crystal System: Indexing Powder Lines *147*	
	7.14	Graphical Method of Indexing *148*	
	7.15	Fibre Diagrams *149*	
	7.16	Determination of Lattice Parameters *149*	
		The Nelson-Riley method *150*	
		Goniometric determination of cell-parameter ratio *150*	
	7.17	Searching Mineral Powder Diffraction File *150*	
	7.18	Selected Powder Diffraction Data for Minerals *152*	
		Mineral file workbook *152*	
		The data card (ASTM/JCPDS) *153*	
	7.19	Single Crystal Methods *153*	
		Stationary crystal: Laue method *153*	
		Scattering and ordered layer lines *156*	
	7.20	Reciprocal Lattice *157*	

7.21 Moving Crystal Method *159*
    Rotation photographs *159*
    Oscillation photograph *160*
    Weissenberg camera *163*
    Lattice line template for Weissenberg indexing *164*
    Precession camera *166*
    Buerger precession method *166*
7.22 Structure Symmetry *167*
    Symmetry of continuum *167*
    Symmetry of discontinuum and omission laws *168*
7.23 Scattering Power of a Set of Atoms *170*
7.24 X-ray Intensities and Atomic Ordering *171*
    Atomic scattering or form factor *171*
    Surface factor *171*
    Structure factor *173*
    Study of halite structure (NaCl, KCl and CsCl) *174*
7.25 Electron Density Distribution: A Fourier Analysis *176*
    Site occupancy derivation *176*
7.26 Electron Diffraction *178*
7.27 Neutron Diffraction *178*

APPENDICES   **181**

    Appendix I *181*
    Appendix II *183*
    Appendix III *187*
    Appendix IV *189*
    Appendix V(a) *190*
    Appendix V(b) *201*

    Author Index *221*
    Mineral Index *223*
    Subject Index *231*

CHAPTER I

General Optics: Interactions of Light with Matter

1.1 PROPERTIES OF LIGHT

Visible light is that part of the electromagnetic spectrum which has wavelengths between 700 nm (red) and 420 nm (violet). Its position in the whole range of electromagnetic spectrum from radiowaves to gamma rays is shown in Fig. 1.1(a). In Table 1.1 the frequency and respective wave numbers of the visible and of the neighbouring colours are shown.

Table 1.1 Visible range electromagnetic spectrum

Colour	Wavelength (nm)	Frequency (Hz)	Wave number (cm^{-1})	Energy eV	Kj mol^{-1}	K cal. mol^{-1}
Infrared	1000	1×10^{14}	1×10^{15}	1.23	120	28.5
Red	700	2.28	1.43	1.77	171	40.8
Orange	620	4.84	1.61	2.00	193	49.3
Yellow	580	5.17	1.72	2.14	2.06	49.3
Green	530	5.66	1.89	2.34	226	53.9
Blue	470	6.38	2.13	2.63	254	68.1
Violet	420	7.14	2.38	2.95	285	68.1
Near ultraviolet	300	1×10^{15}	3.33	4.15	400	95.7
Far ultraviolet	200	1.50	5.00	6.22	600	143

Sun's light spectrum has the approximate form of a blackbody curve (its shape is determined by sun's surface temperature ~6000°C). The spectrum has a broad peak around 2.8 eV, or 560 nm, a yellow-green wavelength. Human eye is sensitive to this wavelength (Fig. 1.1(b)). The dark Fraunhofer lines in the continuous spectrum of the sun are caused by absorption of certain wavelengths of the white light from the hotter regions of the sun by chemical elements present in the cooler chromosphere surrounding it. In Fig. 1.1(c) the extensive lines of the iron and zirconium are shown for comparison. The Fraunhofer lines of different wavelengths of white sunlight are shown in Table 1.2.

D-line of sodium light (589.3 mμ) is often used for monochromatic reflectivity and other optical measurements (*vide* section 4.14).

2 *Fundamentals of Optical, Spectroscopic and X-ray Mineralogy*

Table 1.2 The Common Lines of Light

Fraunhofer line	Colour	Wavelength (mµ)	Source (elemental)
A	Extreme red	760.0	Oxygen
B	Red	686.7	Oxygen
C	Orange red	656.3	Hydrogen
D	Yellow	589.3	Sodium
E	Green	527.0	Iron
F	Blue	486.1	Hydrogen
G	Violet blue	430.8	Iron, Calcium
H	Extreme violet	396.8	Calcium

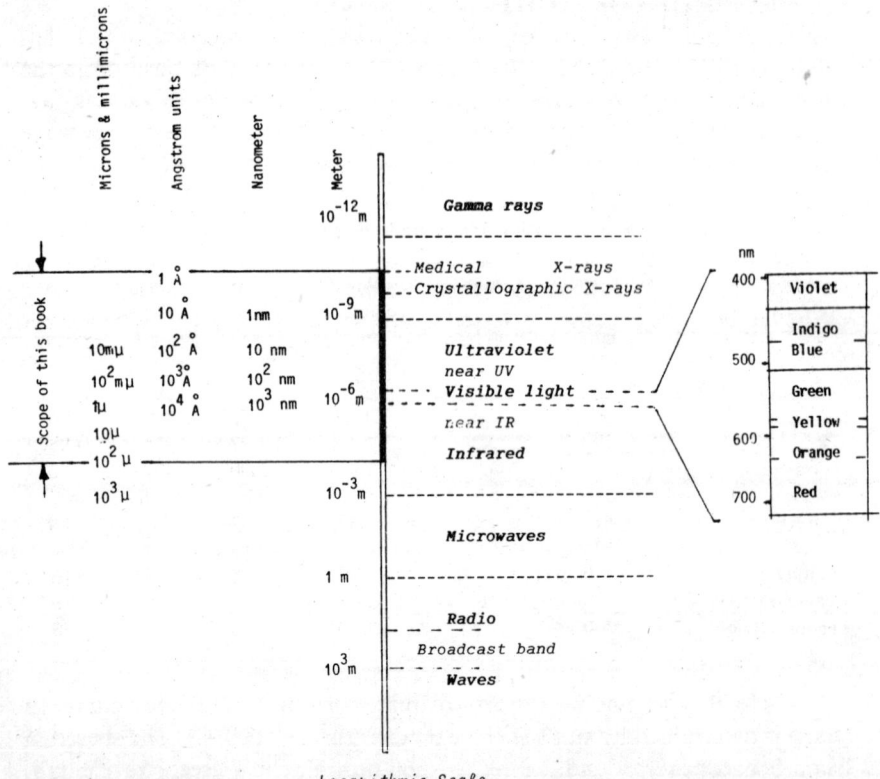

Fig. 1.1(a) Wavelength of the electromagnetic spectrum (1 ev = 1.24 µm) The scope of the book is shown by the dark area in the scale.

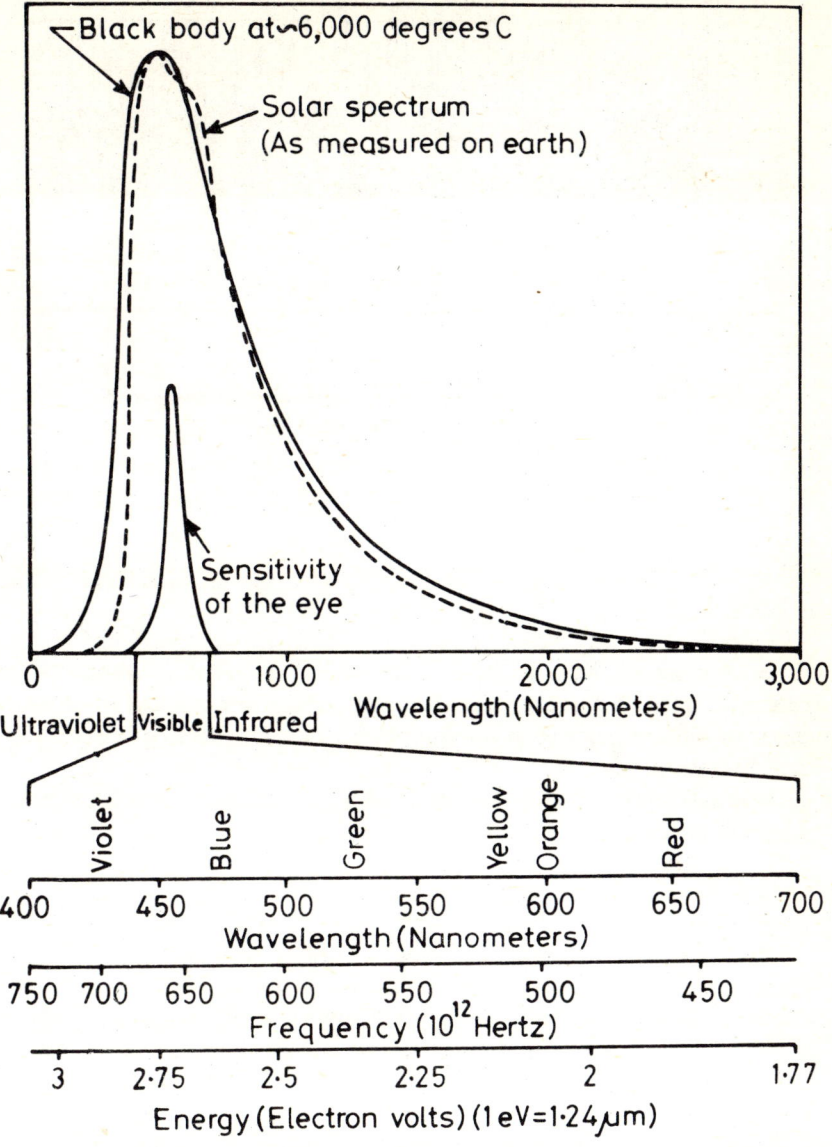

Fig. 1.1(b) Electromagnetic spectrum of the solar radiation and of black body. The range of sensitivity of human eye and the correspondence of energy and frequency with wavelength in the visible range are shown.

In many respects light behaves as a train of waves; in other respects it has particle or corpuscular (photons) properties.

In this context it is important to note the following nomenclature related to the wave-propagation of light.

Amplitude: The maximum displacement of a wave from the line of propagation.

Frequency (v): The number of vibrations per unit time.

4 Fundamentals of Optical, Spectroscopic and X-ray Mineralogy

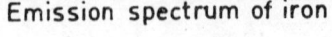

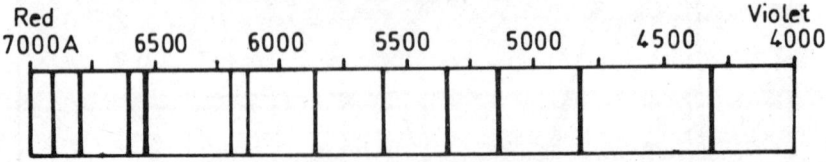

Fig. 1.1(c) Absorption spectrum of the Sun (Fraunhofer lines). Emission spectrum of iron and absorption spectrum of zirconium in the visible region are shown.

Period: It is the time interval necessary for a wave to go through a complete oscillation.

Phase: Two points on a wave are said to be in phase when they move in the same direction and are in the same relative position with respect to the crest or trough. They are in opposite phase when their spatial positions are identical, but are moving in opposite directions with reference to the line of propagation.

Phase difference: This is the portion of wavelength by which one wave train fails to match the other.

Ray: The path followed by light in moving from one point to another. It is the direction of the propagation of energy and is perpendicular to the electrical field.

Wavelength (λ): The distance between two corresponding points in the waves; normally measured between two successive crests or troughs.

Wave front: It is a surface passing through a connected series of points in the waves moving in the same phase. Normally, in isotropic medium the wave front is perpendicular to the direction of transmission. In anisotropic medium, the wave front is perpendicular only in certain directions.

Several problems relating to the optics of minerals, as well as of other objects, can be understood by considering the wave properties of light. In vacuo, light has a velocity of 3×10^{10} cm/sec (in air it is nearly the same), but this velocity changes in solid media due to interactions of the quanta or waves with the matter. One of the common manifestations of this interaction lies in the observation of refractive index of the medium.

1.2 REFRACTIVE INDEX

A stick immersed in a glass of water would look bent at the interface between water and air. This is due to the diminished velocity of light in the medium of water as compared to air. The ratio of the velocity of light in transparent medium and the velocity in air is called the index of refraction (n). The larger is this index, the slower is the velocity in the medium. The index of refraction is related to the angle of incidence (i) and the angle of refraction (r) as

$$n = \operatorname{Sin} i / \operatorname{Sin} r,$$

known as Snell's Law. The refractive index increases as the wavelength of the light decreases.

At the critical angle of incidence the light wave gets reflected totally. According to Snell's Law, the greater is the refractive index, the smaller is the critical angle. The facets of a gem are so cut that the internal angles between these facets make total reflecting surfaces. When light enters through the upper facets, it gets internally reflected and is re-emitted through one of these facets. Thus, the gem looks sparkling and becomes precious*.

Velocity of a monochromatic (single wavelength) light is dependent on its wavelength. Ordinary white light is composed of VIBGYOR (420 nm to 700 nm) wavelengths (Table 1.1). Each component i.e. 420 nm, ... 700 nm would have different velocities and different indices of refraction while travelling through a single transparent medium. Refractive indices for red are smaller than those for blue. Consequently, they have different angles of refraction and critical angles. This causes the 'split-of-light' and consequent multicoloured reflections from cut gems.

This splitting of light to seven colours is seen in a prism and this phenomenon is called *dispersion*. Newton first displayed the dispersion of light by using a trigonal prism cut from a transparent quartz.

Crystals of cubic symmetry have a regular arrangement of atoms which is identical in all directions. Hence they have single refractive index, and are called *isotropic*. Crystals of lower symmetry possess regular repeat patterns in different directions. They have different refractive indices in different directions and are called *anisotropic*.

1.2.1 Polarisability of Ions and Refractive Index
When the electric field (of the incident light wave) interacts with the molecules in a mineral, the electron distribution and the molecular geometry of the ions in the lattice get distorted. This results in polarisation of light[†]. This can be explained with $CaCO_3$ molecules in calcite (Fig. 1.2).

---

*Here the refractive index of the mineral serves as a major property to make it a gem, the other being hardness, colour etc.

† This is effected the cumulative effect of atomic and electronic polarisation. Atomic polarisability is due to geometrical distortion, while the displacement of electrons bring about electronic polarisability. When molecules in a mineral have a permanent dipole moment, the applied field orientates the molecules and the mineral acquires a net polarisation. Highly polarisable molecules respond strongly to the application of the field. Molecules that have intense transitions in the optical and lower frequency region of the spectrum are highly polarisable. Since refractive index is measured at optical frequencies, it is related to electronic polarisability.

6 *Fundamentals of Optical, Spectroscopic and X-ray Mineralogy*

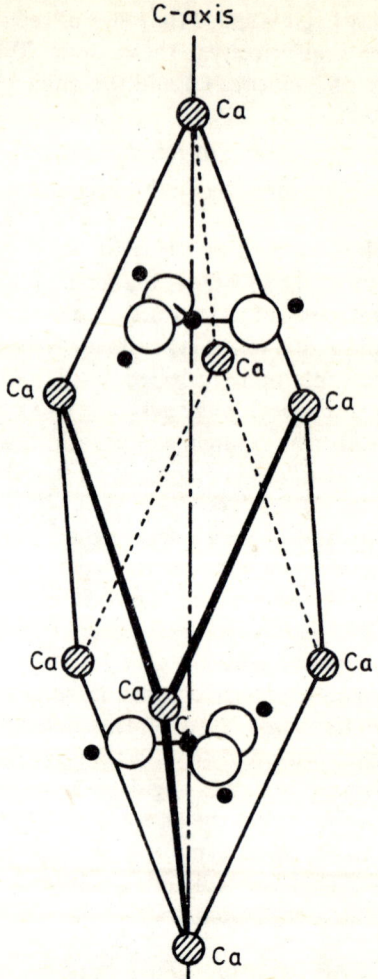

Fig. 1.2 The structure of calcite.

In calcite, equilateral triangles of carbonate ions have the three apices occupied by oxygen ions and carbon at the centre of the triangle. The C-O bonding is predominantly covalent and forms strong polarisation along the plane of the bonds. Therefore, light waves vibrating on this plane enhance the polarisation already present. Light waves moving parallel to the *C*-axis will, therefore, have their vibrations strongly polarised along the plane of carbon ions, in the *C*-perpendicular directions. Consequently this will cause slower velocity and hence higher refractive index along the C-axis.

Light travelling along the plane of carbonate ions will have waves vibrating parallel to *C*. This will reduce polarisation and increase the velocity, i.e., the index of refraction becomes lower.

Extraordinary waves always vibrate in a plane having the *C*-direction and the ray-path but perpendicular to the vibration direction of ordinary rays. Extraordinary ray becomes ordinary ray along the direction of *C*-axis but as the direction moves farther away the velocity of the extraordinary ray increases till it reaches a maximum at 90° to the *C*-axis. This is due to its optically negative character, discussed in Section 2.5.

1.2.2 Lustre

Lustre of a mineral depends on the way in which light is reflected from the surface of a mineral (i.e. *R*). Reflection is again dependent on the refractive index, *n*, of the mineral. Normally, the greater is the index of refraction, the brighter is the lustre.

Lustres, in general, are classified into two broad classes (a) non-metallic and (b) metallic.

Non-metallic Lustre

The cause and nature of non-metallic lustres are due to the interaction of light with dielectric semiconducting and poor conducting substances. The nature of transmission and the relationship of refractive indices with lustre will be discussed later.

The electrons in non-metallic minerals are localised at specific sites if the bonding is primarily ionic, covalent or Van der Waals in nature. The energy of separation between the ground states and the excited states is much more than the energies of the photons of visible light (Fig. 1.3(a)). Hence they get transmitted. This produces non-metallic lustre.

In non-metallic mineral, the energy gaps between the ground and excited states are much higher than the incident photon energies.

Metallic Lustre

In minerals having metallic bonding, (e.g. native metals) or a high degree of covalent bonding (sulphides, sulphosalts etc.), there are a large number of energy levels and transition energies which cover the energies of the photons of visible light (Fig. 1.3b). Hence, the photons are absorbed readily but the oscillation of the free-electron surface in metals causes re-emission of light and consequently there is bright metallic lustre on a fresh surface.

The minerals of metallic lustre have reflectivities of 20–50% (generally over 30%), with high refractive indices. This has been discussed in section 4.14 on reflection optics.

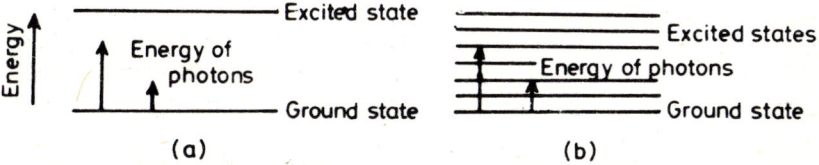

Fig. 1.3 Electronic causes for lustres of minerals (a) When the gaps in electronic energy between the ground and excited states are more than the energy of impinging photons, many photons are not absorbed and they are transmitted causing nonmetallic lustre, (b) In cases where the excited states are numerous and close in energy few photons get transmitted and they look opaque with metallic lustre.

In minerals with metallic lustre, the excited states cover all the energies of the photons of visible light. Hence all photons are absorbed and light is re-emitted by the oscillating free-electron surface.

1.2.3. Lustre and R.I.

The relationship between refractive index, n, and lustre is expressed by the following Bohr equation for the reflectivity, R, (i.e. percentage of light reflected from a polished surface normal to the incident ray) as:

$$R = \frac{(n-1)^2 + n^2 k^2}{(n+1)^2 + n^2 k^2}$$

where k is the absorption coefficient. For transparent crystals k is equal to zero.

The Bohr relationship between reflectivity R and refractive index n for all classes of lustre is presented in Fig. 1.4.

The lustre of a mineral is in general greater when its R. I's are higher. According to A.S. Povarennykh (*vide* Battey and Tomekieff, p. 488, 1964) the lustre can be classified as:

Semivitreous
($n = 1.3 - 1.5$)
↓
Vitreous
($n = 1.5 - 1.8$)

Subadamantine
($n = 1.8 - 2.2$)
↓
Adamantine
($n = 2.2 - 2.7$)
↓
Superadamantine

($n = 2.7 - 3.4$)

Submetallic
($R = 8 - 20\%$)
↓
Metallic
($R = 20 - 50\%$)
↓
Supermetallic,
metallic splendent
($R = 50 - 95\%$)

Usually, minerals with covalent bonding have higher refractive indices compared to those with ionic bonding. Ionically-bonded minerals normally show subvitreous to vitreous lustre. The covalent bonded minerals show lustres sub-adamantine to adamantine (Fig. 1.4). Opaque minerals possess bonding between covalent to metallic. Minerals with pure metallic bonding have lustre with high R (and n values) and with k values between 0.5 to 1. Metallic splendent (supermetallic) lustre should have higher values for these three.

The relationship between R, n and k with bonding is as follows:

	R	n	k
Ionic bonding	low	low	low
Covalent bonding	moderate	high	low
Metallic bonding	high	high	high

Adamantine lustre is found in minerals with refractive indices of $1.9 - 2.7$.

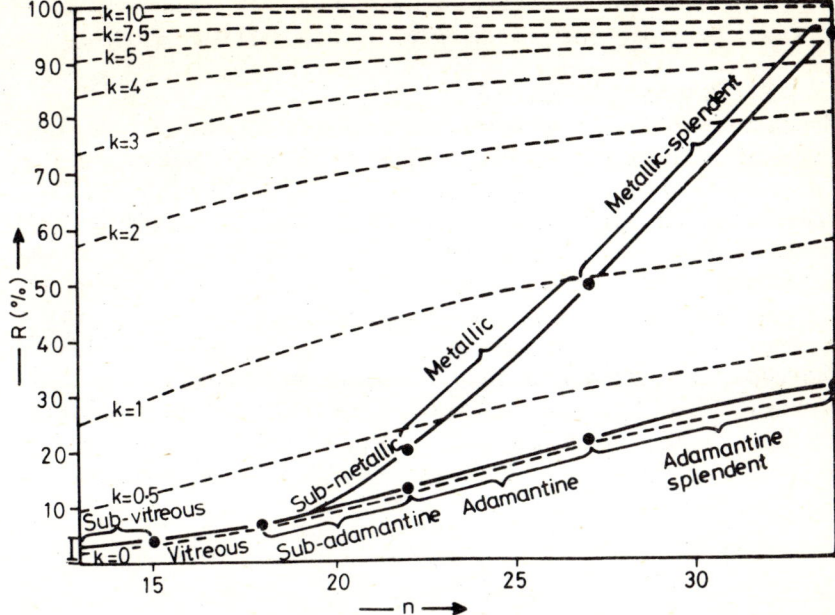

Fig. 1.4 Bohr relationship between reflectively (R) and refractive index (n) for all classes of lustre. (after Bloss, 1971).

Factors like covalent bonding (diamond), heavy elements like lead (cerussite), titanium (rutile) also contribute to this lustre.

Vitreous lustre is found in minerals of predominantly ionic bonding of an atomic number less than iron in the periodic table. Their refractive indices range between 1.3-1.9, e.g. quartz, rocksalt etc.

Resinous lustre is shown by semi-transparent minerals of refractive indices greater than 2, e.g., sphalerite.

Submetallic lustre is found in semi-opaque oxides with refractive indices of 2-3, e.g., cuprite, chromite, etc.

Metallic lustre is found in minerals having R.I's greater than 3, for e.g. gold, galena pyrite, etc.*

Pearly lustre is developed due to the presence of many successive reflecting planes within the mineral, e.g. mica, talc, gypsum, brucite etc.

Earthy or dull lustre is generated by the scatter of light produced by very fine grained and porous minerals, for e.g., clay minerals or clays etc.

About 70% of all mineral have a vitreous lustre viz., silicates, phosphates, carbonates, nitrates, sulphates etc. and the remaining 30% possess an adamantine or submetallic (rarely typically metallic) lustre (sulphides, native elements etc.)

Forms and shapes often affect lustre. Quartz has a vitreous lustre on crystal faces but on fractured surfaces it is greasy. Perfect cleavage surfaces of many minerals show pearly lustre. Fibrous minerals with vitreous lustre often look silky viz. asbestos etc.

---

*Metallic lustre is also a characteristic of those materials with R.I. less than 1.

Like the refractive index, the reflectivity of a mineral is characteristic of it. The reflectivity property is discussed in chapter IV (section 4.14).

1.3 DETERMINATION OF R.I.

Common minerals and crystalline substances have R.I's between 1.45 – 1.80. Single or mixtures of immersion liquid media are available for R.I. determination of solid grains. Liquids with known R.I.'s (determined at 20°C in sodium, D-line, spectrum) are obtained in bottles. They range from 1.331 (methyl alcohol) to 2.92 (selenium melt), and mixtures of R.I. upto 3.17 can be prepared (Table 1.3). On storage they might change in colour and the R.I's need to be determined by a refracto-meter before they are used for mineral studies. D-line is usually used because the dispersion curves of liquid and solid intersect at 589.3 mμ (Fig. 1.5). Some of the commonly mixed liquids used in 1.373 to 1.880 range are listed in Table 1.3(b). Refractive indices of minerals can be determined under microscopes in the form of loose fragments of grains or in thin sections. Some of the methods are described below.

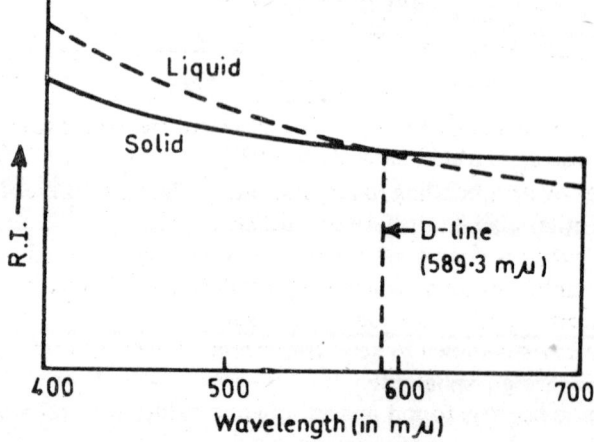

Fig. 1.5 Dispersion curves of liquid and solid with the point of intersection coinciding with the D-line.

A) FRAGMENTS OF GRAINS

The R.I. of mineral fragments relative to an immersion oil may be measured under polarising microscope (discussed in chapter II) by the central illumination or oblique illumination method.

i) Central Illumination Method

In this method the intensity of light is reduced by narrowing the aperture of the substage diaphragm of the microscope. As the field darkens, a narrow zone of brightness develops at the contact surfaces between the grains and oil. From a focussing position when the microscope tube is raised slowly, this bright band moves towards the medium of higher (H) refractive index; conversely, when the tube is lowered the band moves towards the lower (L) R.I. material (Fig. 1.6(a) and (b)). This phenomenon is known as the "Becke-

Table 1.3(a)   Refractive Indices of Immersion Media at 20°C

	Liquid		Liquid
1.00027	Air	1.496	Toluene
1.331	Methyl alcohol	1.498	m-Xylol
1.333	Water	1.501	Benzene
1.535	Ethyl ether	1.513	Trimethylene bromide
1.359	Acetone	1.526	Monochlorobenzene
1.362	Ethyl alcohol	1.530	Clove oil
1.372	Ethyl acetate	1.547	O-Nitrotoluene
1.375	Hexane	1.553	Nitrobenzene
1.387	Heptane	1.560	Bromobenzene
1.391	Trimethylpentane	1.569	Benzyl benzoate
1.393	Ethyl Valerate	1.585 / 1.600	Cinnamon oil
1.400	n-Butyl alcohol	1.598	Bromoform
		1.619	Cinamic aldehyde
1.409	Amyl alcohol		
1.411	n-Decane	1.625	α-Monochloronaphthalene
1.423	Methyl cyclohexane	1.697	α-Iodonaphthalene
1.445	Chloroform	1.738	Methylene iodide
1.448	Kerosene	1.810	Tetraiodoethylene
1.466	Carbon tetrachloride	1.843	Phenyldiiodoarsine
1.473	Glycerine, petroleum oil	2.003	Sulfur (20%) and arsenic disulfide in arsenic tribromide
1.480	Castor oil	2.11	Arsenic tribromide arsenic disulfide and selenium
1.486	Isoamyl phthalate	3.17	Arsenic selenide
1.492	Tetrachloroethane		

Note: Most of the heavier liquids are poisonous to human system.

Table 1.3(b)   Name and R.I's of Different Mixed Liquids Used in Optical Studies

Liquids	Range of R.I.
Ethyl acetate and amyl acetate	1.373 – 1.396
Amyl acetate and xylene	1.396 – 1.490
Para-cymene and monobromobenzene	1.490 – 1.559
Monobromobenzene and bromoform	1.559 – 1.598
Bromoform and α-bromonaphthalene	1.598 – 1.658
α-Bromonaphthalene and methylene iodide	1.658 – 1.740
Solutions of sulphur in methylene iodide	1.740 – 1.780
Solutions of sulphur, Sn I_4, As I_3, Sb I_3 and Iodoform in methylene iodide	1.780 – 1.880

line" effect. The same principle may be applied to determine the relative refractive index of a mineral with respect to Canada balsam ($n = 1.54$). By using oils of progressively higher or lower R.I.'s a matching of the R.I. of the oil with a mineral may be obtained (when the Becke line is replaced by a reddish-orange and a bluish band at the contact edges). By knowing the R.I. of the liquid the R.I. of the mineral is determined.

ii) Oblique Illumination Method

In this method, half of the light field on the stage is darkened by inserting an obstruction between the mirror and the stage. Normally, the condensing lens is removed. The mineral grains would now show partial illumination on one side while the other side will be shaded (Fig. 1.6c). If the R.I. of the mineral is higher than that of the oil, the shaded side of the grain is away from the darkened half of the field, and if the index of the mineral is lower than that of the oil, the shaded side will be toward the darkened half. When the R.I's of the liquid and the mineral match the contact planes appear lined by bluish bands.

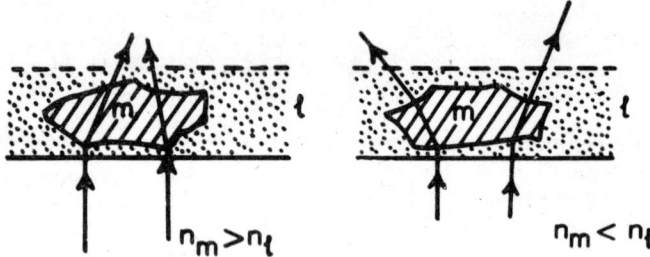

Fig. 1.6(a) Schematic diagram exhibiting the Becke line effect. n_m = Refractive index of mineral; n_l = Refractive index of liquid.

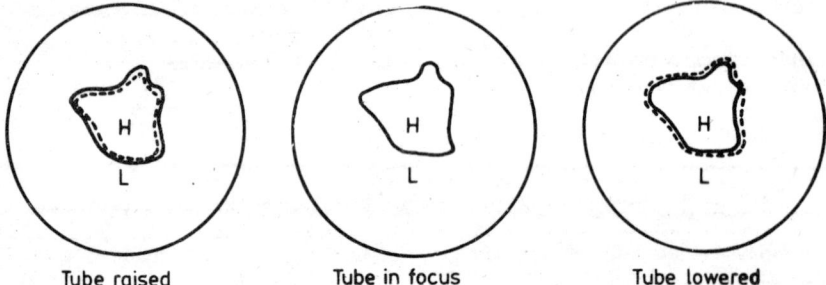

Fig. 1.6(b) Becke line (dotted) test of a mineral in a liquid or canada balsam.

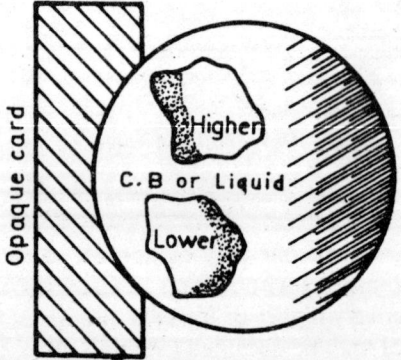

Fig. 1.6(c) Oblique illumination.
Higher = mineral grain, n > liquid
Lower = mineral grain, n < liquid

iii) Variation Method

The R.I. of any substance is a function of temperature and the wavelength of light. The R.I. changes in liquid are more marked than in solids. Therefore, if change, either in temperature (T) or wavelength (λ), is made a matching of the R.I's of the standard (liquid) and sample (solid) can be achieved. This is called the single variation method. But when both the variables, T and λ, are changed, the method is called the Double Variation Method.

a) In the *single variation method,* a liquid of slightly higher R.I. than that of the solid is selected and the temperature of the liquid is increased by gradual heating (using hot stage wth control temperature cell) until the Becke line disappears. The temperature (t) is read and the R.I. is calculated as follows:

$$n_{\text{solid}} = n_{\text{liquid}} - \frac{dn}{dt}(t - t_o)$$

n_{Solid} = The R.I. of the liquid at increased temp. t, when the Becke line disappeared.

n_{liquid} = R.I. of the liquid at room temp. t_o (i.e. about 25°C).

$\frac{dn}{dt}$ = Temperature coefficient of the R.I. of the liquid.

b) The *double variation method* is a combination of thermal and wavelength variation methods.

The immersion liquids employed in this method should have high temperature coefficients of R.I. and high dispersion.

In this procedure a few drops of the chosen immersion liquid are put on the prism of the refractometer and hot water is circulated through the refractometer and the cell. At each successive temperatures, say, 20°C, 30°C, 40°C etc. a match is made by changing the wavelength of the incident light (Fig. 1.7). The matching points specifying λ's for consecutive T°C's will give

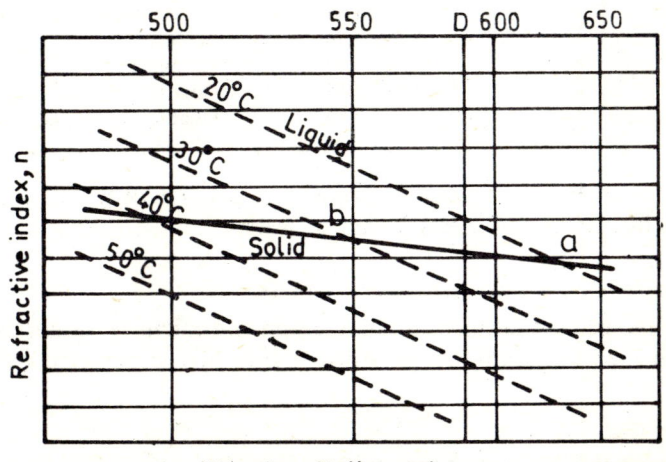

Fig. 1.7   Principle of the double variation method.

the trend (mostly linear) of R.I. change with wavelength (i.e., the dispersion). The standard n value for D-line (589.3 mμ) can thus be determined (Fig. 1.7).

B) THIN SECTIONS

The relative refractive indices of minerals can be determined with respect to other minerals or Canada balsam in contact, by using the Becke line effect as described above.

A measure of the refractive indices of minerals can be made (specially with respect to the R.I. of Canada balsam, $n = 1.54$) by studying the relief of the minerals.

Minerals with high R.I.'s stand out as bodies of high relief with bold outline. This property diminishes with progressive decrease in R.I.'s. On this basis the R.I. characters may broadly be classed as:

High R.I.................... bold outline, high relief, rough surface.
Moderate R.I................ moderate outline, moderate relief and surface.
Low R.I...................... weak outline, low relief and smooth surface.

It should, however, be noted here that moderate to high relief can also be shown by minerals with R.I. approximately less than that of Canada balsam (i.e., the medium).

A further subdivision of the relief-measure with corresponding R.I. ranges and examples are given in Table 1.4.

Table 1.4 Relief and Refractive Index

Relief	R.I.	Example
Extreme	> 1.90	Rutile
Very high	1.76 – 1.90	Almandine garnet
High	1.67 – 1.76	Epidote
Moderate	1.58 – 1.67	Actinolite
Low	1.50 – 1.58	Quartz
Very low	< 1.49	Fluorite

The lowest R.I. of a common mineral is 1.43 (fluorite). Less common cryolite has R.I. (1.34) nearing that of water.

1.4 VISION AND COLOUR SENSATION

For appreciation of the mechanism of vision and colour sensation by human eyes, an elementary knowledge of the human eye becomes pertinent.

Human Eye

Light enters the eye through the cornea, a lens of aqueous material, and makes an image on the retina at the back of the eye.

There are two types of light-sensitive cells in the retina. These are *rods* and *cones**, named after their shapes. The rods and cones are connected to the fibres of the optical nerves, by which the signals produced by the rods and cones are transmitted to the brain. There are 137 million light-sensitive cells in the eye, but only one million fibres in the optical nerve. Human eye is maximum sensitive near the centre of the visible (~ 550 nm) and extremely insensitive to the red and blue ends of the spectrum (Figs. 1.1(b) and 4.8). Red ($\lambda = 700$ mμ), green ($\lambda = 546.1$ m μ) and blue ($\lambda = 435.8$ m μ), are the principal colours, and the remaining colours are derived from combinations of these. Human eye has these three colour-sensitive elements.

Colour blindedness or Daltonism (named after the famous chemist, Dalton who was colour-blind) is about 9 p.c. among men but 0.5 p.c. among women. Most of these people can not distinguish between red and green, but perceive normally other combination colours.

1.5 COLOUR OF MINERALS

To understand the cause of colour of minerals it becomes important to understand the physics of colour generation. Basically, a substance is coloured if it absorbs a particular waveband from the incident radiation; colour can also be generated by it if it scatters one frequency more effectively than others.[1]

Besides internal reflection and scattering (responsible for iridescence in some felspar and opal) the most common cause of colour in minerals is due to electronic transitions brought out by absorption of light in the visible region 700 nm (red) to 420 μm (violet).[2] When red is absorbed (because of the presence of low-lying excited states) the mineral appears blue; when blue or violet is absorbed it appears red[3]. The most intense colours are, however, due to electric dipole transitions.

Generation of colours is mainly due to five main mechanisms: (a) excitation of free atoms and ions, and vibrations in molecules, (b) crystal-field effects, (c) transition between states of molecular orbitals, (d) transitions in the energy bands of solids and (e) effects due to physical optics.

Surprisingly, all these mechanisms involve wavelengths which are narrowly sensitive to human eye. Because of this reason, the interactions of electromagnetic radiation with electrons in matter are important in this wavelength region. On the otherhand, waves of lower energy mainly stimulate the motions of atoms and molecules, causing thermal vibrations or heat.

---

* The diameter of the rod is ~ 0.002 mm (i.e. 2 microns) and its length is about 0.06 mm (60 microns). The cones are a little longer (about 0.005 mm in diameter and 0.07 mm long). The cones have bleachable pigments (like rhodopsin in rods), one each for electromechanical response to red, green and blue. The brain blends these colours to make scores of other hues.

[1] The blue colour of the sky is due to Raleigh scattering, while of the sea by the Raman effect.

[2] The energies of photons over this range vary from 1.7eV to 3.0 eV.

[3] In living systems low-lying energy levels are not common. Therefore, the abundant colours tend towards yellow-red rather than blue. Chlorophyll, however, is specially constructed; therefore, green is ubiquitous.

16 *Fundamentals of Optical, Spectroscopic and X-ray Mineralogy*

Waves of higher energy can ionize atoms and permanently damage molecules.

The causes of colour can be classified into 14 categories of five broad classes (*vide* section 3.12). All but one of the colour causing mechanisms (vibrations of the ions in molecules), can be traced to changes in the site of electrons in matter. Electronic transitions are the most important causes of colour because the energy needed to excite an electron commonly falls in the range that corresponds to visible wavelengths of light. The causes of colour classified by Nassau (1980) is presented in Table 1.5. The electronic processes causing colours of minerals are discussed in Chapter III.

Table 1.5 Causes of Colour in Minerals and Some Common Objects

Electronic transitions and Vibrational transitions	Electronic excitations (free atoms and ions)	Incandescence, flames, lighting, gas discharges.
	Vibrations (molecules)	Blue green tint of pure water and ice.
Crystal-field effects	Transition-metal compounds	Most pigments, some phosphors and lasers, fluorescent material
	Transition-metal impurities	Ruby, emerald, some fluorescent and laser materials.
	Color centres	Amethyst, smoky quartz, fluorites, amethyst glass.
	Charge transfer	Magnetite, blue sapphire.
Transitions between molecular orbitals	Conjugated bonds	Organic dyes, plant and animal colors, lapis lazuli, fireflies, dye lasers.
	Metallic conductors	Gold, silver, copper, iron, brass.
Transitions in materials having energy bands	Pure semiconductors	Silicon, diamond, galena, cinnabar.
	Doped semiconductors	Coloured diamonds (yellow, blue etc.), light-emitting diodes, semi-conductor lasers, some phosphors.
	Dispersive refraction	"Fire" in gemstones, chromatic aberrations, the rainbow.
Geometrical and Physical optics	Scattering	Blue of the sky, red of sunsets, star sapphire, moonstone.
	Interference	Oil film on water, lens coatings, some insect colours.
	Diffraction grating	Opal, liquid crystals, some insect colors.

1.5.1 Chromophores in Minerals

Most of the silicate minerals are coloured because of the presence of transition metal and lanthanide ions in their structures, which absorb radiation through internal 3d or 4f electron transitions. These are known as chromophores.

From the compositional point of view, colours can be considered as inherent or *iodochromatic,* when the colour absorbing chromophore element is essential for the natural composition of the mineral. Examples: pyroxenes, amphiboles, biotites etc. When chromophore occurs as an impurity i.e. not an essential constituent, the mineral is *allochromatic.* Ex., blue quartz (with Ti), ruby (with Cr^{3+}) etc.

Some of the colouring ions or chromophores are listed in (Table 1.6).

Table 1.6 Chromophores in Allochromatic Minerals

Chromophore	Colour	Minerals
Fe^{2+}	Green, Yellowish green	Microcline, olivine (peridot)
Fe^{3+}	a) Pink to red	Calcite, quartz, andradite, grossularite, kyanite, microcline
	b) Greenish yellow	Corundum, epidote, beryl (aquamarine), chrysoberyl
$Fe^{2+} + Fe^{3+}$	a) Blue	Sapphire (corundum)
	b) Yellowish green	Corundum
Ti^{3+}	a) Blue	Corundum, kyanite
	b) Pink	Titanaugite, fitzroyite (Ti-phlogopite)
$Ti^{3+} + Fe^{2+}$	Blue	Sapphire (synthetic)
Cr^{3+}	a) Red	Ruby (corundum), kammererite (chlorite)
	b) Green	Uvarovite (garnet), emerald (beryl)
	c) Red-green	Alexandrite (red/green)
Ni^{2+}	Bluish green, green	Garnierite, bunsenite
Mn^{2+}	Pink	Rhodonite, rhodochrosite, beryl (morganite) spessartine
Mn^{3+}	Pink	Piemontite, viridine, rubellite (tourmaline)
Cu^{2+}	Blue	Malachite, Egyptian blue, turquoise
	Green	Malachite, dioptase
Co^{2+}	Blue	Spinel
V^{4+}	Pink	Erythrite
V^{3+}	Green	Apophyllite
	Blue	Zoisite

References and Selected Readings

Battey, M.H. (1972) Mineralogy for Students, Oliver and Boyd, Edinburgh, p. 323.

Battey, M.H. and Tomkieff, S.I. (1964) Aspects of Theoretical Mineralogy in the USSR, New York.

Bloss, F. Donald. (1971) Crystallography and Crystal Chemistry, Holt, Rinehart & Winston, New York.

Brookes, C.A., O'Neill, J.B. and Redfern, B.A. (1971) Anisotropy in the hardness of single crystals, *Proceedings of the Royal Society,* London, 322A, 73-88.

Gay, P. (1967) An Introduction to Crystal Optics, Longmans, Green & Co., London.

Genkins, F. and White, H.G. (1957) Fundamental of Optics, McGraw Hill, New York.

Greenway, D.L. and Harbeke, G. (1968) Optical Properties and Band Structure of Semiconductors, Pergamon Press, Oxford.

Johnson, K.H. (1973) Scattered-wave theory of the chemical bond. *Advances in Quantum Chemistry, 7,* 143-85.
Kerr, P.F. (1958) Optical Mineralogy, 3rd Ed., McGraw-Hill Book Company, New York, p. 442.
Kostov, I. (1968) Mineralogy, Oliver and Boyd, England, p. 587.
Nassau, K. (1980) The causes of colour. *Scientific American, 243*(4), 106-123.
Pauling, L. (1960) The Nature of the Chemical Bond. 3rd Ed., Cornell Univesity Press.
Plendl, J.N. and Gielisse, P.J. (1962) Hardness of non-metallic solids on an atomic basis. *Physical Review, 125,* 828-32.
Plendl, J.N. and Gielisse, P.J. (1963) Atomistic expression of hardness. *Zeitschrift fur Kristallographie, 118,* 404-21.

CHAPTER II

Study in Polarised Light

2.1 POLARISATION OF LIGHT

Light as well as other electromagnetic waves result from vibrating electric charges. These generate alternating electric (E) and magnetic (H) vectors (field) at right angles to each other and to the direction of propagation of the wave (Fig. 2.1).

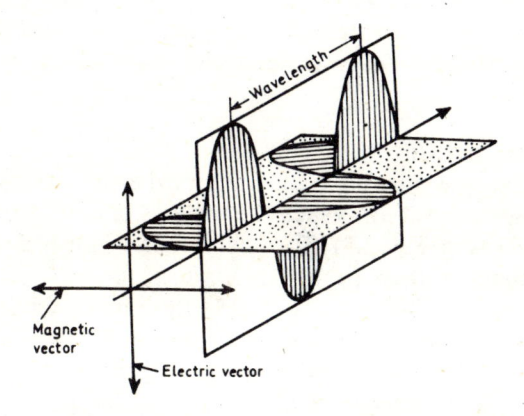

Fig. 2.1   Propagation of a plane-polarised light with magnetic and electric vectors shown.

The propagation of electromagnetic waves is characterised by the vibration of both the electric and magnetic fields. In ordinary light there are no preferential directions of vibration.

In polarised light both the direction and magnitude of electric vector vary regularly with time. If the field lies in one plane, we get *plane polarised* light with mututally perpendicular magnetic vector and the electric vector. Ordinarily a smooth reflecting surface like a glass plate or a varnished table top would reflect light polarised in the plane of the incident and reflected light.

When the magnitude of the field (i.e., amplitude) remains constant, but the direction changes regularly, we get *circularly polarised* light (Fig. 2.2 (a)). If both the direction and magnitude change, so that the ends of the representative vector move around in an ellipse, the light is said to be *elliptically polarised* (Fig. 2.2b). Plane polarised light comprises a right- and a left-circularly polarised beam. Any factor which retards one of these beams

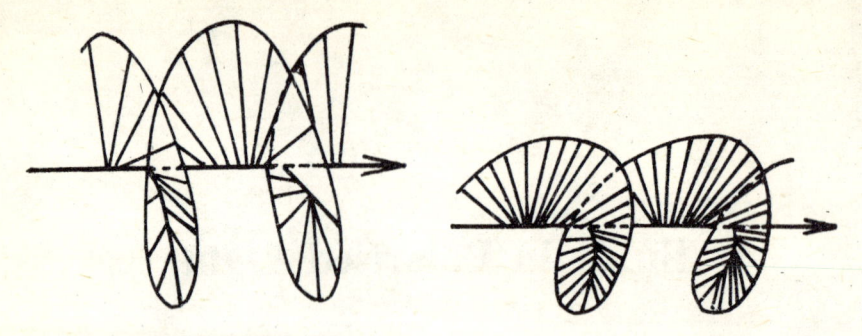

Fig. 2.2(a) The wave fronts circularly polarised light.

Fig. 2.2(b) The wave fronts of the elliptically polarised light.

more than the other will affect the rotation of the original plane of polarisation.

When a crystal structure has a certain element of symmetry which may be described as an n-fold alternating axis of symmetry ($n = $ 1, 2 or 4), it shows optical activity. Crystals of poor symmetry with inversion centres are optically active.

2.2. NICOL POLARISER

The nicol polariser (invented in 1928 and named after the inventor, Nicol) is made by a calcite crystal (of which the width is one-third the length) cut along a diagonal plane (Fig. 2.3). The interfaces are polished and cemented together by Canada balsam ($n = 1.54$), which has R.I. in between the refractive indices of ordinary ($n_\omega = 1.658$) and extraordinary ($n_\epsilon = 1.537$) ray of the calcite crystal. Since n_ϵ is near to that of Canada balsam, the extraordinary ray suffers less deviation on crossing through the interface and reach the other face of the prism. The ordinary ray, because of its high R.I., travels obliquely and meets the interface with Canada balsam at an angle greater than the critical angle. This results in a sideway reflection of the ordinary ray.

The extraordinary ray, polarised in one plane, thus emerges singly on the other side of the prism. The result is a single plane polarised light (with a little lateral displacement of the beam). Two such nicols are used in a polarising microscope. One is set with an alignment perpendicular to the other, so that when they are inserted (called crossed nicols) no light can emerge out.

Calcite nicol polarisers perform well in visible region but the cementing canada balsam causes absorption of wavelengths in the UV range.

2.2.1. Polarisation by Absorption

Strong absorption in one direction of a mineral may cause polarisation of light; e.g., tourmaline crystals strongly absorb light waves vibrating

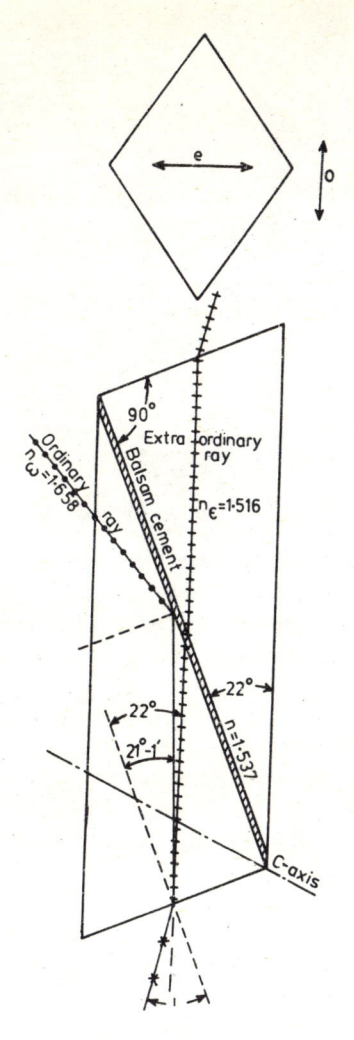

Fig. 2.3 The polarization and deviation of light in a nicol prism.

perpendicular to the crystallographic axis. The transmitted light is, therefore, polarised parallel to C-axis of tourmaline (Fig. 2.4).

2.2.2 Commercial Polaroids
Commercial polaroids are substances showing *dichroism*. In these an array of small dichoric crystals (e.g. iodo-quinine sulphate) are brought into alignment by embedding them in a plastic sheet of cellulose nitrate or polyvinyl alcohol and stretching it in one direction. The individual crystals then tend to turn with their long axes parallel to the stretch axis, known as the axis of polaroid. As a result of dichroism the transmitted light is polarised and strongly coloured.

Polaroid sunglasses are simply polaroid sheets with their polarised axis vertical. Since the glare of light is mostly polarised in the horizontal direction,

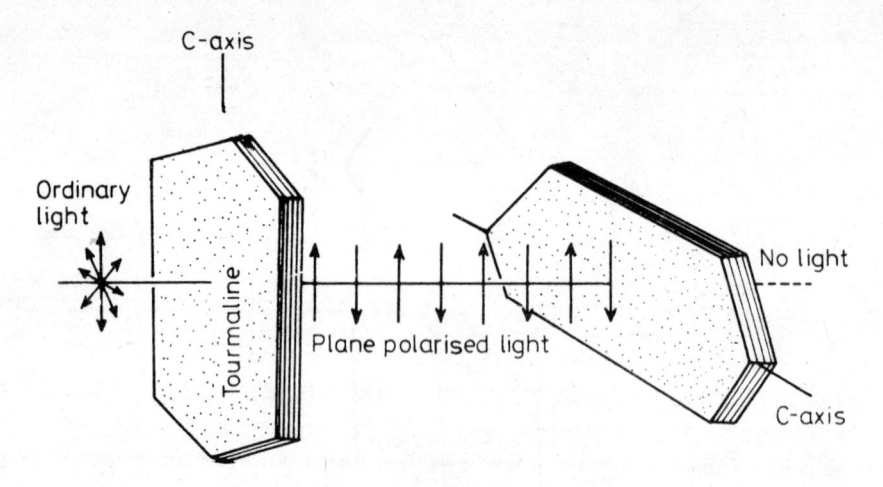

22 *Fundamentals of Optical, Spectroscopic and X-ray Mineralogy*

Fig. 2.4   Polarization by absorption.

it gets cut off by the polaroid sunglasses and the intensity of the transmitted light gets greatly reduced. Ships have polaroid films incorporated in the porthole transparent sheets in order to reduce the glare of the reflecting water surface. Polaroid cameras, however, have nothing to do with the polarisation of light.

2.3 OTHER USES OF POLARISED LIGHT

The utility of polarised light is enormous. Sugar solution has the property of rotating the plane of polarisation. Quantitative estimates of the concentration of sugar solutions are made with a saccharimeter, which determines the degree of rotation of polarisation, which again, is proportional to the concentration of sugar in the solution.

Size and shapes of viruses can also be more easily determined by polarised light than by ordinary light because the viruses are optically active.

The blue skylight reflected at high angles to the sun's rays is partly polarised and some insects, bees and ants, and some aquatic lives like horse-shoe crab (*Limulus*) make use of the electric vector of sky polarisation for navigation. The water flea (*Daphnia*) tends to swim in the direction perpendicular to the electric vibration direction. In polar regions where magnetic compasses are rendered useless polarisation of sky light is sometimes used for navigation. This polarisation is a result of the scattering of the sun's rays by the molecules in the air. Raleigh's well-known inverse-fourth-power law relating wavelength and scattering intensity accounts for the blue colour of the scattered light. At high altitudes the scattering effect is less and the degree of polarisation of sky light is less.

Beneath the surface of water in oceans (or even in a pond) light is found to be polarised with the dominant direction of the electric vibration being horizontal. The main cause of this submarine polarisation is due to the scattering of light by suspended fine particles in water. The particle effect on polarisation has been observed by biologists and oceanographers working under the waters off Bermuda and in the Mediterranean Sea.

2.4 THE POLARISING MICROSCOPE

The polarising microscope, commonly known as 'petrographic microscope' has two nicol prisms (the polariser, below the *stage* and the analyser above the *objective*), a rotable stage with a substage *condenser,* an auxiliary lens, *Bertrand lens,* and a slot in the microscope tube for insertation of *accessory plates*. A set of *objectives* and *occulars* change the magnification of the field of view (Fig. 2.5).

The *magnifying power,* M, is the ratio of the angle subtended at the eye by the image of an object when the image of the object is at a *standard optical distance* of human eye i.e. 25 cm. For simple lens, $M \simeq \frac{25 \text{ cm}}{|f|}$, where f is the focal length in cm.

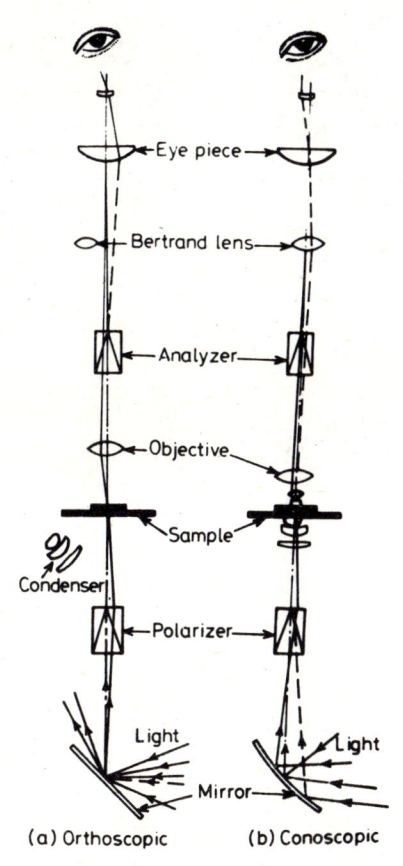

Fig. 2.5   View of the optic path in a polarising microscope
a) orthoscopic set up, b) conoscopic set up

When only the lower nicol is in optical train the study is said to be in *plane polarised light.* When the upper nicol is also inserted the study *under crossed nicols* is done. With the auxiliary lens in place and the substage condenser 'in' the optical system is called *conoscopic* (convergent light); otherwise it is *orthoscopic.*

2.5 WAVE PROPERTIES

In plane polarised light the electric vector, E, is restricted to one plane only (the magnetic vector, H, is perpendicular to this).

If E's of two waves (a and b) of plane polarised and monochromatic light are superimposed, the two waves interfere with each other and a resultant of the waves is generated (Fig. 2.6).

When they are exactly in phase with each other, interference results in the vector addition of the amplitudes (Fig. 2.6b). When the E's are out of phase by ½ λ cancellation in wavelength results (Fig. 2.6c). Any other difference in phase will give intermediate wavelengths (Fig. 2.6a).

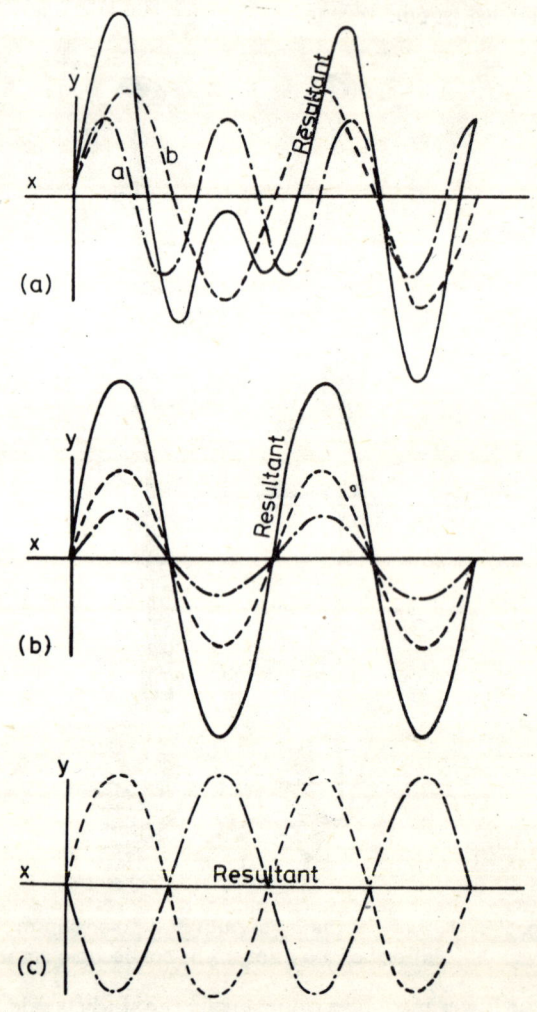

Fig. 2.6   Interference or resultant of waves of polarised light having phase differences: (a) Phase difference *not* equal to one or ½ λ (b). Phase difference equal to 0, amplitude difference exists. (c) Phase difference is ½λ, amplitude the same, causing no light to emerge.

Isotropic: The substance in which light travels with the same velocity in all directions is called isotropic. In this case, the wave surface is a sphere and the wavefront is a circle. Examples, glass and all minerals crystallising with isometric symmetry.

Anisotropic: The substance in which light travels with different velocities in different directions is termed anisotropic. In this the wave surface is an ellipsoid and the wavefront is an ellipse. Examples. all minerals crystallising with symmetry other than isometric.

In an uniaxial crystal having one optic axis (*vide* section 2.7) two sets of rays are formed. One set travels with uniform velocity in all directions and is known as *ordinary (O) ray*. The other set has different velocities in different directions and is called the *extraordinary (E) ray*.

Wavefront

The natue of wavefronts (i.e. velocity surface) in two anisotropic crystals, which are optically 'positive' and 'negative', are discussed below.

At a given instant the wavefront of ordinary ray would be spherical, while that of extraordinary would be ellipsoidal (sections of it would be elliptical except along two planes whereon they are circular). If the velocity of ordinary ray is greater than that of the extra-ordinary ray, i.e. $O > E$ the mineral is *optically positive*. The reverse (when $O < E$) is true for *optically negative* minerals. The velocities are inverse of refractive indices; hence $O = 1/n_\omega$ and $E = \dfrac{1}{n_\epsilon}$. Often, n_ω and n_ϵ are shown as ω and ϵ respectively.

Examples of negative (calcite) and positive (quartz) are given below:

(i) Calcite (negative)

a) Ordinary-ray (O): All waves of ordinary rays vibrate normal to the propagation direction and also to the C-crystallographic direction. Velocity surface is a sphere (Fig. 2.7a).

b) Extraordinary ray (E); All waves from extraordinary rays vibrate normal to the vibration direction of the corresponding ordinary wave in a plane

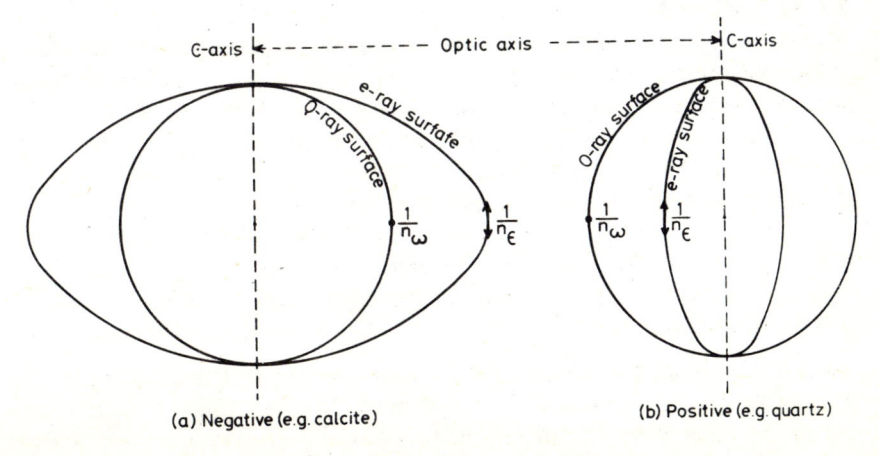

(a) Negative (e.g. calcite) (b) Positive (e.g. quartz)

Fig. 2.7 Wave surfaces, (a) negative and (b) positive uniaxial crystals.

parallel to the C-crystallographic direction. The velocity surface is an oblate spheroid of revolution (i.e. on ellipse rotated about its minor axis).

The combined velocity surface would be composed of a sphere enclosed in an oblate spheroid, the minor axis of the latter is the same as the radius of the sphere.

(ii) Quartz (positive)

Opposite is the case for quartz, for which the velocity surfaces consist of a sphere which encloses a prolate spheroid of revolution with the semi-major axis as the axis of a revolution (Fig. 2.7b). The maximum velocity is along the C-axis and the minimum velocity is normal to C.

The *indicatrix* is a geometrical figure relating the values of refractive indices to their directions of vibration and to the direction of propagation of light through the crystal. For positive crystals it is a prolate spheroid of revolution; for negative crystal it is an oblate spheroid of revolution (Fig. 2.8). This has been discussed in section 2.9.

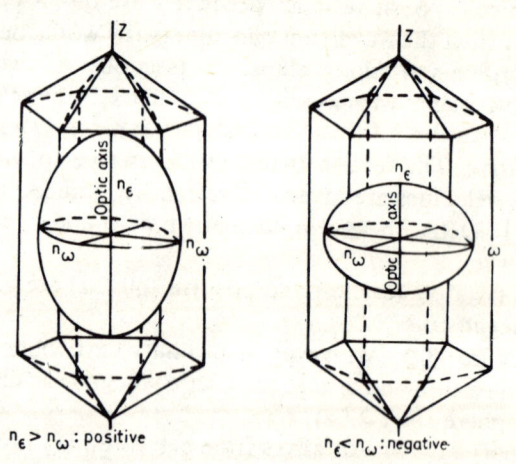

Fig. 2.8   The uniaxial indicatrices of positive and negative crystals.

2.6 PLEOCHROISM

This is generated by differential absorption of polarised light in different directions in minerals. This is mostly due to the presence of transition metal ions in low symmetry or distorted coordination sites in the crystal structures.

Pleochroism may also be generated by electron transitions between neighbouring ions in a structure, facilitated in certain crystallogrphic directions through suitably oriented overlapping d-orbitals.

Pleochroism in minerals, as is commonly known to occur in the visible region, is seen under a polarising microscope. Many silicates (e.g. olivines, etc.) however, show pleochroism beyond the optical region and are studied by spectrophotometers, aided by polarisers. This has been discussed in chapter III.

2.6.1 Polarisation Colour and Pleochroism for Mineral Identification

Primarily colour is generated by absorption. In cubic minerals there is no change in absorption with direction. In uniaxial (tetragonal and hexagonal symmetry) minerals the absorption along O and E are determined. When differences in absorption colours are observed it is called *dichroism*.

In biaxial minerals (of orthorhombic, monoclinic and triclinic symmetry) light waves vibrating parallel to the three optic directions show different degrees of absorption and generate tri-chroism or *pleochroism*. The colours are noted against each optic directions (X, Y, and Z) and the relative degrees of absorption are stated in terms of absorption colours. These are explained with biotite and hornblende colours as examples.

Examples:

Vibration direction	Absorption colour
(a) Biotite:	
α or X	yellow
β or Y	dark brown
γ or Z	dark brown
$\therefore$ X < Y = Z	
(b) Hornblende:	
α or X	pale yellow
β or Y	yellow green
γ or Z	green
$\therefore$ X < Y < Z	

Estimation of absorption can be made by use of absorption spectroscopy.

Green colours in β and γ directions are usual for some minerals (viz. amphiboles etc.) which has Fe^{2+} in octahedral coordination. To account for the red colour in α-direction, a study of the crystal structure and polarised spectra of hypersthene becomes a necessity. This has been discussed in section 3.11.2. The red brown pleochroic colours are thus explained in the case of minerals such as hypersthene, staurolite, augite, biotite and hornblende.

The blue colours found in kyanite, cordierite and glaucophane are, due to $Fe^{2+} - Ti^{4+}$ charge transfer, as discussed in section 3.12.3.

2.7 ANISOTROPISM

Minerals crystallising in the cubic system are optically isotropic, and have only one index of refraction. Because atoms are equally spaced in three mutually perpendicular directions, the impinging light rays are equally affected in all the directions. Any distortion in the arrangement causes deviation from isotropism.

A change in R.I. (which is inverse of the light velocity), in any particular direction will give an indication of the degree of distortion of the electron shells surrounding the atoms (the *electron polarisation*), This is expressed quantitatively by the moment of the dipole induced by electric vector of light ray acting upon them.

In *anisotropic* minerals, because of unequal spatial distribution of atoms in different directions, a light ray gets split into two polarised rays vibrating in mutually perpendicular directions.

Minerals possessing tetragonal, hexagonal (also trigonal) crystal symmetry have a spindle-like distribution of atoms and so they have two R.I.'s; ϵ for the rays vibrating along the principal symmetry axis and ω for rays vibrating in a perpendicular plane. Along the principal axis no splitting occurs and this is called the *optic axis*.

Birefringence is the difference between n_e and n_ω when $\epsilon > \omega$ the crystal is called *optically positive* (e.g. quartz), when $\epsilon < \omega$ it is optically negative (e.g. calcite). Such minerals with symmetry offering only one optic axis are said to be *uniaxial* (Fig. 2.8). A plane perpendicular to this optic axis behaves as isotropic (in orthoscopic set up). This axis may also be loosely termed as the direction of 'isotropism'.

In minerals crystallising in orthorhombic, monoclinic and triclinic systems, there are three mutually perpendicular principal optical directions; viz., X, Y and Z. Consequently there are three principal R.I.'s designated as the smallest (α), intermediate (β) and greatest (γ) respectively. In such minerals no distinction exists between ordinary and extraordinary rays, and *all rays are extraordinary*.

In a plane having α and γ there occur two directions with *R.I.'s* equal to β, along which 'isotropism' occurs. The angle between these two directions, known as *optic axes,* is called the *optic axial angle* (2V) and the crystals are optically *biaxial*.

The relationship between R.I.'s and the optic axial angle (2V) (*vide* Section 2.15) is:

$$\tan V = \frac{\gamma}{\alpha}\sqrt{\frac{\beta^2-\alpha^2}{\gamma^2-\beta^2}}$$

2.7.1 Optics and Structures

Optical anisotropism depends on the type of crystal structures. In layered structures along the planes of densely packed atomic planes the R.I.'s are higher than along perpendicular directions; and all minerals with layered structures are optically negative, eg: Muscovite, $KAl_2[AlSi_3O_{10}](OH)_2$; $\beta \simeq \gamma = 1.59$ and $\alpha = 1.55$, optically negative with $2V = 47°$.

But in minerals having higher content of OH groups viz. $Mg(OH)_2$, called brucite, and $Al(OH)_3$ (gibbsite) are positive; because the oxygen atoms are polarised under the influence of the hydrogen atoms along the direction perpendicular to the layers of the structure.

Carbonates and nitrates with their CO_3^{2-} and NO_3^{1-} groups, are optically negative because of high atomic polarisation in the plane of these groups of ions in the crystal structure. The higher index of refraction lies in the plane of the complex group, e.g. calcite, magnesite, aragonite, etc.

Minerals with chain-like structures are commonly optically positive. In enstatite ($MgSiO_3$, $\gamma = 1.667$, parallel to vertical C-axis) chains of silicon and oxygen atoms lie along the C-axis. Along the C-direction the polarisation (deformation) of the highly refractive ions (e.g. O^{2-}) is greater and hence the interaction with the passage of light is greater and R.I. is higher. This is also the case with Cinnabar, HgS, which has screw-like ... Hg–S–Hg–S... chains which run parallel to its 3-fold axis. Calomel, HgCl, also has Cl-Hg-Hg-Cl molecules preferentially oriented parallel to the 4-fold axis. Cinnabar and Calomel are optically positive.

Structures having close packing in more than two directions should exhibit weak or no birefringence. The reason being that light travels more slowly in the direction of densest layering. But a full explanation demands a very complicated mathematical modelling and analysis.

2.7.2 Interference Colours

Let us study a biaxial mineral section under crossed nicols. A section has only two mutually perpendicular vibration directions. The polarised light on striking the lower part of the section gets resolved into mutually perpendicular polarised rays, vibrating along the vibration directions of the mineral. These rays travel through the length of the tube and reach the analyzer, which has only one polarising direction. The two rays again get resolved in the plane of polarisation of the analyzer. The resolution of the polarised rays (r_1 and r_2) in the analyzer direction is determined by the angles they make with the analyzer plane (Fig. 2.9a). The interference colour is generated as the resultant of the two superimposed polarised light emerge out of the analyzer. The interference colour, therefore, is also a measure of the amount of retardation ($\triangle$) between their velocities as well as their R.I.'s. The relationship between the indices of refraction for fast (n_f) and slow (n_s) rays emerging from the mineral and the retardation is expressed by equation:

$$\triangle = t\,(n_f - n_s)$$

where t is the thickness of the mineral section.

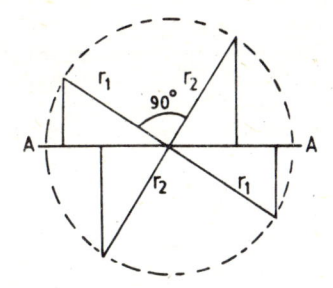

Fig. 2.9(a) Resolution of polarised rays r_1 and r_2 in the plane of the analyser (A-A) causing interference colour.

Interference colour of a mineral section depends on three factors: (i) birefringence, (ii) orientation of the section with respect to the major optic directions, and (iii) thickness of the section.

2.7.3 Orders of Interference Colours

The interference colours, as observed with quartz wedge, are generated by relative retardation. For convenience, orders of colours are demarcated at a retardation of 550 mμ (i.e., red colour). The first such red (rather reddish violet) coloured band is called first-order red, and the next red marks the border of the second order of the coloured bands and so on.

Through a little practice one can learn to identify the order of a colour by its brightness etc. Higher order colours pose a difficulty as they get blurred with shades of pinks and greens and ultimately change to white. Compared to the first order white the higher order white is non-uniform with coloured granulation. On insertion of a gypsum plate, this white does not change even on stage rotation; while the white of first order will change colours to the second order blue (rise) or first order yellow (fall); and rotation will bring a remarkable change in these colours.

2.8 BIREFRINGENCE CHART

The chart is a graphical representation of the relation $\triangle = t\,(n_f - n_s)$ and helps determining directly any one variable, if the other two variables are known. In this chart (fig. 2.9b) the interference colour orders, as obtained under crossed nicols on gradual insertion of qurtz wedge, are drawn vertically from left to right. Each colour corresponds to a specific retardation, which is marked in *nm* at the bottom. Thickness also controls the birefringence and it is marked in *mm* on the left. Lines of equal birefringence $(n_f - n_s)$, radiate from the lower left and reach the top and right side of the chart.

Birefringence of a mineral changes from zero to a maximum value; the latter is its characteristic property. Grains showing maximum interference colour in the chart have maximum retardation and are chosen for specific optical studies. The order of interference colours and the range of respective birefringence maxima are presented as follows.

Order	Colours	Max. birefringence
First	Grey, white, yellow, red	0.000–0.018
Second	Violet, blue, green, yellow, orange, red	0.018–0.036
Third	Indigo, green, blue, yellow, red, violet	0.036–0.055
Fourth and above	Pale pink and green	>0.055

2.9 OPTICAL INDICATRIX

Optical indicatrix is a solid-geometrical representation of the relationship between the crystal symmetry and refractive indices. The refractive indices in different directions, drawn as radii, define the surface of an ellipsoid.

2.9.1 Uniaxial Indicatrix

In uniaxial minerals the two rays *O* and *E*, generated by splitting of transmitted light, travel with vibration planes perpendicular to each other and have different velocities. The *O*-ray travels with equal velocity in all directions, while the *E*-ray travels with different velocities in different directions.

Along the optic axis, which coincides with the *C*-axis, the two rays travel at the same speed and under crossed nicols, therefore, sections cut normal to the *C*-axis (i.e. basal section) appear isotropic.

In all other directions their velocities differ. The difference becomes maximum in a direction perpendicular to the optic axis. Therefore, maximum interference colours are exhibited in such sections.

When *E* is slower than *O*, the mineral is said to be optically positive, when *E* is faster it is called negative. In negative minerals, therefore, the refractive

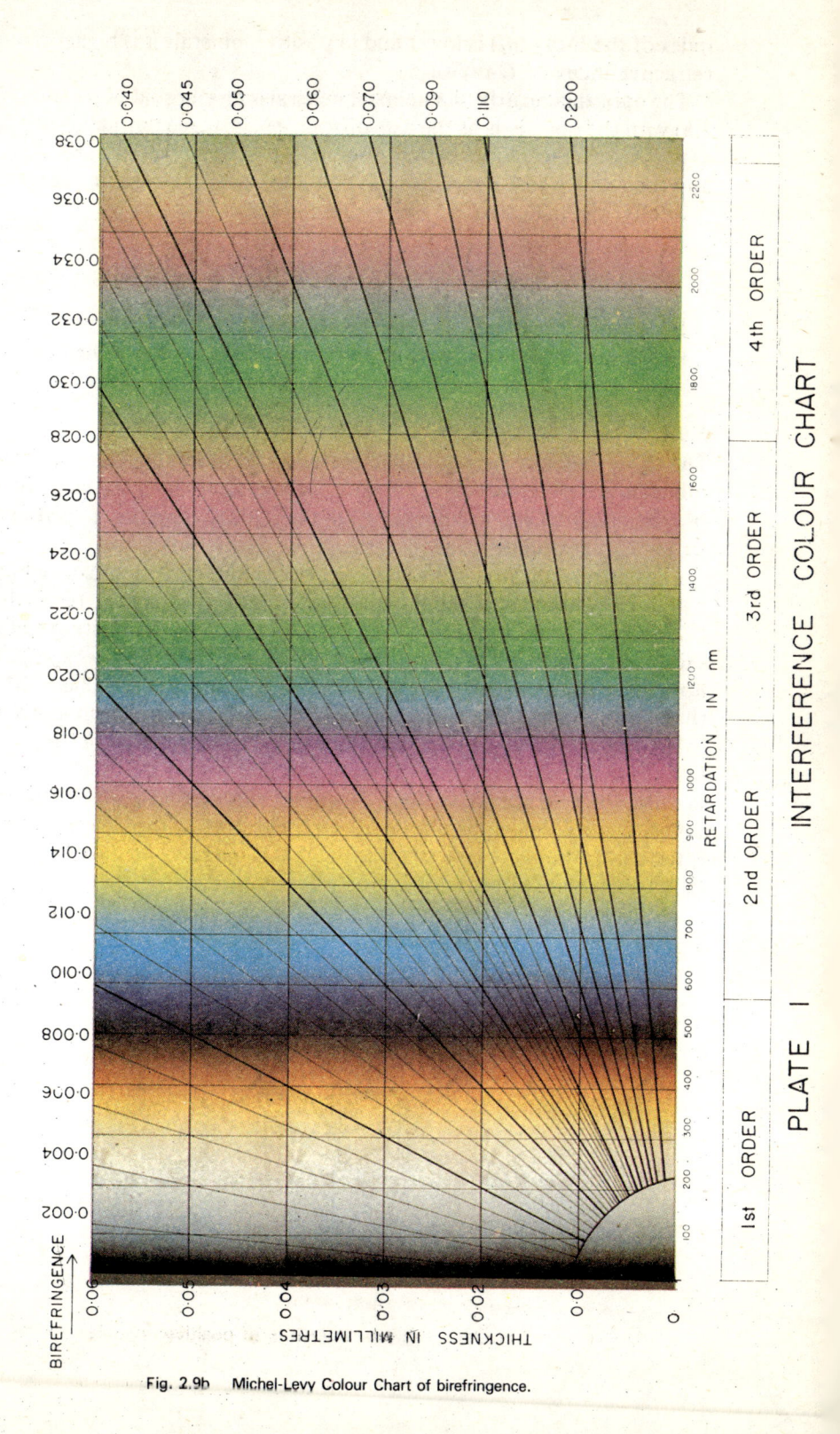

Fig. 2.9b   Michel-Levy Colour Chart of birefringence.

index of the *E*-ray (n_ϵ) is lower and in positive minerals it is higher than the refractive index of *O*-ray (n_ω).

The optical indicatrix of a uniaxial mineral is an ellipsoid of rotation (Fig. 2.8) with the optic axis as the axis of rotation. A plane normal to the optic axis intersects the indicatrix to form a circle. All other sections are elliptical, and the semiaxes of a section containing the optic axis are n_ω and n_ϵ.

Briefly, the indicatrix may be represented as:

$$\frac{x^2+y^2}{n^2_\omega} + \frac{z^2}{n^2_\epsilon} = 1$$

In isotropic crystals $n_\omega = n_e$, the spheroid is a sphere.

In negative crystals, $n_\omega > n_\epsilon$ and the spheroid is oblate; while in positive crystals, $n_\omega < n_\epsilon$, the spheroid is prolate (Fig. 2.8).

2.9.2 Biaxial Indicatrix

The three mutually perpendicular preferred vibration directions in biaxial minerals are designated as *X*, *Y* and *Z*; these correspond to the directions of minimum (n_α) and maximum (n_γ) of the refractive indices; *Y* corresponds to some intermediate value n_β.

The optical indicatrix of a biaxial mineral is, therefore, a triaxial ellipsoid with minor, intermediate and major semiaxes of n_α, n_β and n_γ respectively.

All plane sections passing through the center point of this figure are ellipses except two, which are circles with a radius of n_β. The *optic axes* are the two normals to these circular sections. They also lie in the *XZ* plane, which, therefore, is called the *'optic plane'*; the *Y* direction is the *optic normal*.

Between the optic axes the acute angle is called the *optic axial angle*, and marked as *2V*. The principal vibration axis that bisects the optic axial angle is called the *acute bisectrix* (Bx_a). The other vibration axis which bisects the obtuse angle is called the *obtuse bisectrix* (Bx_o). If *Z*, the slowest vibration direction, is the acute bisectrix, the mineral is optically *positive* (Fig. 2.10); if *X* is the acute bisectrix, it is optically *negative*.

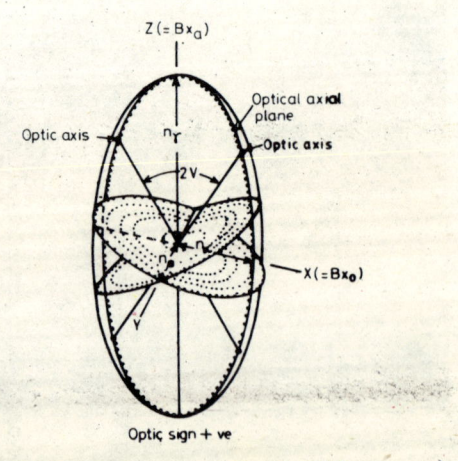

Fig. 2.10   Biaxial indicatrix of positive crystals

In orthorhombic minerals the three principal vibration directions (X, Y and Z), coincide with three cyrstallographic axes a, b and c. In monoclinic minerals, only one of the principal vibration directions coincides with a crystallographic axis—always b (two fold) axis. In triclinic minerals there is no necessary coincidence.

2.10 EXTINCTION ANGLE

It is the angle between the vibration plane and one of the crystallographic planes (i.e. the crystal face or trace of cleavage planes etc.), usually, $Z \wedge c$.

An extinction angle is said to be *parallel* when in the grain any of the crystallographic axial planes, made parallel to the crosswires, makes the grain dark i.e. in position of extinction.

At the position of extinction when a cleavage plane or a crystal face makes angles with the cross-wires the extinction is called *inclined* or *oblique*. When this direction, marked by crosswires, bisects the angle between two sets of cleavages or faces the extinction is called *symmetric*.

Uniaxial crystals show parallel extinctions with respect to the basal and prismatic planes and symmetrical extinctions with respect to pyramidal planes.

The extinction angles can serve as distinguishing features for groups of mineral viz. pyroxenes, amphiboles, plagioclase felspars etc.

2.11 ACCESSORY PLATES

Optic signs and the length-fast or-slow characters of minerals are determined by use of accessory plates with known slow-or fast-ray directions as shown in Fig. 2.11. Gypsum plate (called 'sensitive tint' or first order red) and mica plate ($\lambda/4$) are commonly used for minerals with low to moderate birefringence.

a) *Gypsum plate:* Usually a cleavage plate of gypsum gives first order red interference colour (retardation 550 mμ). Such a plate causes samples with very low retardation to exhibit a second order blue interference colour as a result of addition. By subtraction, first order yellow interference colour results. Because of this noticeable colour change as a result of slight changes in retardation, the gypsum plate is sometimes called the *sensitive tint* plate (Fig. 2.11a).

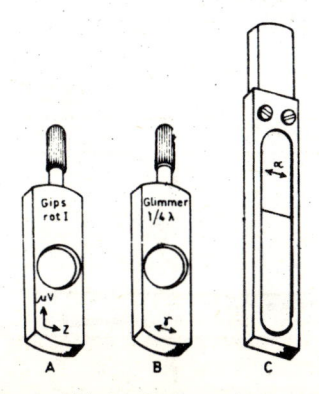

Fig. 2.11 Accessory plates: (A) the Gypsum (1st order red) plate; (B) the Mica (¼λ) plate; (C) the Quartz Wedge.

b) *Mica plate or $\frac{1}{4}\lambda$* plate causes interference colours to 'add' or 'subtract' by 147 mμ which is $\frac{1}{4}$ th the sodium wavelength (589 mμ (hence called $\frac{1}{4}\lambda$). (Fig. 2.11b).

c) *Quartz Wedge* is a quartz plate cut parallel to the optic axis and made to a wedge shape (Fig. 2.11c).

When a quartz wedge is gradually inserted through the 45° angle slot of the microscopic tube and seen under crossed nicols in ordinary light, a succession of broad colour bands, separated by evenly spaced narrower darker bands will be seen running across the direction of insertion. The reason for this effect is as follows:

i) *In monochromatic light:* The monochromatic light on passing through the wedge is split into two vibration directions of the wedge. The two vibrations differ in phase but have the same wavelength. At certain thickness of the wedge where the differences in phases occur as integral numbers of the wavelength (of the monochromatic light used), the two beams interefere destructively and the dark bands are produced. As the wedge thickens the first dark band appears at a relative retardation of λ, the second at 2λ, the third at 3λ, and so on. In between the dark bands are the positions of brightest bands where the two components reinforce each other. Consequently, brightest zones occur at phase differences of $\lambda/2$, $3\lambda/2$, $5\lambda/2$ etc.

ii) *In white light:* In white or ordinary light as the wedge is gradually inserted the observed colour bands under crossed nicols change from grey to white

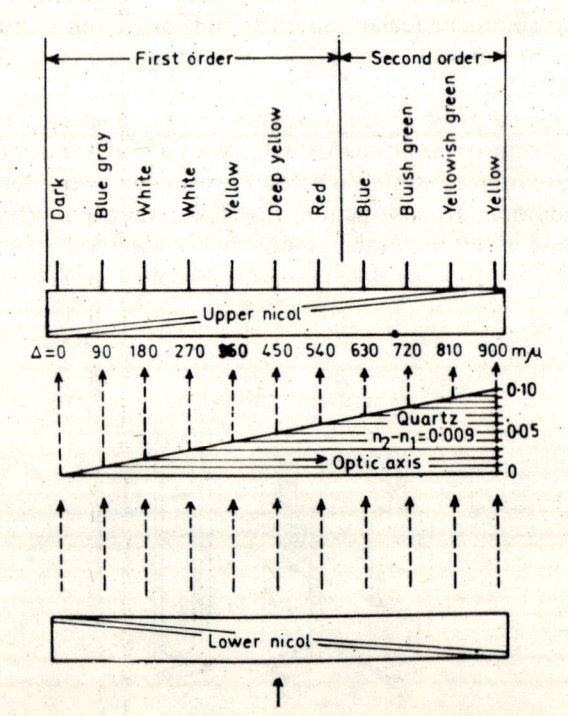

Fig. 2.12   Interference colour bands of quartz wedge

to yellow, orange to red and then is succeeded by the Newton's scale of colours: violet, indigo, blue, green, yellow, orange, red. Four such distinct colour orders can be observed, although becoming progressively paler, before they turn to a pale pinkish hue called the high-order white. Pink/red bands divide the repetitions and each segments are marked sequentially as the first order, second order, third order interference colours. (Fig. 2.12).

2.11.1 Calculation of Phase Difference

The *relative retardation* $\triangle$, (or optical path difference) can be calculated from refractive indices as:

$$\triangle_{(nm)} = (n_s - n_f)t$$

where, n_s and n_f are the R.I.'s of slow and fast vibration directions respectively.

The *phase difference*, δ, is given by

$$\delta = \frac{(n_s - n_f)t \times 2\pi}{\lambda}$$

2.11.2 Use of Accessory Plates

When one of the accessory plates are inserted into the accessory slot, retardation is increased or decreased in alternate quadrants of the optic figure. If it is decreased parallel to the trace of the slow ray of the uniaxial figure, the mineral is said to be *negative* ($n_\epsilon < n_\omega$), while if increased, *positive* ($n_\epsilon > n_\omega$).

In the case of uniaxial negative minerals, on insertion of the length-fast gypsum plate along the second quadrant, the first and third quadrants would appear yellow-green and the second and fourth bluish. The reverse is the case when the minerals are positive. Normally for minerals showing very low birefringence (of the first order grey to yellow) gypsum plate is used.

With the length-fast mica plate along the second quadrant the uniaxial negative minerals will produce dark dots in the first and third quadrants, while positive minerals will produce dark dots in the second and fourth quadrants. (Fig. 2.15)

Quartz wedge is used in minerals showing high birefringence, where many isochromes develop. As the wedge is inserted (along the second quadrant) from the thin edge over a negative uniaxial mineral the isochromes in the first and third quadrant shift away from the centre and those in the second and fourth toward the centre. Opposite shifts take place in case the mineral is positive.

In a non-conoscopic set-up, when the sample and quartz wedge both cause the same phase difference, complete extinction occurs. With a calibrated quartz wedge, therefore, a quantitative estimate of the birefringence of a mineral can be made when the thickness of the sample is known.

The use of accessory plates in biaxial minerals is discussed in section 2.14.

2.11.3 Compensators

These are used for the measurement of birefringence. The Berek Compensator consists of a plane parallel plate of calcite cut perpendicular to its optic axis.

This is inserted in the accessory slot at the 45° position of the microscope tube. This plate can be rotated around an axis in the plane of insertion. The plate, when is in a position perpendicular to the microscope axis, looks isotropic because calcite, when viewed along the optic axis, produces no optical phase difference. By inclining the Berek plate, increasing phase difference is produced, and the birefringence of the sample can be compensated. The magnitude of the rotation angle necessary to accomplish compensation is related to the phase difference, because the compensator is calibrated accordingly.

2.12 DETERMINATION OF FAST/SLOW VIBRATION DIRECTIONS

The mineral grain, in which the fast or slow vibration directions are to be determined, is first brought to an extinction position and then rotated through a 45° angle. In this position one of the vibration planes is brought to a position, which is parallel to the accessory slot, (through which the accessory plates are inserted in the microscope).

On insertion of the plate which has length fast, the order of interference colour will rise over a mineral if the trace of its slow ray (higher R.I. direction) is perpendicular to the axis of the accessory plate and will decrease if its fast ray vibration direction is perpendicular.

If the slow-ray vibration direction is parallel to the length of the crystal or fragment of the mineral, it is said to be of *positive elongation* or length-slow. Conversely, when the fast ray is parallel to the length it is called *negative elongation* or a length fast mineral. For elongated uniaxial crystals the extraordinary wave vibrates parallel to elongation, and the sign of elongation is the same as the optic sign.

2.13 INTERFERENCE FIGURES

When anisotropic minerals are viewed in a conoscopic set-up (i.e. in the convergent light under crossed nicols) interference figures are produced.

These figures consist of a series of concentric bands of interference colours; these isochromatic bands or isochromes, increase in order of retardation from the centre outward. One or more dark bands called *isogyres* are also produced.

The isogyres are of the shape of a cross, bars or segments of hyperbolas. These dark areas are formed due to the loci of points in the interference fields where the vibration directions of the resultant waves are nearly parallel to the planes of vibration of the upper and lower nicols.

Interference figures are useful for a) distinguishing between uniaxial and biaxial minerals, b) determination of optic sign, c) evaluation of birefringence, when the thickness (standard 0.35 mm) is known, d) ascertaining the optic orientation, e) measuring the optic axial angle, $2V$, in biaxial minerals, f) studying the dispersion of the optic axes.

The following types of interference figures are obtained in a conoscopic set up:

Uniaxial minerals: 1) Optic axis figure—sections perpendicular to the optic axis, 2) Optic normal figure—sections parallel to the optic axis, 3) Off-centred figure—sections oblique to the optic axis.

Biaxial minerals: 1) Sections perpendicular to the acute bisectrix, (Bx$_a$)—acute bisectrix figure, 2) Sections perpendicular to the obtuse bisectrix (Bx$_0$)—obtuse bisectrix figure, 3) Sections oblique to the bisectrices—off-centered figure, 4) Sections perpendicular to the optic axes—optic axis figure, 5) Sections parallel to the optic axial plane—flash figure.

2.13.1 Uniaxial Interference Figure

When a uniaxial crystal is cut perpendicular to its optic axis and is observed under cross-nicols in a conoscopic set-up i.e. with the condenser 'in', a centred optic-axis figure is obtained. The thickness and the birefringence of the mineral will determine the number of coloured circles and the sharpness of the isogyres. Higher birefringence and thickness will cause reduction in the spacings of the rings and sharper isogyres. The number of rings also depends on the numerical apertures (N.A.) of the condensing lens and also on the objective lens. N.A. equals $n \sin u$, where u is half the angular aperture and n is the lowest *R.I.* in air.

A dark cross appears superimposed on a set of coloured rings, called isochromes (Fig. 2.13a and b). The centre of the cross represents the emergence of the optic axis, the *melatope*. On rotation the dark figure remains stationary.

The isochromes are formed due to the joining of points of equal relative retardation of the two interfering light rays travelling through the mineral. The *isogyres*, or the dark arms of the cross, as explained earlier, are formed due to zero amplitude generated in the waves along the trace of the preferred vibration directions of the polariser and analyser.

If the optic axis of the mineral is not at right angles to the plane of the section, then the centre of the cross will be off-centred. On rotation the melatope will rotate in the same direction, but the arms will maintain their respective orientations with the cross-wires (Fig. 2.14). When the optic axis is at a high angle with plane of the section, the melatope will be out of the field of vision but on rotation the arms will appear and disappear in the field maintaining their orientations with respect to the cross-wires, as in the previous case.

If the optic axis is so high angled (off-cut) that it is nearly parallel to the microscope-stage, then on rotation a flash figure is obtained. In such cases when the trace of the optic axis is at about 45° with respect to the cross wires, the isochromes forms two conjugate hyperbolas asymptotic to the cross-wire directions.

Optic Sign Determination

The determination of the optic sign from the interference figure, which is centro-symmetric, is done by inserting an accessory plate (gypsum, mica or quartz wedge) and observing the movement of the colour bands. If additive colour is formed parallel to the slow ray of the accessory plate the section is said to be positive. If subtraction colour is formed it is negative. In sections of 1st order grey interference colour, when a length-fast gypsum plate is inserted the colour of the isogyre becomes pinkish (1st order red). In negative minerals, blue colour appears along the direction of the insertion of the

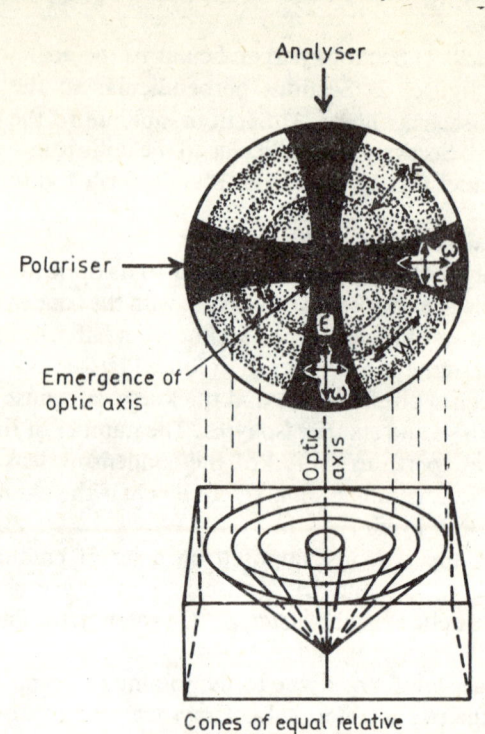

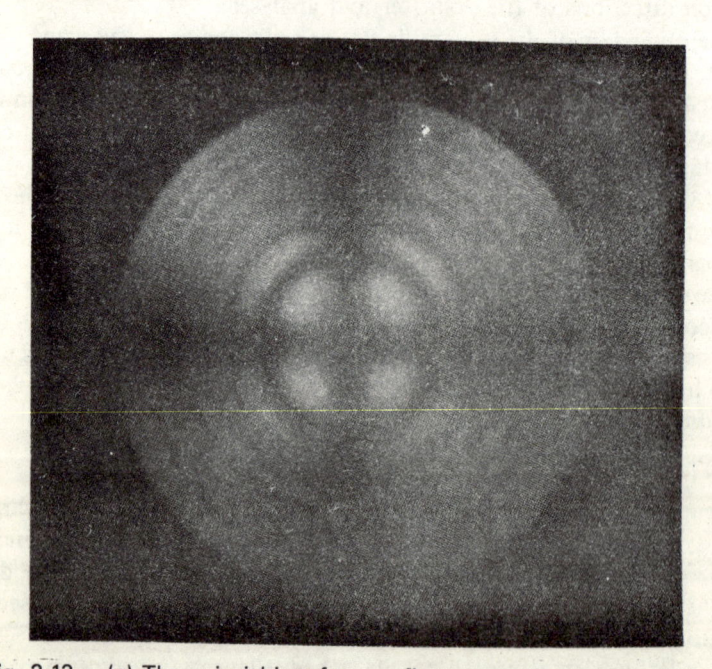

38 *Fundamentals of Optical, Spectroscopic and X-ray Mineralogy*

Fig. 2.13 (a) The uniaxial interference figure.
(b) The uniaxial interference figure showing the colour rings.

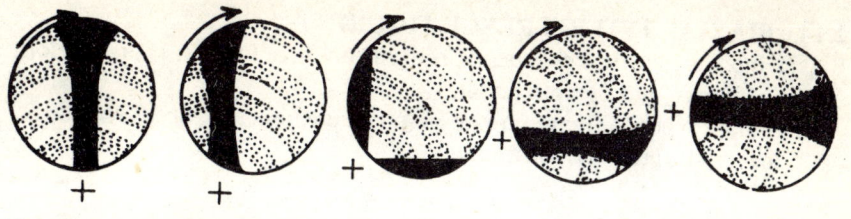

Fig. 2.14 Off-centred uniaxial interference figure rotated clockwise; + marks successive positions of the optic axis.

gypsum plate (Fig. 2.15a). In the case of a positive mineral, black dots appear on the two quadrants at right angles to the length of the accessory mica plate (making an imaginary cross). In sections of higher birefringence mica and quartz wedge are used. The mica plate forms black spots and the gradual insertion of a quartz wedge has a thickening or thinning effect on the coloured rings as shown in Fig. 2.15(b).

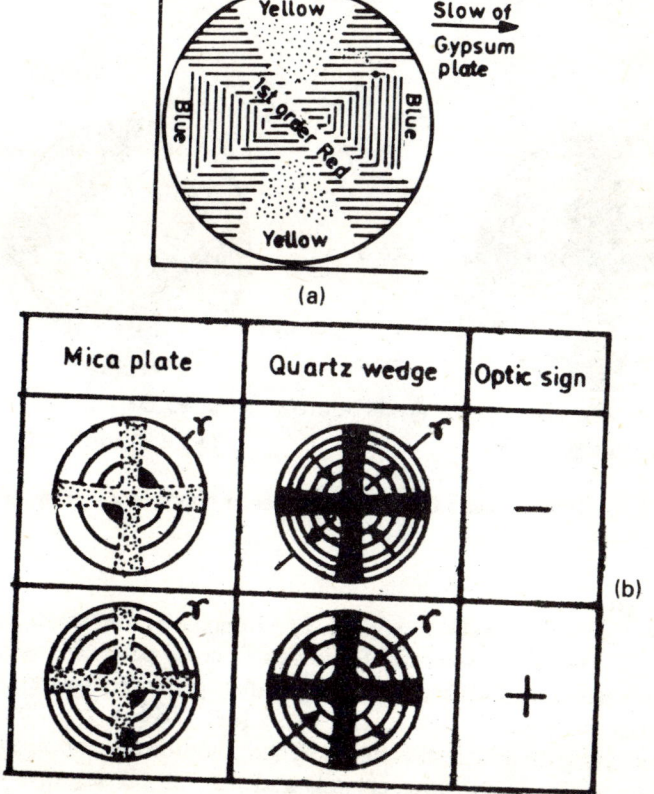

Fig. 2.15 Optic sign determination using length-slow accessory plates (a) Insertion of length-slow gypsum plate over a negative crystal causes increase in retardation along its length. (b) Dark spots appear along the mica plate over negative crystals and colour bands move out in quartz wedge on its gradual insertion. The reverse phenomenon is observed in positive crystals.

2.14 BIAXIAL INTERFERENCE FIGURES

Biaxial interference figures depend primarily on three factors: 1) orientation, 2) birefringence and 3) optic angle.

Four main types of figures are observed: acute bisectrix, obtuse bisectrix, optic axis and optic normal figures.

2.14.1 Acute Bisectric (Bx$_a$) Figure

An acute bisectrix figure is formed when the acute bisectrix (Bx$_a$), either X (in case of negative) or Z (in case of positive minerals), is normal to the stage.

In Fig. 2.16 centred Bx$_a$ figure is shown for a mineral with moderate 2V and moderate birefringence. If 2V is 35° or less, the two melatopes lie within the field in a centred Bx$_a$ figure and remain so throughout a complete rotation of the stage. If 2V is very small, the figure is much like the figure of a uniaxial mineral. If 2V is larger than 35°, the melatopes lie outside the field. But from the nature of the isochromes and the relative thickness of the arms of the cross the optic axial angle can be surmised (*vide* section 2.15).

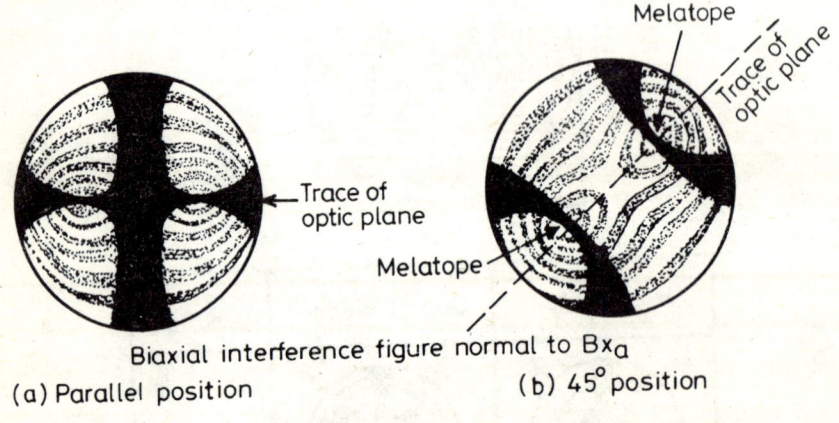

Biaxial interference figure normal to Bx$_a$

(a) Parallel position   (b) 45° position

Fig. 2.16   Acute bisectrix interference figure of a biaxial crystal.

Optic Sign

The optic sign of a biaxial mineral is determined by ascertaining the relative velocity of the acute bisectrix. From nearly cross position (when the two isogyres meet with melatopes nearly parallel to the cross-wires), the stage is rotated through 45 degree clockwise (i.e. NE-SW position), and an accessory plate (length-fast) is inserted; the colour in the concave side of the isogyres will rise for positive crystals and will fall for negative ones.

Whenever the optic plane is oriented NE-SW, retardation will always add between isogyres for positive crystals and subtract for negative ones. It is equally valid for acute as well as obtuse bisectrix figures.

2.14.2. Obtuse Bisectrix (Bx$_o$) Figure

When the angle between the optic axes is a few degrees above 90° the isogyres in an obtuse bisectrix figure would disappear from the field of view on rotation to 45° position. When the obtuse angle is very large complication arises and it is advisable to look for an acute bisectrix figure or an optic axis figure for sign determination and other studies.

2.14.3 Optic Axis Figure

A section cut perpendicular to an optic axis gives the optic axis figure (Fig. 2.17) The ray travelling along it experiences zero birefringernce and a melatope is formed at the centre point of an isogyre. For any optic axial angle (2V) less than 30° the isogyre accompanying the other axis appears in the field. When the axial angle is large only one isogyre appears. The isogyre becomes straight when the optic axial plane is parallel to the analyser or polarisation directions. At 45° position it shows maximum curvature (Fig. 2.17.1). For very small 2V the figure looks much like an uniaxial figure.

When the axial angle is large and the optic axis emerges in the field, the single isogyre remains stationary at the melatope, and on rotation the two arms of the isogyre make small angular sweeps in the opposite direction to the direction of rotation. The line which symmetrically bisects the curved isogyre at the 45° position is the trace of the optic plane (Fig. 2.17.1). The acute bisectrix lies on the convex side of the parabolic isogyre.

When at 45° position the isogyres do not show any curvature, 2V is nearly 90° and the sign cannot be determined. At NE-SW position, retardation will add on the concave side of the isogyre for positive crystals and subtract on the concave side for negative ones (Fig. 2.17.2). Off-centred biaxial interference figures on rotation are shown in Fig. 2.17.3.

2.14.4 Flash Figures

When optic normal i.e. Y is perpendicular to the section, flash figures occur in biaxial minerals. These can very well be confused with obtuse bisectrix figures or uniaxial flash figures.

The dark broad and diffuse cross separates quickly on slight rotation and joins again when X and Z are parallel to the cross-hair. If 2V is not large, at the 45° position the quadrants brighten and the position of the acute bisectrix and the sign can be determined by using the accessory plate.

2.15 2V ESTIMATION

In a section perpendicular to the optic axis, the optic figure consists of a single hyperbolic isogyre with a single melatope which shows a maximum curvature at the centre of the field in a 45° position. The $2V$ angle can be visually estimated by referring to the nature of curvature as in the figure 2.17.1.

The isogyre is straight when $2V$ equals 90°, and bends to about 90° when $2V \cong 0°$ i.e. produces the uniaxial cross. When $2V = 45°$ the isogyre lies half way between.

In a section perpendicular to the obtuse bisectrix, the melatopes are

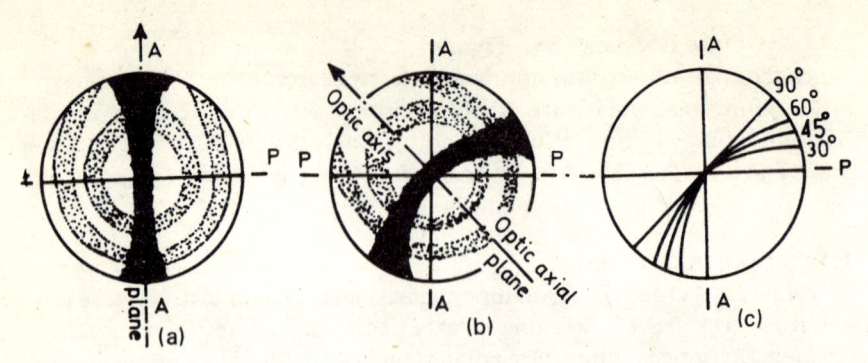

42 *Fundamentals of Optical, Spectroscopic and X-ray Mineralogy*

Figure 2.17.1 The optic axis interference figure of a biaxial crystal. (a) Parallel position (b) 45° position (c) curvature of isogyre, for different optic axial angles. (PP, AA: Planes of polariser and analyser)

Gypsum plate	Mica plate	Quartz wedge	Optical character
green / blue	(dot)	(wedge pattern)	+
blue / green	(dot)	(wedge pattern)	−

Fig. 2.17.2 Optical character determination from optic axis intereference figure of biaxial crystals.

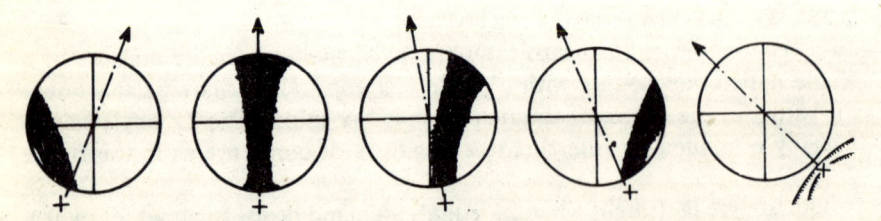

Fig. 2.17.3 Off-centered biaxial figure rotated clockwise; + marks the position of optic axis.

normally beyond the view and sweep swiftly through the field as the stage is rotated.

The bisectrix figure is acute when on rotation the isogyres remain within the limits of the field. If the isogyres disappear within 15° to 30° of stage rotation, the $2V$ is large ($>90°$).

Relation between 2V and R.I.

The optic axial angle can be calculated (approximately) when the indices of refraction have been *accurately* determined.

In uniaxial minerals when R.I.'s of $\epsilon > \omega$ the optic sign is +ve, and if $\omega > \epsilon$ it is $-ve$. In biaxial minerals if $(\gamma - \beta) - (\beta - \alpha)$ is positive, then the optical character of the mineral is +ve, and vice versa.

The $2V$ in biaxial minerals can be determined by employing the following formulae:

Positive *Negative*

$$\sin^2 V = \frac{\gamma^2(\beta^2 - \alpha^2)}{\beta^2(\gamma^2 - \alpha^2)} \qquad \sin^2 V = \frac{\alpha^2(\gamma^2 - \beta^2)}{\beta^2(\gamma^2 - \alpha^2)}$$

$$\cos^2 V = \frac{\alpha^2(\gamma^2 - \beta^2)}{\beta^2(\gamma^2 - \alpha^2)} \qquad \cos^2 V = \frac{\gamma^2(\beta^2 - \alpha^2)}{\beta^2(\gamma^2 - \alpha^2)}$$

$$\tan^2 V = \frac{\beta - \alpha}{\gamma - \beta} \qquad \tan^2 V = \frac{\gamma - \beta}{\beta - \alpha}$$

Mallard's approximation of the above formulae give:

Positive *Negative*

$$\sin^2 V = \frac{\beta - \alpha}{\gamma - \alpha} \qquad \sin^2 V = \frac{\gamma - \beta}{\gamma - \alpha}$$

$$\cos^2 V = \frac{\gamma - \beta}{\gamma - \alpha} \qquad \cos^2 V = \frac{\beta - \alpha}{\gamma - \alpha}$$

$$\tan^2 V = \frac{\beta - \alpha}{\gamma - \beta} \qquad \tan^2 V = \frac{\gamma - \beta}{\beta - \alpha}$$

The error in calculated optic angle increases greatly with a small increase in the error of the R.I. values. This error increases greatly with a smaller value of $2V$. If the R.I.'s are determined with an accuracy of ± 0.001, the error in $2V$ of range $65° - 90°$ is of the order of $2 - 3°$. This error is reduced to a negligible value when the accuracy is in the range of $\pm 0.001 - 0.0002$.

Troger (1952) prepared a nomogram chart using the equation:
$$\tan^2 V = \frac{\gamma - \beta}{\beta - \alpha} = q$$

Using the quotient q, an approximate value of $2V$ can be directly read from the nomogram.

2.16 DISPERSION

Different wavelengths of light (with different refractive indices) have different paths in a mineral. This results in dispersion.

In an isotropic medium the path difference is gradual from red to violet wavelengths. Therefore the velocity surfaces are spheres with different radii.

For uniaxial crystals the refractive indices of slow and fast vibrations differ with different wavelengths. The size and shape of the indicatrix also change with wavelength but because the crystallographic orientation does not change, the dispersion cannot be seen in a uniaxial interference figure.

In biaxial minerals variations in R.I.'s with wavelength change may cause change in the shape of the indicatrix and the angular separation between the two optic axes (i.e. $2V$). This dispersion causing angular change between the optic axes is called *optic axis dispersion*.[†]

In orthorhombic crystals the optic axial angle may vary for different colours of light. This is known as *rhombic dispersion*. An interesting case is orthorhombic brookite (TiO_2), where a change of wavelength causes its 2V to become zero and a further change of wavelength causes its $2V$ to open out again in a plane at right angles to the original one (Fig. 2.18).

In monoclinic crystals, bisectrix dispersion takes several forms, depending on which optical direction is fixed parallel to the b-axis.

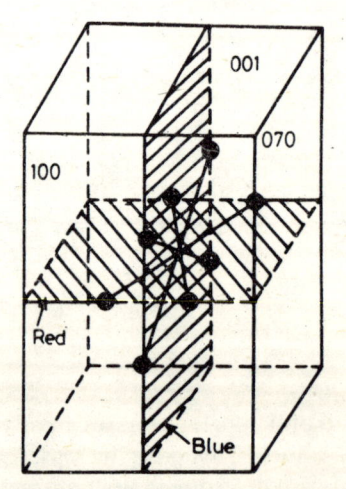

Fig. 2.18 Dispersion of the axial angle in brookite.

[†] Due to dispersion, the optic axial angle measured in white light is not true 2V of the crystal. Some prefer to denote the measured optic axial angle as 2E.

In triclinic crystals, since no optical directions are fixed, additional forms of bisectrix dispersion are generated.

2.17 UNIVERSAL STAGE

In the universal stage (U-stage), the principal planes of the indicatrix can be oriented perpendicular to the microscope axis (Fig. 2.19). The optic axis orientation is determined by seeking the directions of "isotropism". In U-stage the specimen can be rotated around several axes (3 to 5 axes). This stage is used for advanced mineralogical and petrological studies, as in the following cases:

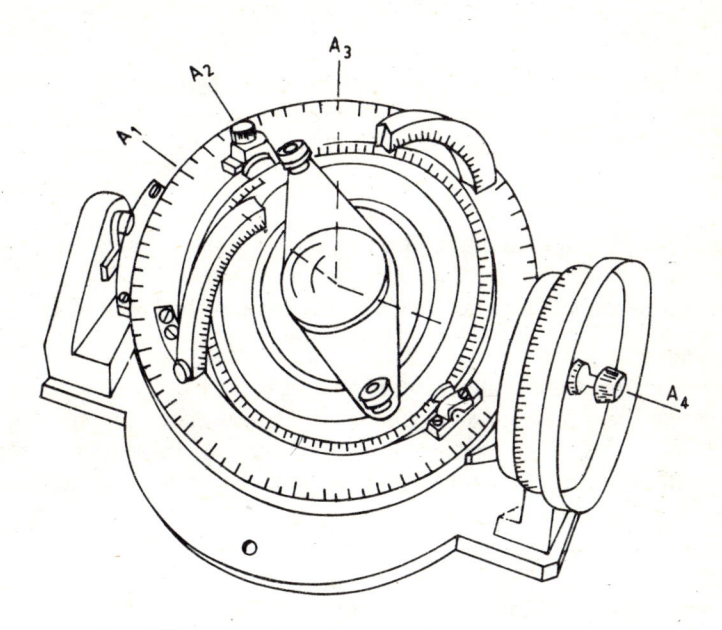

Fig. 2.19 A universal stage with rotational axes shown.

i) The uniaxial optic axis and the biaxial optic axes and the directions X, Y, and Z can be determined in space accurately with reference to any crystallographic directions like cleavage, twinning, etc.

ii) $2V$ can be measured accurately.

iii) Preferred orientation of mineral grains in a rock can be measured. This is done by determining the orientation of the optical indicatrix for a large number of mineral grains in a thin section. The results are plotted on a stereogram to assess the degree of preferred orientation.

Accurate determination of $2V$ by universal stage becomes an important exercise for research workers in petrology because it helps in identifying the problematical mineral (or its species); reporting precise mineralogical data; in modal analysis of rocks; determination of species of minerals, viz. anorthite content of plagioclase or fayalite content in olivines, or species of carbonate minerals; identifying the twin-laws of plagioclase, etc.

2.17.1 The U-Stage Components

The universal stage consists of several (three to five) axes of rotation, two hemispheres and a circular glass plate.

There are two types of universal stage: (a) Four-axis Federov universal stage and (b) five axis Emmons Universal Stage.

The axes, components and operation of a U-stage are described below.

The Axes of Rotation

The axes can be designated as two vertical axes of rotation, A_1 and A_3 and two horizontal axes of rotation, A_2 and A_4. The microscope stage axis is vertical and is called A_5. Normally A_3 is clamped at zero position and not usually used.

A_1	inner stage	} initially vertical
A_3	outer stage	
A_2	N–S axis of tilt of inner ring	} initially horizontal
A_4	E-W control axis	

The universal stage, manufactured by Leitz Wetzlar, Germany, has four axes of rotation. In the five-axis universal stage, there is another extra N-S axis which permits the rotation of the stage on both E-W and N-S axis.

Hemispheres

Glass hemispheres are manufactured with a variety of *R.I.* Depending on the *R.I.* of the minerals to be studied the hemispheres are chosen. Two hemispheres are put above and below the thin section. These help in reducing the high degree of reflection and refraction of light from the tilted mineral section.

A small drop of glycerine is placed on the circular glass plate supplied with the universal stage. The lower hemisphere is placed on the drop. The hemisphere with the plate is placed on the U-Stage and a spring supports this lower hemisphere against gravity.

The thin section is placed on the plate with glycerine in between. On the top of the section a drop of glycerine is put and the upper hemisphere is fixed on it with a screw.

For petrofabric work an upper hemisphere with a square edge is used. Often this is fitted with a slide guide.

If the angle of tilt is greater than 30–40°, or the difference in *R.I.* is greater than 0.10, corrections in readings become necessary. The corrections are made by using the Federov diagram. The observed angle is plotted along a radial line which intersects the concentric line of the particular value of *R.I.* of the mineral. A vertical line is drawn. The radial line drawn through the intersection point gives the corrected or true angle.

Illumination

For accurate measurement of the optic directions (extinction position etc.) a good parallel beam of light is required. For this, special objectives fitted with diaphragms are used. These diaphragms and those placed below the stage produce a good parallel beam of light. This set-up reduces the reflection from the hemispheres.

Adjustment

The U-stage is adjusted by following the steps as:

1) The microscope stage is centred by rotating the A5 and using the centring screws of the microscope.

2) A5 is clamped and A1 is rotated and centred using the centring screws on the U-stage.

3) A4 and A2 are rotated along the axes which lie in the plane of the mineral section. If they do not, the image will appear to move forward and backward on rotation. Adjustment is done by turning the threaded mount of the lower hemisphere.

4) The objective lens is focussed on specks of dust on the top or bottom of the glass sphere. As A_4 is rotated, the dust particles should move north and south parallel to the cross-wire. If this motion is not parallel, the microscope stage (A_5) is rotated until it is. This is the zero position of A_5. The position is noted and clamped.

Rotation about A_2 now would move the specks parallel to the E-W cross-wires.

Rotation on E-W axis would show whether the speck moves parallel to N-S axis or not. If it moves diagonally to the cross-wires approximate rotation is given on the microscope stage to the left or right so that the speck moves parallel to N-S axis.

For centering with vertical axis a point on the thin-section is focussed with the help of basal and objective screw. The inner vertical axis and the microscope stage is rotated to see whether it remains centric on vertical axis or not. Objective screws are operated to adjust the microscope stage.

Limitations of U-stage

There are several limiations of universal stage: (a) The limited space between upper glass hemisphere and objective restricts the angle of rotation (not beyond 50°). (b) The higher magnified objective cannot be used due to the presence of large glass hemisphere. (c) In the absence of using high power objective and condenser interference figure cannot be observed.

Recording

The three dimensional data can be represented on a two dimensional plane surface by the use of a projection, usually on a stereographic net. The method is discussed in detail by Fairbairn (1949), Phillips (1971) and others. The other methods of projections are gnomonic and orthographic.

2.17.2 Measurement of 2V

The following steps are usually followed to measure 2V of a mineral:
(a) The grain with lowest interference colour is chosen.
(b) All the axes of the stage are brought to zero position
(c) With respect to horizontal and vertical axes the stage is made centric.
(d) The grain is rotated on inner vertical axis to an extinction.

(e) East-West horizontal axis is rotated; when the extinction is not disturbed, the vertical plane contains XY, YZ or ZX, but if the extinction is disturbed, a small rotation is given on the E-W horizontal axis to bring the grain into extinction position.

(f) Rotation on the N-S axis is done to see if the extinction is disturbed. If so, the last three stages are repeated till the grain remains extinct on rotation.

In such a position one of the optic symmetry planes, XY, YZ, or ZX is vertical.

(g) To identify if that plane is optic axial plane having X and Z, the stage is rotated 45° anticlock-wise. On rotating along E-W axis, if the grain shows extinction, the plane should contain one of the optic axes. The optic axis in this position lies vertical. These two extinction positions are noted and 2V is measured accordingly on the E-W axis.

(h) If it is the optic axial plane, the nature of the vibration direction is determined with the help of an accessory plate. The coordinates of a second optic symmetry plane are recorded from both E-W and N-S horizontal axes and also from the inner vertical axis. Coordinates of the lines of intersection of the two planes and the optic axial angle may be obtained by plotting this data on a stereographic net.

2.17.3 Petrofabric Analysis

Determination of the directions of optic axes of uniaxial minerals in relation to the rock fabric is done as follows.

(a) A grain showing the lowest interference colour is chosen. This section should be nearly normal to the optic axis.

(b) All the axes of the universal stage are brought to the zero position.

(c) By rotating the inner vertical axis the grain is brought to an extinction.

(d) On the E-W axes the grain is rotated and if the grain remains extinct the optic axis must lie on N-S vertical plane. If it does not remain extinct but shows some interference colour rotation is made on the inner vertical axis till extinction.

(e) The grain is rotated 45° anticlockwise on the microscope stage and again rotated on E-W axis till extinction, when the optic axis lies vertical.

(f) The coordinates of optic axes are noted on E-W horizontal axis and inner vertical axis.

At such a position with optic axis vertical the sign can be determined by rotating the grain a small angle from the extinction position when it gives a first order colour. Using the gypsum accessory plate the fast and slow directions are determined and thence the optic sign.

2.18 PETROGRAPHIC STUDIES

By use of polarising microscope petrographic studies are done following some steps. The optical properties of individual minerals are recorded by studying first in plane polarised light (viz. colour and pleochroism, refractive index, shape and habit, cleavage, fracture, inclusions and alterations etc.). This is followed by noting the properties seen between crossed nicoles (viz. interference colour and its order, zoning and twinning, optic sign etc.).

By determining the refractive indices, the order of birefringence colour and the 2V (determined from refractive indices, *vide* section 2.15 and by use of a universal stage), the *species* of a mineral in a solid solution series can be determined with a fair degree of accuracy.

The grain size of an individual mineral in a rock is described on the basis of the average diameter as:

Grain size	Average diameter
fine	< 1 mm
medium	1 – 10 mm
coarse	1 – 3 cm
very coarse	>3 cm

The mineral *shapes* are classified as equidimenional, acicular or needle-shaped, prismatic or elongated, rhombic or tabular, fibrous, skeletal etc.

The *roundness* of grains is identified within the series: angular-subangular-subrounded-rounded-well rounded.

2.18.1 Microtextures

In petrographic studies of rocks the microtextural characters of minerals are described using some prefixes (Table 2.1) and suffixes (Table 2.2). The connotations of which are listed with examples.

Table 2.1 Prefixes used in describing microtextures

Prefix	Textural connotation	Example
A	Without, not	Aphyric
Allotrio	foreign, alien, without	Allotriomorphic
An	without, not	Anhedral
Auto	of itself	Automorphic
Crypto	hidden, not apparent	Cryptocrystalline
Eu	Well-developed	Eutaxitic
Holo	Completely, entirely	Holocrystalline
Hypo (Hyp)	nearly, incompletely	Hypocrystalline
Hyalo	glassy	Hyalopilitic
Idio	one's own	Idiomorphic
Inter	between	Intersertal
Macro/Mega	large	Macrocrystalline
		Megacrystalline
Ortho	rectangular, straight	Orthopyric
Pan	entirely, all	Panidiomorphic
Phanero	distinct, visible	Phaneritic
Pilo	felt-like	Pilotaxitic
Poiki	spotted	Poikilitic
Pseudo	false,	Pseudoporphyritic
Sub	under, below, less than	Subhedral
Vitro	glassy	Vitrophyric
Xeno	foreign, strange	Xenomorphic

Table 2.2. Suffixes used in describing microtextures

Suffix	Textural connotation	Example
—blastic	a growth, sprout	Crystalloblastic
—hyaline	glassy	Holohyaline
—morphous	form, structure, shape	Amorphous
—oid	similar to	Porphyroid
—phyre	porphyry, porphyritic	Vitrophyric
—pilitic	felt-like	Hyalopilitic

2.18.2 Some samples studies: examples:

Lunar samples, because of their eternal freshness (free from chemical weathering), offer petrographic colours of primary minerals far brighter compared to the terrestrial samples, in which the primary minerals are much affected by later solution actions since their period of crystallisation. Some lunar rocks* are described below as illustrations.

Microphotographs of a coarse-grained high titanium ilmenite basalt (Apollo-17, sample no 70017) consisting of large plagioclase crystals (An = 88-76; upto 5 mm across), that poikilitically enclose clinopyroxene, olivine and ilmenite (Figs 2.20 a, b). The clinopyroxene phenocrysts in turn enclose small crystals.

---

*These samples were available from NASA Johnson Space Centre, through cooperative agreement with the concerned Institutes.

Study in Polarised Light **51**

Fig. 2.20 Microphotographs of ilmenite basalt Apollo-17. Sample No. 70017 with poikilitic plagioclase, clinopyroxenes and opaques. (All views are 3.48 mm wide)

  a) Subhedral rectangular prismatic clinopyroxenes of tan colour enclosing opaque minerals which are ilmenite, armalcolite, troilite and iron (in polarised light)

  b) Under crossed nicols the twinning in the long prismatic crystals of plagioclase and sector zoning in pigeonite to subcalcic augite are seen.

References and Selected Reading

Battey, M.H. (1972) Mineralogy for students, Oliver & Boyd, Edinburgh, 323p.
Bloss, F.D. (1961) An Introduction to the Methods of Optical Crystallography, Holt, Rinehart and Winston, Inc., New York, 294p.
Fairbairn, H.W. (1959) Structural Petrology of Deformed Rocks, Cambridge, Mass.
Kerr, P.F. (1959) Optical Mineralogy, 3rd edition, McGraw-Hill Book Co., New York, 442p.
Phillips, F.C. (1971) The Use of Stereographic Projection in Structural Geology, Edward, Arnold, London, p. 90.
Phillips, W.R. (1971) Mineral Optics, Principles and Techniques, W.H. Freeman and Co. San Francisco, 249p.
Shelley, D. (1975) Manual of Optical Mineralogy, Elsevier Scientific Publishing Co. Amsterdam, 239p.
Troger, W.E. (1952) Ein neues Nomogramm zur Bestimmung des optischen Achsenwinkels. *Heidelberger Beitrage Zur Mineralogie und Petrographie, 3,* 44.
Wahlstrom, E.E. (1969) Optical Crystallography. 4th edition, Wiley, New York, 489p.
Winchell, A.N. and Winchell, H. (1951) Elements of Optical Mineralogy: Part I, Principles and Methods, 5th edition, Wiley, New York, 263p.

CHAPTER III

Optical (Absorption) Spectroscopic Studies of Minerals

3.1 INTRODUCTION

To appreciate the cause of mineral colour and such other important aspects as the distribution of elements in different sites in a mineral or in a group of co-existing minerals, to evaluate of thermodynamic properties etc., a background knowledge in optical. absorption spectroscopy is deemed necessary. This technique for mineralogical investigation is new compared to polarised-light or X-ray studies. The literature on mineralogy since the sixties amply demonstrates its overwhelming importance in studies of minerals, glasses and synthetic compounds, in the study of extraterrestrial materials obtained as moon-samples, meteorites or cosmic dust.

A brief outline of the principle and technique of optical absorption spectroscopy is presented in the later sections with special reference to the first series of transition elements, which are ubiquitously distributed in a large number of minerals. Let us now study the electronic properties of elements and their interactions when in combination.

3.2. ELECTRONIC BUILDING OF ELEMENTS

In the building of elements in the periodic table, three principles are followed. The first of these is called the *'aufbau'*, according to which the electrons are put into orbitals in order of energy. The lowest energy orbitals are filled first before electrons are placed in successively higher energy levels. The order of orbital energies is

$1s < 2s < 2p < 3s < 3p < 4s < 3d ...$

The second principle, known as *Pauli exclusion principle,* states that a maximum of two electrons can occupy an orbital and no two electrons in an atom can have the same set of the four quantum numbers (i.e., n, l, m, s). But two electrons in an orbital may have parallel or opposite (parallel) spins.

The third principle is known as *Hund's rule.* According to this, when a set of degenerate (i.e., same energy) orbitals is to be filled, the first electrons enter different members of the set with parallel spin, and only when each degenerate orbitals contain one electron do the remainder enter the orbitals to form paired spins.

The rules can be illustrated with the electronic structure of nitrogen atom, $1s^2 \, 2s^2 \, 2p^3$ as in Fig. 3.1.

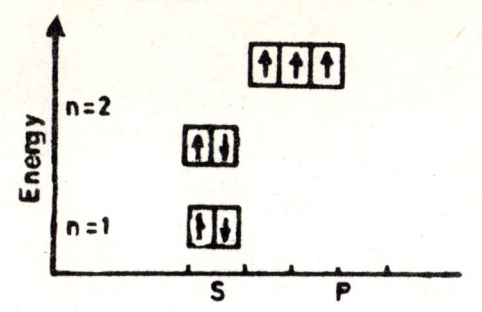

Fig. 3.1   Electronic structure of nitrogen.

A high-spin configuration forms when the electrons get distributed to more orbitals (i.e., less paired). A low-spin configuration with more paired spins occurs in a strong crystal field.

3.3   TRANSITION ELEMENTS

The transition series elements have partly filled d-or f-shells.
The first transition series has the electronic configuration as

$(1s^2)(2s^2)(3s^2)(3p^6)\ 3d^{10-n}(4s)^{1(or\ 2)}$
←– Argon core –→

In an isolated ion of the transition metals, the five $3d$-orbitals are energetically equivalent (called five-fold degenerate), and following *Hund's Rule*, the electrons tend to minimise their mutual electronic repulsions by occupying singly as many orbitals as possible with spins oriented in the same direction. The $3d$-orbitals are shown in Fig. 3.2.

In an octahedral coordination, these five orbitals fall into two groups, t_{2g} and e_g. In tetrahedral environments the subscript g is omitted because a tetrahedron lacks an inversion centre (*gerade*) and the relative energies of t_2 and e orbitals are reversed (in relation to the octahedral configuration).

In t_{2g} group the three d_{xy}, d_{yz} and d_{xz} orbitals have the lobes projecting between the three cartesian axes; while the second e_g group has two orbitals viz., d_{z^2} and $d_{x^2-y^2}$, which have lobes directed along the cartesian axes.

In a crystalline field the degeneracy of the d-orbitals are lost. Coulomb and exchange interactions between electrons cause them to be distributed over as many orbitals as possible so that a maximum number of unpaired electrons with parallel spin can exist (*Hund's rule*).

Again, if in a metal ion one of the d-orbitals is empty and other of equal energy (viz. in e_g group) is half-filled then it is predicted by *Jahn-Teller Effect* that the configuration would distort spontaneously to a geometry (more asymmetric) in which a more stable electronic configuration is achieved by making the occupied orbital lower in energy. In minerals containing Mn^{3+} viz. bixbyite (γ-Mn_2O_3), manganite ($MnO(OH)$) and hetaerolite ($ZnMn_2O_4$) and in minerals with Cu^{2+} viz. malachite and tenorite, the crystal structures are distorted by *Jahn-Teller Effect*.

Electrostatic repulsion of d-electrons varies with direction, and electronic transitions have greater probability in a certain direction. This results in

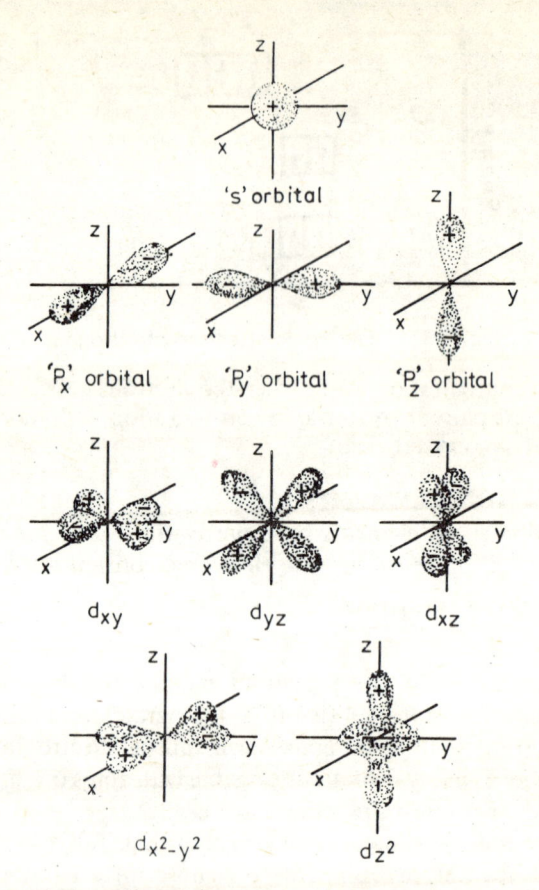

Fig. 3.2 Boundary surfaces of atomic orbitals; the lower five are d-orbitals.

preferential absorption in a polarised light (i.e., pleochroism). Crystal Field Theory (CFT) yields information on the probability of such a transition.

3.3.1 Transition metal ions (Ti, V, Cr, Mn, Fe, Co, Ni)

Ti^{3+}, V^{3+} and Cr^{3+} ions have three d-electrons. They can have only one electronic configuration and d-electrons occupy orbitals singly with spins parallel.

However, ions possessing d-electrons which number four (Cr^{2+}, Mn^{3+}), five (Mn^{2+}, Fe^{3+}), six (Fe^{2+}, Co^{3+}) and seven (Co^{2+}, Ni^{3+}) have a choice of configuration of electrons in the two groups of orbitals.

In a weak field the d-electrons try to maximise the occupancy of all the orbitals in both the groups by singly occupying with parallel spins. Pairing is thereby avoided. But in a strong field (when the crystal field splitting is large) it becomes energetically more favourable for d-electrons to fill low energy t_{2g} orbitals, when pairing occurs. In such a case the pairing energy is less than the Crystal Field Stabilisation Energy (CFSE).

In cases where d-electrons number as much as eight (Ni^{2+}), nine (Cu^{2+})

and ten (Zn^{2+}) only one electronic configuration is possible in each case; and d-orbitals are filled gradually. In a weak crystal field the ions are in high-spin state while in a strong field they are in low-spin state.

3.4. CRYSTAL FIELD THEORY (CFT)

The transition elements of the first series have five 3d orbitals, which are unfilled.

In a free ion of a transition element these five 3d-orbitals are degenerate (i.e., have the same energy), but when the ion is complexed the ligands remove the spherical symmetry of the atom and replace it by a field of lower symmetry. The immediate environment around the ion may be tetrahedral, octahedral etc. In a crystal field the ligands are regarded as point sources of electrical potential. Because of the interaction with the electrical field of the ligands, d-orbital energies loose the degeneracy.

In an octahedral field the $d_{x^2-y^2}$ and d_{z^2} orbitals (i.e., e_g) point directly towards the ligands, whereas the other three (t_{2g}) d-orbitals point in mutually equivalent direction in between them (i.e., between the reference coordinate axes). Electrostatic repulsion, therefore raises the energy levels of the $d_{x^2-y^2}$ and d_{z^2} orbitals, while it lowers the energy levels of the d_{xy}, d_{xz} and d_{yz} orbitals.

In tetrahedral and cubic coordinations, the situation is reversed and $d_{x^2-y^2}$ and d_{z^2} orbitals acquire lower energies with respect to the other three d-orbitals. The separation between these two energy levels is called crystal field splitting and is designated as Δ_o (or 10dq).

In an octahedral environment each electron in one of the lower three d-orbitals stabilises a transition metal ion by $\frac{2}{5}\Delta_o$. Summation of the Δ_o terms gives the *Crystal Field Stabilisation Energy* (CFSE) for a particular ion. This is explained in the following section with Cr^{3+}, showing how CFSE changes with the legand symmetry around the metal.

The relative energies of the crystal field splittings in cubic (Δ_c) tetrahedral (Δ_t), octahedral (Δ_o) are related as:

$$\Delta_c = -\frac{8}{9}\Delta_o$$
$$\Delta_t = -\frac{4}{9}\Delta_o$$

as shown in Fig. 3.3

Example: Cr^{3+} ion

The Cr^{3+} ion in an octahedral field has three 3d electrons occupying singly each of the three t_{2g} orbitals. It, therefore, acquires a very high CFSE of $\frac{6}{5}\Delta_o$ (about 60 Kcal/Cr^{3+} in oxide structures). In tetrahedral environments, the relative energies of the t_2 and e orbital levels (subscript g is omitted because tetrahedrons lack inversion centres) are reversed relative to octahedral coordination. The third 3d electron, therefore, occupies the t_2 orbital and contributes destabilization energy to the Cr^{3+} ion. The net CFSE $\frac{4}{5}\Delta_t$, of Cr^{3+} in tetrahedral coordination would be about 30 percent less than the CFSE of octahedrally coordinated ion. For this reason, Cr^{3+} ions have a high preference for octahedral sites in geochemical and mineral assemblages.

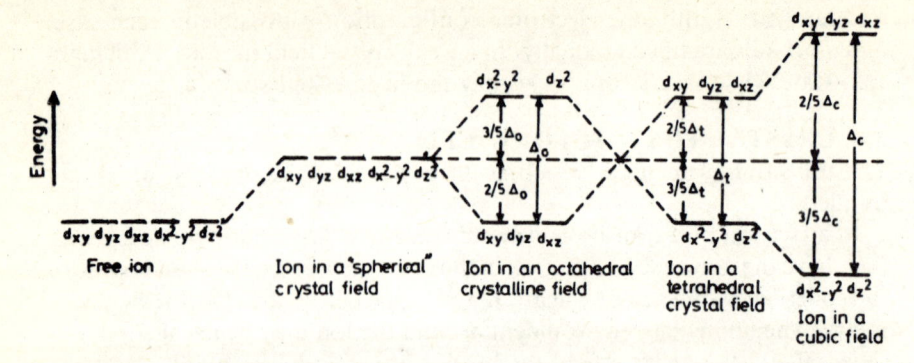

Fig. 3.3 Crystal field splitting. A free ion has d-orbitals of the same energy. A spherical crystal field raises the energy levels. Octahedral, tetrahedral, and cubic crystal fields split the d-orbitals by an amount. Because the energy values lie in the visible and near visible region colours are generated.

Cr^{3+} ion acquires a higher CFSE in a distorted (compressed) site, as offered by an Al^{3+} site, compared to Mg^{2+} site, in an oxide. CFSE of Cr^{3+} in octahedral sites of minerals are of the order spinel > garnet > pyroxene > olivine. No wonder that olivine enclosing chromite (poikilitic) are poor in chromium.

Cr^{2+} ion, $3d^4$, is stabilised in a distorted octahedron because its fourth $3d$- electron occupies the more stable orbital of the e_g group.

3.5. WEAK FIELD AND STRONG FIELD

In a weak crystalline field, the crystal field splitting is small and $3d$-electrons can occupy singly all the t_{2g} and e_g orbitals. In this case ions possess more unpaired electrons and are called *high-spin state*. (Fig. 3.4.1).

But if the field is strong the crystal-field-splitting becomes large, the $3d$-elctrons first fill the energetically more favourable low-energy t_{2g} orbitals. When the electrons number more than three, the fourth, fifth and so on electrons start pairing with the existing t_{2g} electrons rather than going to eg. This happens because the energy for electron pairing is lower than CFSE increase. In this case ions have less unpaired electrons and is called *low-spin state*. Ions having three, six and eight d-electrons (in low-spin state) acquire high stabilisation in octahedral crystal fields, whereas ions possessing zero, five and ten d-electons have zero CFSE in weak octahedral fields.

The energy of spin pairing is greater than the CFSE in silicate and oxide minerals. The distribution of electrons in transition elements in octahedral high-spin configurations are given in Table 3.1.

Transition metal ions will preferentially get into sites which give the lowest CFSE. Distortion in the coordination polyhedra (i.e., symmetry of the environment) reduces the degeneracy of the d-orbitals. Figure 3.4.2 shows the arrangement of ligands and relative energy levels of d-orbitals of transition metal ions in tetragonally distorted octahedral sites. Distortion in the symmetry of the environment affects the energy difference, Δ_o, between the

Table 3.1
Distribution of electrons in d orbitals and CFSE of First Transition Series ions in Octahedral coordination (high spin state) (after Burns, 1970)

Number of d electron	Ions	d_{xy}	d_{xz}	d_{yz}	$d_{x^2-y^2}$	d_{z^2}	Number of unpaired electrons	CFSE
0	Ca^{2+}, So^{3+}, Ti^{+4}						0	0
1	Ti^{3+}	↑					1	$\frac{2}{5}\Delta_o$
2	Ti^{2+}, V^{3+}	↑	↑				2	$\frac{4}{5}\Delta_o$
3	V^{2+}, Cr^{3+}, Mn^{4+}	↑	↑	↑			3	$\frac{6}{5}\Delta_o$
4	Cr^{2+}, Mn^{3+} *	↑	↑	↑	↑		4	$\frac{3}{5}\Delta_o$
5	Mn^{2+}, Fe^{3+} *	↑	↑	↑	↑	↑	5	0
6	Fe^{2+}, Co^{3+}, Ni^{4+} *	↑↓	↑	↑	↑	↑	4	$\frac{2}{5}\Delta_o$
7	Co^{2+}, Ni^{3+} *	↑↓	↑↓	↑	↑	↑	3	$\frac{4}{5}\Delta_o$
8	Ni^{2+}	↑↓	↑↓	↑↓	↑	↑	2	$\frac{6}{5}\Delta_o$
9	Cu^{2+}	↑↓	↑↓	↑↓	↑↓	↑	1	$\frac{3}{5}\Delta_o$
10	Zn^{2+}, Ga^{3+}, Ge^{4+}	↑↓	↑↓	↑↓	↑↓	↑↓	0	0

* also has a low-spin configuration.

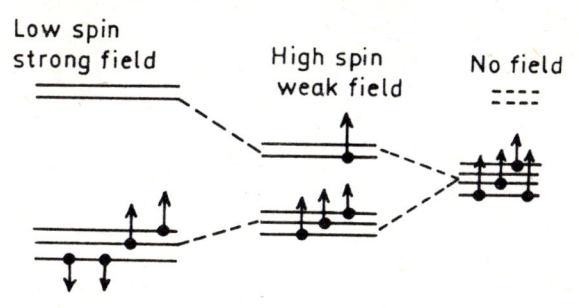

Fig. 3.4.1 Free ion in low spin strong field, high spin weak field, and no field.

lower and upper energy orbitals. This energy difference depends on the number of d-electrons, the coordination number and symmetry of the environment, the metal-anion bond lengths, and the charge of the cation.

The Δ values for many transition metals are of energies which fall in the range of visible and near-visible electromagnetic spectrum. For this reason, many minerals containing these elements are coloured. This has been discussed earlier in Section 1.5.

Crystal field theory is an approximation because it supposes that the energy of ligand-metal bonding is due solely to electostatic effects and ignores the covalent nature of the bonding and the role that π-bonding might be expected to play. These deficiencies are overcome in the broader ligand field theory.

Nevertheless, the CFT is highly successful in depicting the importance of

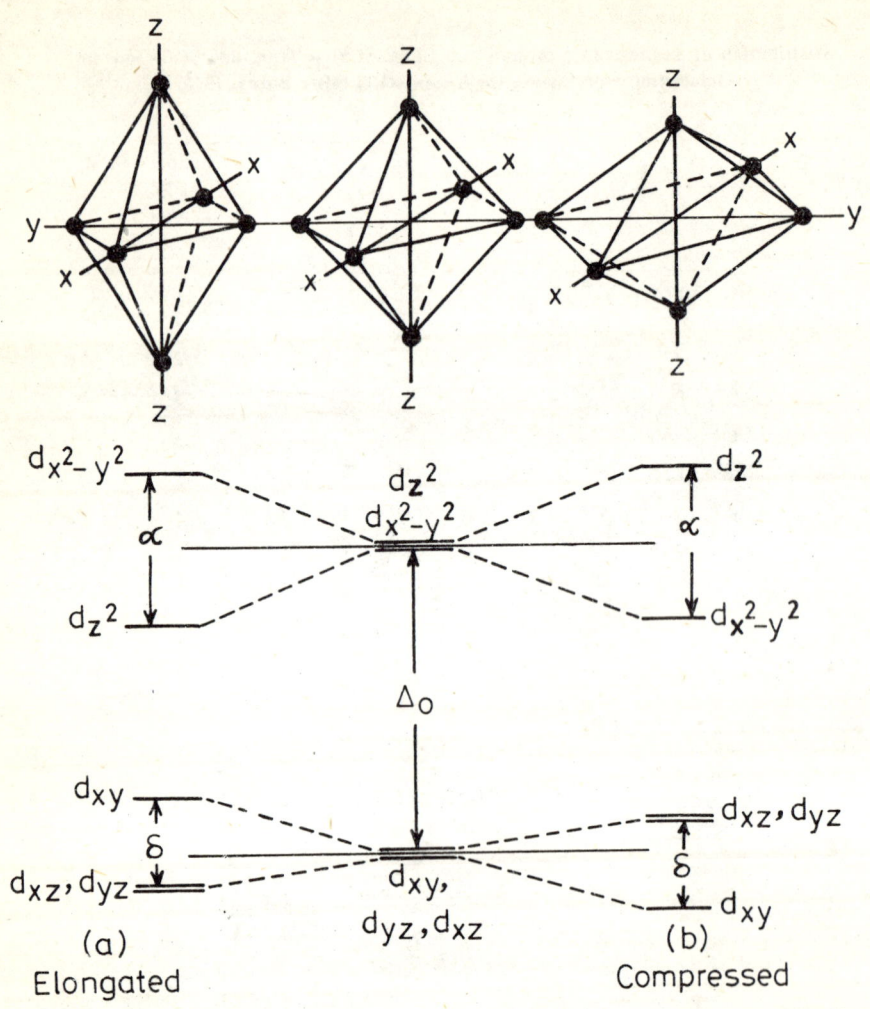

Fig. 3.4.2 (a) Octahedral (transition metal ion) site elongated along the tetrad axis (Z-axis) (b) The same site compressed along the tetrad axis. The arrangements of ligands and energy levels for a regular octahedral site is shown at the middle for reference.

the symmetry of the complex in determining the electric, magnetic and chemical properties of transition metal complexes.

3.6 CFT, MOT AND VBT

Crystal field theory was first developed in 1929 by Bethe, who considered that only the ligands (anions) were responsible for producing a steady crystalline field. This theory has been successfully applied to explain the chemical and physical properties of elements and the compounds of the first transition series of elements viz., Sc, Ti, V, Cr, Mn, Fe, Co, Ni and Cu. These transition elements occur in most minerals and constitute about 18 atomic percent and 40 weight percent of the earth. Study of their distribution in lattice positions in minerals, therefore, makes it as important to understand the thermodynamics of the formation and stability of the host minerals.

In mineralogy and geochemistry, CFT and spectral measurements are used to understand colour and pleochrosim of minerals, distribution of transition metals in different atomic sites and between different minerals in the earth, and even properties of the earth's mantle.

CFT splendidly describes the effects of perturbation of d-orbitals of a transition metal ion in a lattice. It, however, disregards the magnetic and exchange forces in the first approximation.

A transition metal ion in a crystal is subjected to the electric field of the surrounding negatively charged ions or dipolar groups (*ligand*) which, in this theory, are represented by point negative charges. In contrast, the *Molecular Orbital Theory* (MOT) takes into account the effects of the overlap of metal and ligand orbitals on the energy levels of the transition metal compounds. The *Valence Bond Theory* (VBT) is similar to the MO Theory and is also concerned with the covelent bond formation.

Molecular Orbital Theory assumes that electrons occupy polycentric orbitals which are filled in the same way as atomic orbitals. The problem lies in solving the Schrodinger equation for each electron in these molecular orbitals. An introduction to the method can be obtained from Cotton and Wilkinson (1972), and Murrell, Kettle and Tedder (1970). Commonly the overlaps of atomic orbitals are considered. The overlap of metal and ligand orbitals depend on the symmetry properties and is best described by the Group Theory.

Qualitative molecular orbital models and band models have been used in studies of oxides and sulphide minerals and compounds (Goodenough, 1963, 1969, 1972) and thiospinels (Vaughan *et al.*, 1971).

Molecular Orbital Theory is capable of describing bond polarities from the totally covalent to totally ionic, and can be applied to crystalline solids (Tossel, Vaughan & Johnson, 1974).

The energy band theory describes the system in which the outer electrons are delocalised i.e., not associated with a particular atom but free to move throughout the lattice. MOT is capable of describing both localised and delocalised electron behaviour; while CFT and ligand field theory can describe only the localised electrons.

3.7 CFT IN MINERALOGY: A RESUME
Summarising this discussion, we now can see that CFT greatly helps in understanding the following aspects of mineralogy.

1. Colour and pleochroism: Distortion in the environment around transition metal ions cause differential absorption of polarised light in different directions causing pleochroic colours.

2. Fe/Mg ratio in coexisting ferromagnesian silicate minerals: The relative orders of CFSE of Fe^{2+} in different structural environments of Fe^{2+} or Fe/Mg ratio in some minerals as compared to others.

3. Metal-ligand distances and magnetic properties: The number of unpaired electrons in transition metal ions and the strength of the crystal field in a mineral controls its magnetic property.

The internuclear distances between the metal ion and the surrounding ligands depend on the electronic configuration. High crystal field leads to shorter internuclear distances and lesser magnetic property. The high-spin pyrrhotite ($Fe_{1-x}S$) has greater Fe–S distances as compared to those Fe-S distances present in low-spin pyrite (FeS_2). This explains why pyrite is diamagnetic while pyrrhotite is paramagnetic.

4. The geochemical behaviour of transition metals in igneous, metamorphic, edimentary and aqueous environments is well-explained by the application of crystal-field theory.

3.8 FACTORS CONTROLLING THE CRYSTAL FIELD SPECTRA

The crystal field spliting parameter, Δ, can be measured from the absorption spectra. As discussed earlier in the colour of minerals, the energy to excite an electron from one d-orbital to a higher energy state corresponds to the visible or near infra-red region of the electromagnetic radiation. The absorption of this radiation generates the colour in transition metal bearing minerals or compounds. The splitting parameter, Δ, may be related to the factors as discussed now.

1. The valence state of the cation: Crystal field splittings are larger for higher valence states. The Δ values are, therefore, lower in divalent ions than in the trivalent ions of the same metal (i.e. $\Delta M^{3+} > \Delta M^{2+}$).

2a. The coordination number (N) of the cation: Increase in N results in a decrease in Δ. e.g. tetrahedral ($N=4$) > octahedral ($N=6$) > cubic ($N=8$) $\geq$ dodecahedral ($N=12$).

  b. *The coordination symmetry:* Δ depends on the symmetry of the coordinating oxygen ligands; the CFSE for tetrahedral coordination is 50-60% less than those of the octahedral coordination. Octahedral Δ_o > Cubic Δ_e >> tetrahedral Δ_t. Distortion from symmetric polyhedra (viz., regular octahedron, tetrahedron etc.) generates additional bands.

3a. Δ is dependent on the ligands and the degree of covalent bonding.

  b. In order of increasing Δ, the ligands may be arranged in a series (called *spectrochemical series*) as:

$I < Br^- < Cl^- < F^- < OH^- \leq$ carboxyions $\leq H_2O < NH_3 < SO_3^{2-} < NO_2^- << CN^-$

The series starts with the generation of weak crystal fields (around transition metal ions) followed by low-spin configurations at the central series while the end of the series generate strong crystal fields (often leading to low-spin states). When an ion crosses over from a high-spin to low-spin state, the internuclear distance is greatly reduced due to ligand configuration.

4. *Oxygen-bonding distance, R:* The Δ is inversely proportional to the fifth power of R, the cation-oxygen bond length ($\Delta \propto 1/R^5$). Increase in pressure reduces R and the absorption bands move to shorter wavelengths ('blue shift') and Δ increases. High temperature has reverse effects.

3.9 CFSE IN Fe DISTRIBUTION IN SILICATES

CFSE of ions determines the thermodynamic properties viz., lattice energies and heats of hydration etc. Determination of CFSE by optical spectroscopy has revealed that major Fe-Mg silicate solid solution systems (viz. olivine, pyroxene and amphibole series) hardly conform to the ideal-solution behaviour. A necessary criterion for a solution to be ideal is that the heat of mixing of the components is zero. But CFSE measurements of ions (Fe^{2+}, Mg^{2+}) in different sites in the lattice are found to be different. The distribution of Fe^{2+} ions in silicates are governed by the relative orders of CFSE of the Fe^{2+} ion in ferromagnesian silicate crystal structure, and the following order of decreasing relative enrichment is predicted in Mg-rich phases: olivine $M_2 \geqslant$ olivine $M_1 >$ garnet $>$ orthopyroxene $M_2 >$ orthopyroxene $M_1 >$ pigeonite $M_2 >$ pigeonite $M_1 >$ cummingtonite $M_4 >$ cummingtonite $M_1, M_2, M_3 \geqslant$ (actinolite M_1, M_3) $>$ diopside $M_1 \geqslant$ (actinolite M_2).

Thus, Fe^{2+} - Mg^{2+} ratios of co-existing phases would be expected to decrease in the order:

Olivine $>$ garnet $>$ orthopyroxene $>$ pigeonite $>$ cummingtonite (actinolite) $>$ diopside.

This order predicted by CFT has largely been found valid by analyses of coexisting mineral samples (Burns, 1970).

3.10 OPTICAL ABSORPTION SPECTROSCOPY

When electromagnetic radiation passes through a substance certain parts of its wavelengths are absorbed. This induces electronic, vibrational and rotational transitions.

The energies of these transitions, which are equivalent to the amount of light absorbed at each energy, are shown in a spectrum obtained by an absorption spectrometer.

The energy or *position* of the absorption and the *intensity* and width of the absorption bands are characteristic of the electronic transition and the bonding involved. The absorption spectrum (also called crystal field spectrum) provides information on the electronic configuration (oxidation state, spin state), the crystal field splitting parameter (Δ), covalence of the metal-anion bond and the distortion of the coordination site.

Spectral measurements of transition metal-bearing minerals in the visible (0.4-0.77μm) and nearby infrared (0.77-3.0μm) and ultraviolet (0.2-0.4μm) regions are made to ascertain the coordination of ligands about a transition metal ion and the nature of bonding.

The position of an absorption band is measured on a wavelength scale which is calibrated in any of the units viz. $\overset{\circ}{A}$, μ or μm. The relationships between them are:

$1\mu m = 10^{-6} m = 10^{-4} cm = 10^{3} m\mu = 10^{4} \overset{\circ}{A}$.

The spectra are also recorded on a wavenumber scale, which is inversely proportional to wavelength in cm.

$1\mu m = 10^{-4} cm$ or $10,000 cm^{-1}$
$ = 10,000 Å$

Thus, $2\mu m = 5,000 cm^{-1}$.
and, $4,000 Å = 25,000 cm^{-1}$ and $20,000 Å = 5,000 cm^{-1}$
Again, in terms of energy values,
$1 cm^{-1} = 2,859 cal = 1.24 \times 10^{-4} eV$
Therefore, the wavelength range 4,000 to 20,000 Å (or 25,000 cm^{-1} to 5,000 cm^{-1}) corresponds to 71.5 to 14.3 Kcal and 3.1 to 0.6 eV.

The intensities of absorption are related to the probability of electron transitions between the various $3d$ orbital energy levels. The greater the probability, the higher the intensity of an absorption band. The probability of transitions is deduced from some selection rules.

3.10.1 Selection Rules:

1) The *Laporte selection rule* says that transitions between orbitals of the same type and quantum number, such as two $3d$ orbitals, are forbidden. The transition of a *d*-electron to a *p*-orbital is, therefore, allowed.

Transitions may occur and absorption bands gain intensity when transition metal ions occur in a co-ordination, which lacks a centre of symmetry, viz. tetrahedral etc.

In centro-symmetric octahedral sites, weak absorption bands may occur through the mechanism of *vibronic coupling*, as a result of coupling of the wave-functions of vibrational transition to electronic transitions in metal ions lying away from the centre of symmetry of the site. Intensities of such transitions increase with temperature, which increases vibrational amplitudes.

For this reason, absorption bands in non-centrosymmetric tetrahedrally coordinated Co^{2+} in cobalt spinel ($CoAl_2O_4$) are 100 times more intense than those in centrosymetric octahedrally coordinated Co^{2+} in bieberite ($CoSO_4 \cdot 7H_2O$).

2) Electronic transitions between the ground-state and excited states are forbidden when there is a change in the number of unpaired electrons (or *spin-multiplicity*). For this reason, transitions in the $3d$-electrons in high spin Mn^{2+} and Fe^{3+} are spin-forbidden, because transition from ground to excited state changes the number of unpaired electrons from five to three.

But *spin-orbit coupling* often relaxes this rule and causes weak absorption. The magnetic fields produced by an electron spinning about its own axis and rotating in an orbital interact to induce a probability of transition to an otherwise spin-forbidden transition between the half filled five $3d$-orbitals.

Hematite, and topazolite[*], containing Fe^{3+} show weak absorption in the

---

[*] a grossular garnet with composition $Ca_3(Al, Fe^{3+})_2(SiO_4)_3$.

visible region (emitting a red internal reflection). Mn^{2+} bearing rhodonite and rhodochrosite emit a pink colour due to the relaxation or break down of the spin-multiplicity selectron rule.

3.10.2 Positions of Bands

From the positions of the absorption bands the parameters that are determinded are (a) the crystal field splitting parameter, Δ, which contributes to thermodynamics properties, and (b) the Racah parameters, B and C, which provide a measure of the degree of covalent character of cation-anion bonds. Peaks for tetrahedral Fe^{2+} and Fe^{3+} cations occur at lower energies (longer wavelengths) than octahedral cations.

3.10.3. Intensities

The intensity of an absorption band can be represented by the Beer-Lambert relationship as follows:

$$\log_{10} I_o/I = E.c.d.$$

where $\log_{10} I_o/I$ is the optical density or absorbance. I_o and I are the incident and emergent light respectively, d is the thickness of the sample; c is the concentration of absorbing species (moles litre^{-1}); E is the molar extinction coefficient (in litre mole^{-1} cm^{-1}).

The molar extinction coefficient can be calculated when other values are determined. This can be illustrated with an example of olivine $(Mg_{0.88} Fe_{0.12})_2 SiO_4$, having an absorption hand at 10,500 Å. For a specimen of 0.055cm thickness its absorbance is 1.26 log.units. The molar volume of this olivine, $Fe_{88}Fa_{12}$ is calculated as 44.1 cc. mole^{-1}. Therefore the concentration, c, becomes $(1000 \times 0.12) 44.1 = 2.79$ moles litre^{-1}. The molar extinction coefficient, E, therefore is $1.26/2.76 \times 0.055 = 8.2$ litre mole^{-1}cm^{-1}.

The intensities of absorption bands depend on (a) The symmetry of the coordination site. Cations in the non-centrosymmetric environments (e.g. tetrahedral sites; in olivine distorted octahedral M_1 site) produces absorption bands one to two orders of magnitude more intense than centrosymmetric sites (e.g. M_2 site in olivine); (b) spin-forbidden transitions (as in Fe^{3+}) are typically two orders of magnitude less intense than spin-allowed transitions (in Fe^{2+}, Cr^{3+}, Ti^{3+}); (c) Increase in P and T causes increased covalency and vibronic coupling, resulting in greater intensities.

3.10.4 Widths:

Atoms are in continuous thermal vibrations in their mean positions. As they vibrate in a lattice the metal-ligand distances vary and change about a mean energy corresponding to the mean positions of the atoms. When there is a little variation in the energy separation between ground and excited states with increasing Δ, the electronic transition between these states will lead to a sharp absorption band. But if the energy separation varies over a wider range, the electron transitions will result in a broad band.

The widths can be correlated with energy level diagrams for transition metal ions as a function of increased intensity of a crystal field. Crystal field splitting, Δ, is sensitive to subtle changes of metal-oxygen distance, so that vibrations along bands leads to fluctuating metal-oxygen distances and therefore, Δ.

The other causes of line broadening are:

1) Overlapping of absorption bands of two or more similar sites in a crystal structure (e.g. Fe^{2+} in M_1 and M_2 sites in olivines and pyroxenes) or superposition of two or more bands of transition metal ions in a single low-symmetry site.

2) Overlapping of bands formed due to distortion of the coordination site.

3) Splitting of energy levels of the excited state during an electronic transition due to Jahn-Teller distortion (e.g. E_g state of Ti^{3+} and $5E_g$ state of Fe^{2+} in hexahydrates).

4) Temperature effect. Higher temperatures make the atoms vibrate with larger amplitudes about their mean positions in the structure. This causes Δ to vary over a wide range and causes line-broadening. Lower temperature sharpens the lines.

3.10.5 Instrument used

Optical absorption spectra of minerals can be measured in the range 4,000 - 25,000 Å using double-beam spectrometers (Beckman, Cary make). Minerals or rock sections of dimensions 2mm × 2mm, and thin polished (preferably on both sides) are put in the sample holder fitted with variable aperture windows. The spectra are recorded on a chart recorder.

To obtain polarised spectra a polarising microscope equipped with a 3- or 4-axes universal stage attachment (Fig. 2.19) is used. The universal stage attachment enables a crystal to be oriented accurately in any desired manner and helps accurate measurement of polarised absorption spectra along any crystallographic or vibration axis in the mineral. A thin polished slide of the sample is placed between the glass hemispheres of the universal stage. The microscope with the universal stage is placed horizontally in the sample cabinet in such a way that the radiation from the light source impinges on the Nicol-prism polariser before it reaches the sample, followed by the detectors. Another microscope is set in an identical alignment in the reference beam path with the deviation that no sample is kept on the glass slide.

In the output the ordinate shows the absorption scale and the abscissa represents the energy or wavelength.

3.11 STUDY IN POLARISED SPECTRA

Identical absorption spectra are obtained for light propagation along all directions in a cubic crystal. But, as discussed earlier, in crystals having lower symmetry coordination sites the electric vector has different properties in different directions. Therefore, different spectra would be obtained in light polarised parallel to each crystallographic axis in a non-cubic mineral. This is illustrated with some examples below.

3.11.1 Polarised spectra of olivines

Olivines are orthosilicates with individual SiO_4 tetrahdra which are linked together by divalent cations like Mg and Fe^{2+} in octahedral coordination with

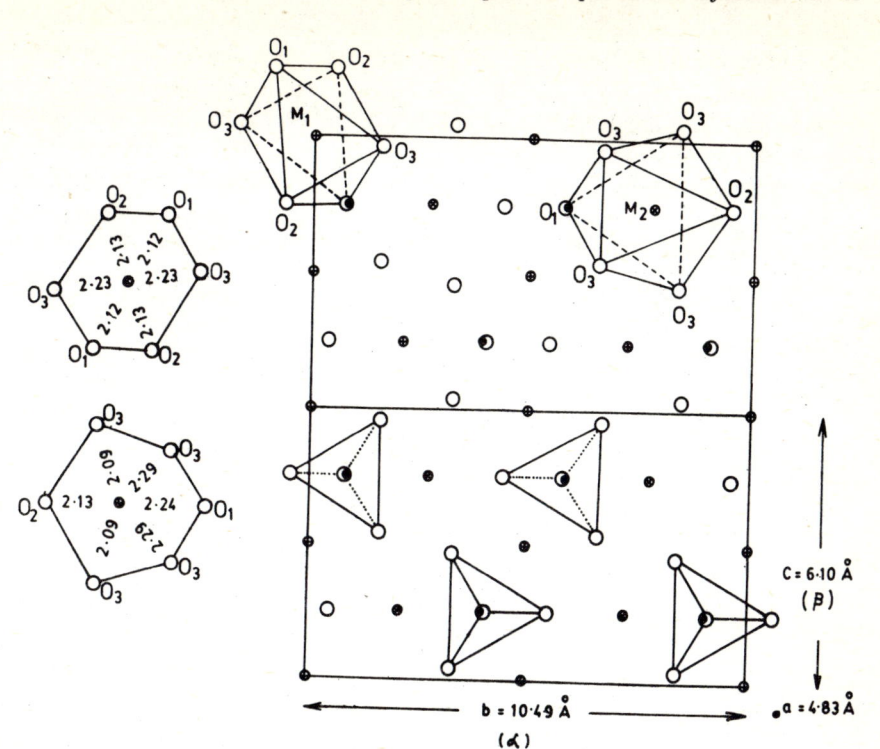

Fig. 3.5(a) Crystal structure of olivine

the oxygens of the tetrahedra. There are two such octahedral (i.e. six coordinate) positions, M_1 and M_2, which are distorted from regular octahedral symmetry.

The centro-symmetric M_1 site is elongated along the O_3-O_3 axis (i.e., tetragonally distorted). Although the point symmetry of the M_1 position is C_i, the local symmetry of the M_1- coordination site is approximately D_{4h}. The M_2 coordination site is non-centrosymmetric and irregular and is approximately of C_{3v} symmetry with the 3-fold axis parallel to the a-axis. Average metal-oxygen distances in the site in forsterite, Fo_{90} Fa_{10}, are: at M_1 site, 2.10Å; at M_2 site, 2.14Å. The values increase in fayalite, Fo_2 Fa_{100}, to 2.16Å for the M_1 site and 2.19Å for M_2 site (Hanke, 1965).

The absorption bands (Fig. 3.5 b) occur beyond the visible region and along the three crystallographic directions; the absorption spectra show differences. The dominant band at 10500Å is assigned to Fe^{2+} in the non-centro-symmetric, trigonally distorted M_2 site. The other two less intense bands at 9000Å and 12000Å are from Fe^{2+} in the centro-symmetric tetragonally distorted M_1 site.

Therefore, the crystal field transitions in Fe^{2+} in olivine (periodot) take place mostly in the infrared, but they do extend in the visible. Such absorption of red light is responsible for the yellow-green colour of olivine and other ferromagnesian silicates.

In the forsterite-fayalite series the profiles of the absorption spectra do not change significantly, except in the shift towards longer wavelengths (lower energies), with the gradual substitution of Mg^{2+} by larger Fe^{2+} ions in both M_1 and M_2 sites. This indicates that there is no significant cation ordering in Fo-Fa series. But in fayalite-tephroite series (Fe-Mn olivines) the spectra show pronounced decrease in intensity of the 10500Å absorption band relative to the 9000Å and 12000Å bands, indicating Mn^{2+} preferentially replacing Fe^{2+} in the M_2 site.

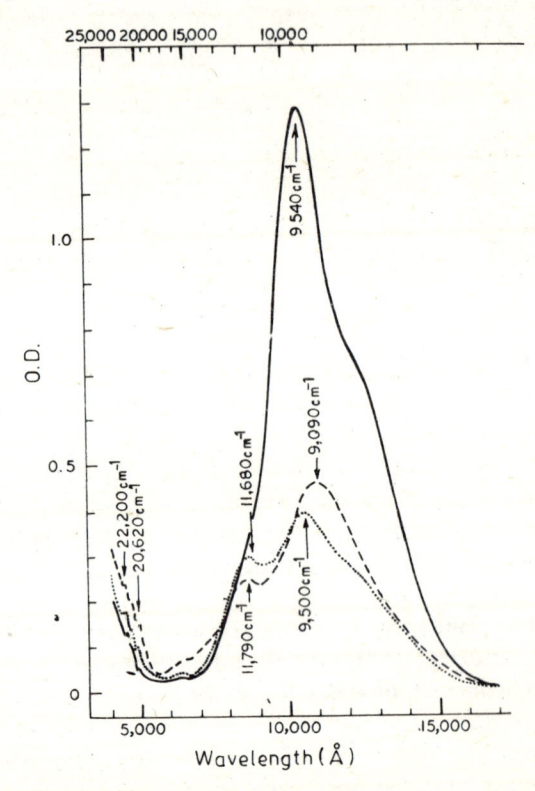

Fig. 3.5(b) Polarised absorption spectra of olivine (Fa_{12}) ....α spectra; ---β spectra; ——γ spectra. (Burns 1970, p. 80).

3.11.2 Polarised spectra of orthopyroxene (hypersthene)

In orthopyroxenes Fe^{2+} ions are concentrated in M_2 positions and other cations such as Mg^{2+}, Al^{3+}, Fe^{3+} etc. occupy M_1 positions (Fig. 3.6a). The polarised absorption spectra of an orthopyroxene ($Fs_{14.5}$) in α,β and γ directions are shown in Fig. 3.6b.

The green colour observed in polarised light along β and γ directions are well explained by Fe^{2+} located in six-fold coordination in the orthopyroxene structure. The absorption spectra show that the intense charge transfer bands in the UV region are more pronounced in the aluminous orthopyroxene than in the non-aluminous one. The red colour in α direction is due to absorption of blue and green radiation i.e. transmission of red light.

Optical Spectroscopic Studies of Minerals 67

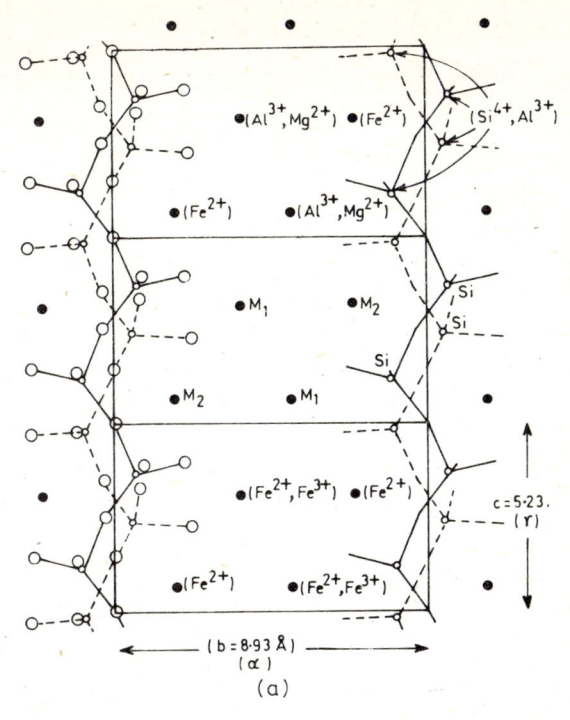

Fig. 3.6(a) The atomic structure of orthopyroxene (projected on <100>)

Various charge transfer processes are possible involving Fe^{2+} ions in M_2 positions and Al^{3+} ions in M_1 positions in the pyroxene structure. Electronic transitions between neighbouring ions are facilitated in certain crystallographic directions through suitably oriented overlapping d-orbitals, which generate pleochroism. In polarised light such transitions become possible by the favourable alignment of the lobes of t_{2g} orbitals belonging to Fe^{2+} ions in M_2 positions, Al^{3+} ions neighbouring M_1 positions and Al^{3+} ions in adjacent (Si, Al) O_4 tetrahedra (Burns, 1970).

3.11.3 Importance in characterisation of 'dubious' ions in minerals
Absorption spectral studies have resolved many mineralogical problems, which otherwise (by chemical or x-ray) could not (or hardly) be resolved. A few examples are stated below.

Reason for reverse pleochroism in biotite has been found to be due to tetrahedral Fe^{3+} ions.

The purple-violet pleochroic colour in titanaugites is due to $Fe^{2+} - Fe^{3+}$ charge transfer and Fe^{3+} in tetrahedral coordination. It was earlier wrongly suggested that this colour is due to Ti^{3+} ions.

In piemontite (Mn-epidote), and many other Mn-bearing silicate minerals (Mn-andalusite, manganophyllite, etc.) the colour is due to Mn^{3+} alone and *not* Mn^{2+} or Mn^{4+} ions, as thought earlier.

The orange colour of the glassy globules found in the lunar soil (regolith) is due to the presence of Ti^{3+} ion (not Fe^{3+}, as earlier suspected). Presence

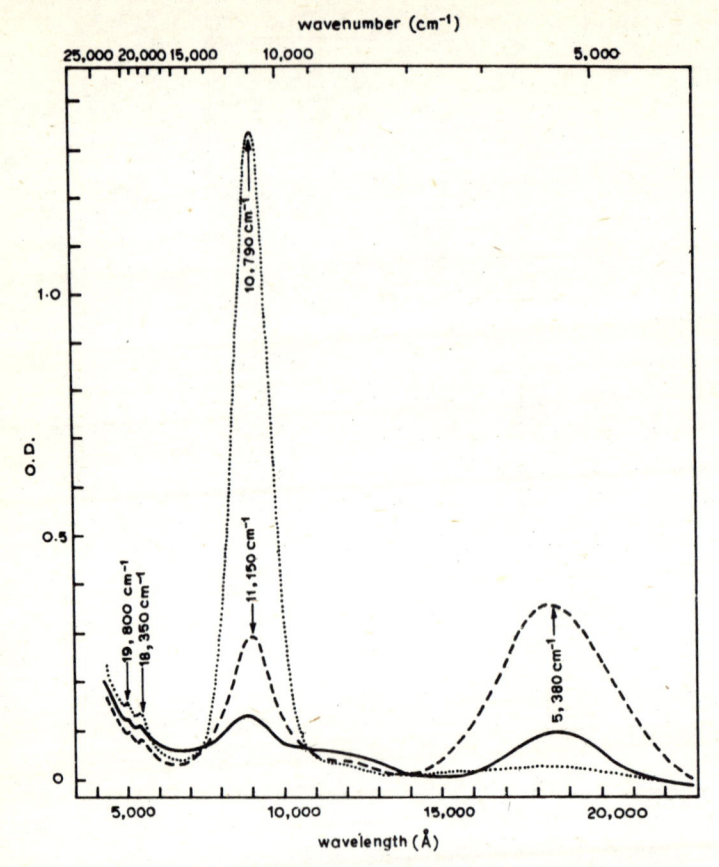

68 *Fundamentals of Optical, Spectroscopic and X-ray Mineralogy*

Fig. 3.6(b) Polarised absorption spectra of orthopyroxenes....α-spectra, ---β-spectra,—γ-spectra of bronzite, Fs$_{14.5}$ (Burns 1970, p. 88).

of Ti^{3+} in lunar pyroxenes has also been confirmed by absorption spectral measurements.

3.12 COLOURS OF MINERALS: FUNDAMENTAL PRINCIPLES

As stated earlier, colour is formed due to a selective absorption of light by matter. Since energy is quantized, certain discrete energy levels are allowed for electrons in an atom. When electromagnetic radiation interacts with a material, the wavelengths, whose energies correspond exactly to the energy differences between the electronic energy levels in the material, will be absorbed, and electrons are excited from one level to another.

The energy of absorption is related to the wavelength and wavenumber as

$$E = h\nu = \frac{hc}{\lambda} = hc\tilde{\nu}$$

where E is energy, h is Planck's constant, c is the speed of light, ν is the frequency, λ is the wavelength and $\tilde{\nu}$ is the wave number (expressed in cm^{-1}).

Transition elements are the major causes of colour in minerals and cause colours by various crystal fields and charge transfer transitions.

The absorption character of a substance determines its colour. All colourless ions have absorption bands in the ultraviolet region. But when the perturbation, caused by bonding with increase in covalent character, causes an increase in the wavelength of the single absorption band, the absorption colour passes through the visible spectrum. The colour of electronegative atoms goes through the sequence: lemon yellow, yellow, orange, red, purple, and so on. Using a colourless ion, therefore, a measure of the ionic vs covalent character can be made from the sequence of colours (Pitzer and Hildebrand, 1941). The spectral colours and their complementary colours are shown below.

Table 3.2 Spectral colours and their complementary colours

Wavelength Å	Spectral colours	Complementary colours
4100	Violet	Lemon yellow
4300	Indigo	Yellow
4800	Blue	Orange
5000	Blue-green	Red
5300	Green	Purple
5600	Lemon-yellow	Violet
5800	Yellow	Indigo
6100	Orange	Blue
6800	Red	Blue-green

Colour deepens with increase in covalent bonding, passing through yellow (90-80 percent covalent) and orange to red and black. We know, therefore, that in the case of electronegative atoms, with increase in covalent character the absorption bands shift from the UV to the visible region. For electropositive atoms the opposite is true. Because in the process of covalent bond formation the electropositive atoms gain electrons from a donor and the absorption bands shift toward violet.

The colourless cupric ion, which has an absorption in the infrared becomes blue on hydration ($CuSO_4$-colourless; $CuSO_4$, $5H_2O$-blue). The colour deepens in a more covalent complex ($[Cu(NH_3)_4]^{++}$ — deep blue). Similarly, the nickelous ion becomes green on hydration, blue and violet in more covalent complex ions.

3.12.1 Electronic processes causing colours in minerals

The electronic processes causing colours are mainly of four types: (1) electronic transitions with d and f orbitals of the transition metal and lanthanide elements respectively, (2) charge transfer, (3) electron transfer induced by structural imperfections and (4) band gap transitions.

Electronic transitions

In transition elements the $3d$-electrons have small energy splittings due to the crystal field of the surrounding ligands. The intensity of the colour is usually low because the d-d transitions are forbidden by the selection rules in operation and the intensity is due to a vibronic transition.

The same chromophore ions may generate different colours due to a difference in the ligand environment i.e, the crystal field around it. This is illustrated with Cr^{3}-ion below.

Cr^{3+} in ruby and emerald:
Cr^{3+} replacing Al^{3+} in small amounts in different minerals generate different colours viz. red in corundum (Al_2O_3) (ruby) structure and green in beryl ($Be_3 Al_2 Si_6 O_{18}$) (emerald) structure. Similarly, green colour is formed in Cr-garnet (uvarovite) and pink in Cr-chlorite (Kammererite) etc.

The visible and infrared absorption of beryl is shown in Fig. 3.8. In the visible region (0.4 to 0.8 μm) the absorption is caused by Fe^{3+}, Cr^{3+}, while the absorption in the region of 0.8 to 6.5 μm is due to molecular vibration; and absorption beyond 4.5 μm is due to lattice vibrations (Wood and Nassau, 1968).

The Cr^{3+} in ruby as well as in emerald is surrounded by six oxygen ions in an octahedral configuration, and the bond length is about 0.19 nm. But there is a significant difference in the nature of the chemical bonding. In emerald the bonding is less ionic by a few per cent, consequently, the magnitude of the electric field surrounding Cr^{3+} ion is reduced as compared to a ruby.

Each Cr^{3+} ion has three unpaired electrons in the 3d-shell, whose lowest possible energy is ground state designated as 4 A_2; the three excited states are designated as 2 E, 4 T_2 and 4 T_1 (Fig. 3.7). Electrons can be excited to both of the 4 T levels from the ground state but selection rules forbid a direct transition to the 2 E states from the ground level. To understand the enegy-state notations *vide* chapter III and appendix I.

In a ruby, because of the selection rules the excited electrons from 4 T levels can return to the 4 A_2 ground state only through the intermediate 2E level. The initial transitions from 4 T to 2 E release small amounts of energy corresponding to infrared wavelengths, but the drop from 2 E to the ground state gives rise to strong red emission.

This red fluorescence of ruby is best observed by illuminating it with ultraviolet or violet light. The red light of a ruby laser is derived from the red fluorescence of the synthetic ruby (corundum doped with Cr^{3+} and free from iron, which quenches this fluorescence).

In emerald, which is less ionic in bonding than ruby (63 p.c. ionic), the electric field surrounding Cr^{3+} is smaller. As a result, the two 4T levels lie at slightly lower energies; the position of the 2 E band is essentially unaltered. This brings about a shift in the absorption band covering the yellow and red part of the spectrum. This results in absorption of most of yellow and red; and consequently in emerald transmission of blue and green is greatly enhanced. The visible to infrared absorption spectrum of emerald are shown in Fig. 3.8, which depicts the regions of molecular vibrations and lattice vibrations as well.

Alexandrite (Cr^{3+} replacing Al^{3+} in $Be Al_2O_4$) however, shows an intermediate transmission between ruby and emerald. In it the crystal field around Cr^{3+} ion is weaker than ruby but stronger than emerald, resulting

Optical Spectroscopic Studies of Minerals 71

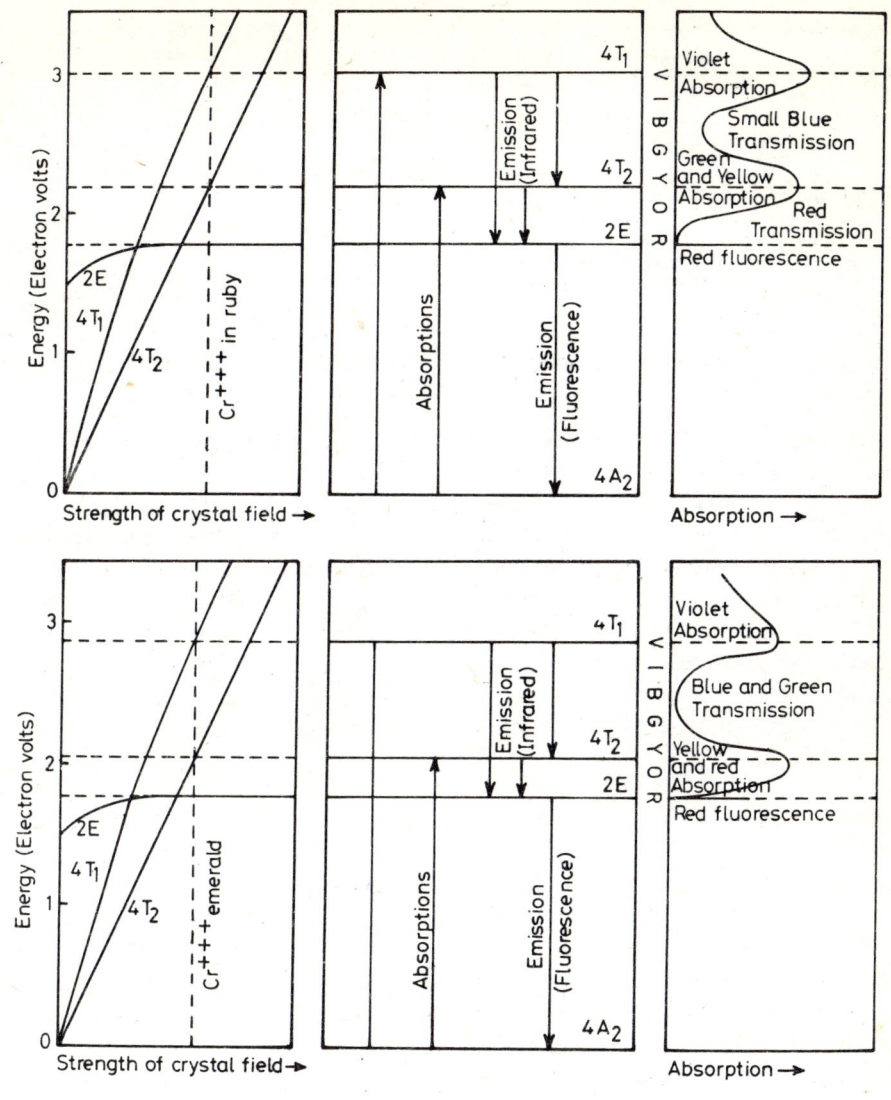

Fig. 3.7 Energy level diagrams of Cr^{3+} in ruby and emerald, showing their absorption and transmission colours (after Wood and Nassau, 1968).

in a balance of red and green transmission bands. It, therefore, appears blue-green in blue rich sunlight while it looks red in lamp light.

Charge-transfer (CT)
However, more intense colours are generated due to charge-transfer transitions (inter-element electron transitions) in which electrons flip between the orbitals of the metal and the ligand. Eg. manganate (VII) ion, MnO_4^-, in intense permanganate solution. Here transitions possess a large *transition dipole moment*.

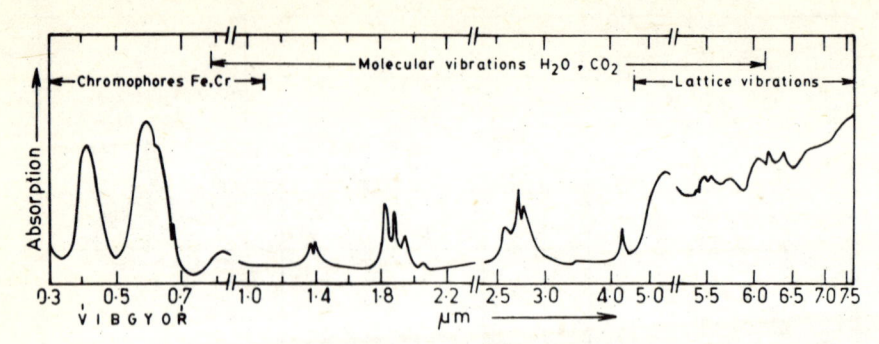

72 *Fundamentals of Optical, Spectroscopic and X-ray Mineralogy*

Fig. 3.8   The visible to IR absorption spectrum of emerald (after Wood and Nassau 1968).

Charge transfer is common among silicates containing transition metals which can exist in two or more oxidation states (e.g. Fe^{2+} - Fe^{3+}; Mn^{2+} - Mn^{3+}, Ti^{3+} - Ti^{4+}). The electron transfer takes place as simultaneous reduction ($Fe^{3+} + e^- \rightarrow Fe^{2+}$) and oxidation ($Fe^{2+} \rightarrow Fe^{3+} + e^-$) between two ions. Charge transfer is facilitated when there is local inbalance of charge accompanying isomorphous substitution e.g. Fe^{2+} and Mg^{2+} by Fe^{3+} and Al^{3+}. The blue pleochroic colours in alkali amphilboles, some tourmalines, kyanite, glaucophane, benitoites, cordierite etc. are due to charge transfer between neighbouring Fe^{2+} and Fe^{3+} ions in the structure. Red brown colours in hypersthene, hornblende, biotite and staurolite are also due to this transition.

Structural defects

Deviation from stoichiometry results in excess positive ions accompanied by negative vacancies or excess negative ions accompanied by positive vacancies. These result in colouring of minerals.**

Structural defects (lattice defects and colour centres) in minerals (halides etc.) can also impart colours. Electrons filling anion vacancies (F – centres) and electrons filling interstices (F'-centres) are the most common causes of colours in fluorites, halites, calcite etc. The quantum state of such trapped electrons lie in the forbidden energy region. Structural defects caused by irradiation (gamma, neutron-or x-ray) can also cause colour[†] viz., smoky quartz (by gamma irradiation). Many sulphides, arsenides etc. owe their colours to the electronic transitions from valence band to the conduction band, called band-gap transitions. The absorption edge often extends into the visible region. In cinnabar, HgS, the absorption edge covers visible blue end and low-frequency red is transmitted. Dispersion of light by colloids may cause

---

**   Excess potassium in KCl crystals gives violet colour, excess Zn in ZnO and excess Li in Li F give yellow and pink colours respectively.
Such irradiated crystals usually contain both F and V centres. In the latter excess anions accompany positive-ion vacancies, trapping holes.

†   Coloured halos formed due to structural damage by impacts of particles from radio-active inclusion are found in zircon, biotite etc.

colour. The rich red ruby colour is found in glass having colloids of gold, which scatter the blue components of transmitted light.

Band gap transitions have been discussed in section 3.12.7 (also vide table 3.5).

3.12.2 Colour Centres:

These centres are generated a) by the presence of impurity ions in the intersititial positions, b) by the presence of vacancies, produced by structural deviations with *excess positive ions* or metal ions accompanied by negative vacancies or trapped electrons.

Examples: excess of Li in Li F → pink
excess of K in KCl → violet
excess of Zn in ZnO → yellow

The quantum states of trapped electrons lie somewhere in the forbidden-energy region. The trapped electrons get excited to a higher energy state, in the forbidden region, with the absorption of energy from the impinging white light this produces colour in the crystal. The range of energies to which these electrons are excited is called the *F-band*. The F absorption band is asymmetric with a tail on the short wave-length side.

The narrow tail-side, called *K-band*, is due to electronic transitions to quantum states having higher energies but still lying below the bottom of the conduction band.

Often the electrons in the F-band on heating become "free" electrons by absorbing thermal energy. These electrons can combine with other F centres with single electrons. Thus centres with two electrons, called *F' centres*, are formed. When an F' centre absorbs light it becomes an F centre again. This is optically observed as losing colour or *bleaching*. The energy of this light is less than that necessary to excite electrons trapped in F centres.

In crystals having an *excess of anions*, with corresponding positive-ion vacancies or trapped holes, colour centres are produced. These are called *V-centres*. The absorption spectra of V-centres show several peaks, which are marked as V_1, V_2 etc. bands.

When high-energy radiations like X-rays, γ-rays or neutrons impinge upon crystals, colour centres may be produced. In this process first recoil electrons are produced by the incident radiation. These electrons interact with the valence electrons and produce electron-hole pairs. At such a stage both V and F centres are present. These coloured radiation-damaged crystals can easily be bleached or faded by heating or exposure to visible light. The excited holes and electrons in these cases combine with each other on further excitement.

The colour centres in fluorites are often *trapped holes and electrons*. The colour of smoky quartz is due to hole colour centres in impure natural quartz with Al^{3+} in Si^{4+} site and charge-compensated by H^+ or Na^+, K^+ etc. This makes it easy to remove an electron from a neighbouring oxygen ion. Such quartz when exposed for a few minutes to strong X-ray or γ-ray, or for a long geological period to a weaker radiation, the radiation expels one electron from a pair of electrons in an oxygen atom adjacent to an Al^{3+} ion, thereby leaving an unpaired electron in the orbital. This forms a hole colour centre, since an election is missing. The absent electron or a hole and the remaining unpaired electron have a set of excited states much like that of an excess electron. This causes light absorption, producing a smoky colour*

The colour in amethyst is also due to hole centres but the adjacent impurity ion is iron rather than aluminium.

Heating often stabilises the structure (or decolourises) by bringing the displaced ions to their original positions. Amethyst when heated therefore changes colour to become greenish or yellowish quartz; yellow flourites may become colourless.

The colour of bluish-green amazonite (microcline) is due to the substitution of $2K^+$ by Pb^{2+} in $KAlSi_3O_8$ with the generation of K^+ lattice vacancy.

A list of colour centres causing characteristic colours of some common minerals is given in Table 3.3.

Table 3.3 Colour Centres in some common minerals

Mineral	Composition	Colour	Colour Centres
Halite	NaCl	blue	Cl^- vacancy
Fluorite	CaF_2	purple	F^- vacancy
Quartz (smoky)	SiO_2	brown	Al^{3+} for Si^{4+}
Quartz (amethyst)	SiO_2	purple	Fe^{3+} for Si^{4+}
Microcline (amazonite)	$KAlSi_3O_8$	blue-green	Pb^{2+} for $2K^+$

3.12.3 Molecular Orbital Transitions:

In a crystal field electronic transitions take place within an ion. But in a large number of cases the electronic transitions take place between a group of atoms. This situation is explained by the Molecular Orbital Theory (MOT).

In minerals the molecular orbital transitions of participating electrons fall under the general heading of Charge Transfer (CT). Two types of this transition are common: (a) Ligand to metal (in silicate minerals, mostly oxygen to metal; in sulphide minerals, sulphur to metal) and (b) metal to metal charge transfer.

Oxygen-metal charge transfer (O-M): In this process the electrons from oxygen $2p$ orbitals are transferred to $3d$ orbital of a metal. In vanadinite (VO_4^{3-}) and crocoite (CrO_4^{2-}), oxygen → metal ($O^{2-} \rightarrow V^{5+}$, $O^{2-} \rightarrow Cr^{6+}$) charge transfer transitions are responsible for the strong absorption of the violet, blue and green wavelengths resulting in orange to orange-red colours.

O-M charge transfer in transition metal ions (Fe^{2+}, Fe^{3+}, Ti^{3+}, Ti^{4+}, Cr^{3+}) in octahedral mineral sites, occurs in the UV range and causes general opacity and higher reflectivity. For octahedrally coordinated cations, oxygen

---

* Heating smoky quartz to 400°C may supply adequate energy to the removed electron to return to the original oxygen ion causing destruction of the colour. Thus, the colour of smoky quartz fades at about 400°C.

to metal charge transfer (CT) energies are calculated to decrease in the order $Cr^{3+} > Ti^{3+} > Fe^{2+} > Ti^{4+} > Fe^{3+}$.

Aquamarine and heliodore are beryl with Fe^{3+}. Fe^{3+} ions occupy the same AlO_6 site as does Cr^{3+} in emerald. If the $(Fe^{3+}O_6^{9-})$ cluster is analysed by MO theory, the absorption edge in aquamarine and heliodore seems to be due to oxygen $\rightarrow Fe^{3+}$ charge transfer. The optical absorption spectra of aquamarine has been treated in section 3.12.6. The mineral examples of oxygen $\rightarrow$ metal charge transfer are presented in Table 3.4.

Table 3.4 Oxygen to metal charge transfer in minerals

Beryl (heliodore)	$Be_3Al_2Si_6O_{18}$	$O \rightarrow Fe^{3+}$	Yellow
Crocoite	$PbCrO_4$	$O \rightarrow Cr^{6+}$	Orange
Vanadinite	$Pb_5(VO_4)_3Cl$	$O \rightarrow V^{5+}$	Orange-brown

Metal-metal charge transfer: This involves transfer of an electron between metals or between metal ions of differing valence (intervalence CT). $Fe^{2+} \rightleftharpoons Fe^{3+}$ and $Fe^{2+} \rightleftharpoons Ti^{4+}$ are common examples. Energies of this reversible electron-hopping process often correspond to energies of visible wavelengths (i.e. between 1.7 to 3eV; vide Table 1.1). The blue colour in sapphire ($Fe^{2+}, Ti^{4+}/Al_2O_3$), kyanite ($Fe^{2+}, Ti^{4+}/Al_2SiO_5$), glaucophane [$Na_2(Mg, Fe^{2+})_3 (Al, Fe^{3+})_2 Si_8 O_{22}(OH)_2$], cordierite [$(Mg, Fe^{2+})_2 (Al, Fe^{3+})_4 Si_5O_{18}$] are due to such transitions.

The energy of transition, however, depends on the bond-distances. As the distance increases the probability of electron transfer decreases and absorption intensity diminishes. The energy decreases in the order of bonding thus: corner-sharing > edge-sharing (oct - tetr) > edge sharing (oct-oct) > face sharing (oct-oct). When the metal-metal axis is parallel to the polarised light absorption maxima takes place.

Blue saphire is a corundum (Al_2O_3) in which $2Al^{3+}$ is couple-substituted by Fe^{2+} and Ti^{4+} with balanced charge. Some varieties may also contain Fe^{3+}. The possible metal $\rightarrow$ metal charge transfers, therefore, are: $Fe^{2+} \rightarrow Fe^{3+}$ and $Fe^{2+} \rightarrow Ti^{4+}$. The absorption spectrum of blue sapphire (Fig. 3.9) indicates that the $Fe^{2+} \rightarrow Fe^{3+}$ CT transition absorbs yellow, orange and red producing the blue colour, which makes it a gem.

In $Fe^{2+} \rightarrow Ti^{4+}$ CT the transition involves a higher energy and a net change in spin of the system as:

$$Fe^{2+}(d^6) + Ti^{4+}(d^0) \rightarrow Fe^{3+}(d^5) + Ti^{3+}(d^1)$$
$$\text{(one pair spin)} \qquad \text{(no paired spin)}$$

In $Fe^{2+} \rightarrow Fe^{3+}$ CT the total spin of the system remains constant and involves lower energy than the visible red.

Some cases of metal-metal CT are presented in Table 3.5.

3.12.4 Bond length in CT colour

In charge transfer process the variation in bond distance in different planes causes change in absorption. With the same bonded ions, an increase in bond distance will cause decrease in absorption. There are two geometries for metal-

Table 3.5 Metal-Metal charge transfer in some minerals

Mineral	Composition	Transition	Colour
Beryl (aquamarine)	$Be_3Al_2Si_6O_{18}$	$Fe^{2+} \rightarrow Fe^{3+}$	ω = yellow ϵ = blue
Corundum (sapphire)	Al_2O_3	$Fe^{2+} \rightarrow Ti^{4+}$	ω = dark blue ϵ = light blue
Cordierite (iolite)	$(Mg, Fe^{2+})_2(Al, Fe^{3+})_3$	$(AlSi_3O_{18})$ $Fe^{2+} \rightarrow Fe^{3+}$	α = colourless β = blue γ = violet blue
Kyanite	Al_2SiO_5	$Fe^{2+} \rightarrow Ti^{4+}$	α = dark blue
Vivianite	$Fe_3(PO_4)_2 \cdot 8H_2O$	$Fe^{2+} \rightarrow Fe^{3+}$	β = pale green γ = pale green

metal interactions in a corundum structure. Metal coordination polyhedra share faces in a direction parallel to C and edges in a direction approximately perpendicular to C. Metal-metal distances in sapphire (with $Fe^{2+} \rightarrow Ti^{4+}$) and the corresponding absorption colours show the inverse relationship.

Along face (//C) the Fe^{2+} - Ti^{4+} distance is 2.65 Å and the absorption in the direction of ω = dark blue, while the greater distance, 2.79Å along edge ($\perp C$) cause ϵ = light blue.

3.12.5 Other Molecular transitions causing colour
Lapis lazuli $(Ca, Na)_8 (Al, Si)_{12} O_{24} (SO_4, S, Cl)$. H_2O is blue, due to electronic transitions in the S_3^- molecular ions in that mineral.

3.12.6 Combination of transitions
Optical absorption spectra of aquamarine (Fe^{2+}, Fe^{3+}/$Be_3Al_2Si_6O_{18}$) and blue sapphire (Fe^{2+}, Ti^{4+}/Al_2O_3) are presented in Fig. 3.9: to illustrate cases of minerals which simultaneously involve molecular vibration, spin-allowed transition, spin-forbidden transition bands and charge transfer. The spectra are taken in ordinary and polarised (C// and $\perp$) positions. The peaks are attributed as follows:

Aquamarine:
1) ~ 10,000cm^{-1} : molecular vibration in trapped H_2O
2) 12,300cm^{-1} : spin-allowed crystal field absorption in Fe^{2+}
3) 25,000cm^{-1} : spin-forbidden band due to crystal field in Fe^{3+}
4) Absorption edge in UV : $O^{2-} \rightarrow Fe^{3+}$ charge transfer
5) 16,100cm^{-1} : $Fe^{2+} \rightarrow Fe^{3+}$ charge transfer in $E//c$ (ϵ polarisation); this causes blue colour.

Blue sapphire:
1) 10,000 and 11,300 : $Fe^{2+} \rightarrow Fe^{3+}$ charge transfer (metal-metal)
2) 13,000 and 17,000 : $Fe^{2+} \rightarrow Ti^{4+}$ charge transfer (metal-metal)

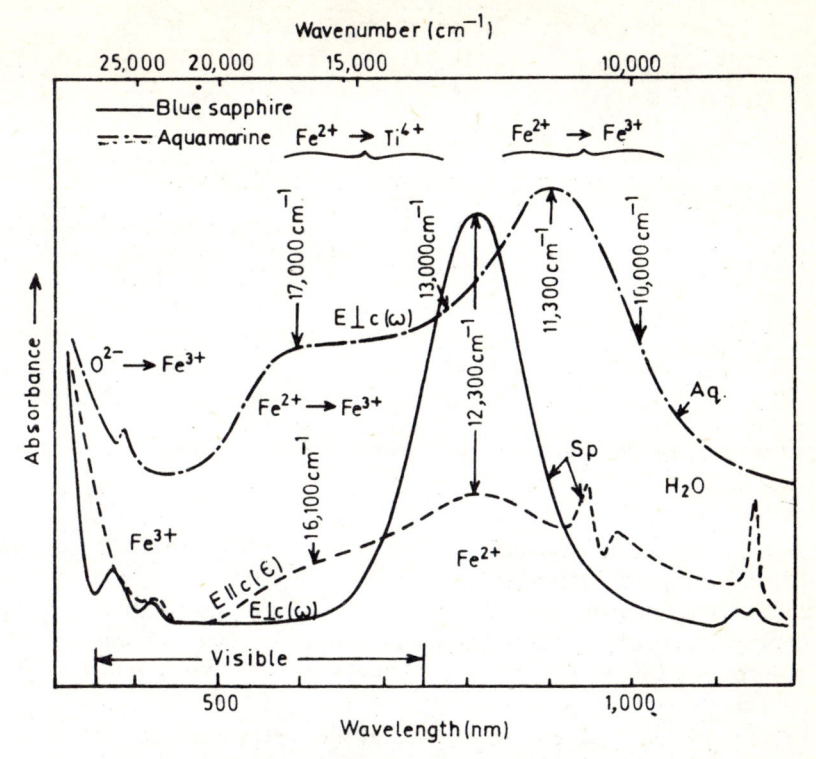

Fig. 3.9 Optical absorption spectra of Aquamarine Fe^{2+}, $Fe^{3+}/Be_3Al_2 Si_6 O_{18}$ and blue sapphire (Fe^{2+}, Ti^{4+}/Al_2O_3). Sapphire shown in polarisation absorption [E// c(ε) and E⊥ c(ω)].

3.12.7 Metals, semiconductors and insulators

In metals and semiconductors, the enormous amount of mobile electrons (10^{23} electrons/cm³), generate special optical and electrical properties. The reflective property of metal is due to the jumping back to the original state by the excited electron after absorption of light (*vide* section 4.4.1).

The variation in colour of metals, copper (red) or silver (white), is due to the differences in the number of states available at particular energies above the Fermi level. Because the density of the states is not uniform, some wavelengths are absorbed and transmitted more efficiently than others. Thinly hammered gold (strained) leaf transmits a green colour. Colloidal gold (unstrained) is purple-red.

Semiconductors are distinguished from metals by having the band of energy levels in two parts, separated by a gap of forbidden energies. All the lower levels form a valence band and are filled at ground state; while all excited states lie in a separate conduction band, which in the ground state is entirely empty (Fig. 3.10).

A pure semiconductor is coloured when the magnitude of the band-gap energy falls in the visible range. The red colour of HgS (Cinnabar) is due to its band gap energy of 2.1 *eV*. All photons with energies higher than this are absorbed while the ones with longest wavelengths transmitted give a red

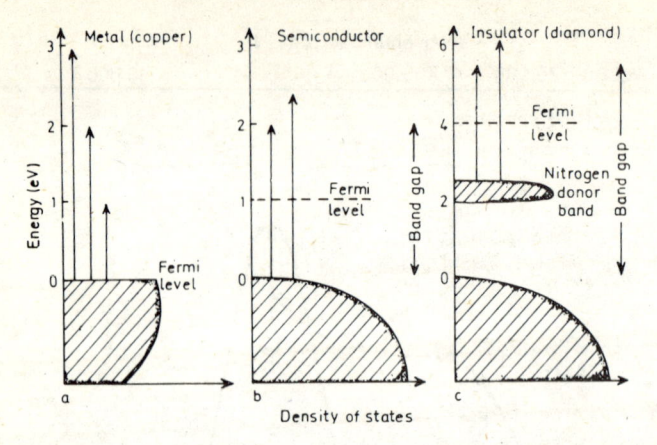

78 Fundamentals of Optical, Spectroscopic and X-ray Mineralogy

Fig. 3.10 Energy levels and density of states in metals, semiconductors and insulators.

colour. For proustite (Ag_3AsS_3) where the band gap (1600 cm^{-1}) is in the visible range (orange), all wavelengths of visible light from orange though violet are absorbed. Only the lower-energy wavelengths are transmitted.

Greenockite (CdS) is yellow-orange because its band gap is about 2.6 eV. It absorbs energies greater than 2.6 eV and gives out blue and violet light (vide Table 1.1).

In an insulator, a large band gap makes the energy required to promote an electron from the filled to the empty levels prohibitive. Hence it transmits all wavelengths and looks colourless.

Diamond (C) with large band gap of 5.4 eV, which is higher than the highest energy of visible light, can absorb no wavelength of visible light, hence it is transparent and colourless.

The minerals with colours due to band gap transitions are listed in order of energies involved (Table 3.6).

Table 3.6 Mineral colours due to band gap transitions

Mineral	Composition	Colour	Band gap Energy (eV)	Wavelength (cm^{-1})
Galena	PbS	Greyish black	0.37	3,000
Molybdenite	MoS$_2$	Greyish black	~1.0	8,000
Proustite	Ag$_3$AsS$_3$	red	~2.0	16,000
Cinnabar	HgS	red	2.1	16,900
Orpiment	As$_2$S$_3$	bright yellow	2.5	20,000
Greenockite	CdS	bright yellow	2.58	20,000
Sphalerite	ZnS	light coloured	3.6	29,000
Diamond	C	colourless	5.4	43,500

Some dopants or impurities can introduce a set of energy levels in the gap between the valence band and the conduction band. The energies of light required to excite electrons from filled impurity levels to the empty levels above the band gap may be in the visible region.

The impurities in diamond thus generate colours. As shown in the Fig. 3.10, yellow colour in a diamond is due to nitrogen impurity. All wavelengths high in energy than 2.5eV (20,000 cm^{-1}) are absorbed. This absorption is in the blue and violet, therefore yellow colour is observed.

A freshly cut metal surface shines with mirror like metallic lustre. Let us study the effect of the incident light on such a surface. The highly mobile free electrons of the metal on being impinged upon by the oscillating light photons, move back and forth and quench the incident energy. But the oscillating surface electrons themselves give rise to a radiated light field, and so almost all light is reflected. However, in some brightly colored metals like gold and copper there are true absorption bands in the visible region, and both these metals absorb blue light and get hotter in the process.

Doping also changes the electrical properties and hence solid state electronic devices are possible. In light-emitting diodes and semiconductor lasers, an electric current populates excited states and the electrons emit radiation while returning to the ground state. Doped semiconductors also act as phosphors which give off light when stimulated electrically or otherwise. Thus fluorescent tubes have phosphors and in the picture tube of a coloured television set, three phosphors are distributed over the surface with emissions at red, green and blue wavelengths.

3.13. CHARACTERISTIC ABSORPTION LINES OF SOME GEM MINERALS

In Table 3.7 certain characteristic absorption lines of some gem minerals are presented.

3.14. OPTICAL PROPERTIES OF MINERALS DUE TO EXCITATION OF CRYSTAL ENERGY

Each electron shell of an atom is characterised by a certain quantum of energy; the inner ones are of lesser energy than the outer ones. The energy field of the crystal lattice, however, influences the energy-level patterns of atoms and a series of energy levels or bands are formed. These bands are often separated by forbidden zones, where electrons hardly exist (Fig. 3.11).

Colours are generated mostly by electrons in the valence bands. Absorption of energy from an impinging electromagnetic radiation (light) may promote the electrons from the less energetic inner band to the higher-energy outer band and absorption of light occurs. But when the amount of energy absorbed is great the electron may pass into a conduction band. In this band the electrons are free to move about through the crystal under the influence of external electric fields.

In insulators and semiconductors the energy separation between the conduction and valence bands is termed the "band gap" or "forbidden energy

Table 3.7 Characteristic absorption lines of some gem minerals
(absorption colour ranges are shown in brackets)

Mineral	Absorption lines
Andalusite	4550Å (brown), 5525Å (green), 5495, 5775Å (strong absorption in blue-violet region)
Apatite	Yellow variety gives 5800, 5200Å, while blue ones give 5120, 4910 and 4640 Å
Beryl	For ε-ray 6830 and 6800 Å (red) only, for ω-ray absorption occurs at 6830, 6800 (red doublet), 6370 and 4775 Å (blue)
Bronzite	A sharp 5060 Å for iron
Chrysoberyl	Yellow and brown varieties give peaks at 4400 Å (due to Fe^{2+}). Alexandrite shows broad bands in the yellow-green and violet portion of the spectrum.
Cordierite	4920, 4560, 4360 Å (blue-violet)
Corundum	Ruby (with Cr^{3+}) has absorption bands at 6962, 6928 Å (red doublet), 6680, 6592 Å (orange), 4765, 4750, 4685 Å (blue). Sapphire has at 4500Å, while the presence of iron gives the lines 4710, 4600Å.
Diamond	4155 Å (deep violet), or 4530, 4660, 4780Å. A short exposure of a colourless diamond in a cyclotron or nuclear-reactor gives green colour which changes to yellow or brown after heat-treatment. Long exposure turns colourless diamonds black. High energy electron bombondment can give blue colour. Irradiated yellow diamonds give absorption bands at 5920, 4980, 5040Å
Diopside (Cr-variety)	5080, 5050Å
Epidote	4500Å (strong, violet)
Fluorite	green variety shows a strong band at 4270Å
Peridot	4930, 4730, 4530 Å (all due to Fe^{2+})
Rhodonite	5480, 5030, 4550Å
Sphalerite	6900, 6670, 6510Å
Spinel	Cr-spinel (yellow-green) 5400Å. Fe^{2+}-spinel (blue) 4580, 4780Å, weak likes in orange, yellow green
Topaz	A doublet at about 6830 Å may be observed due to the pressure of Cr^{3+}.
Zircon	6535A (red), 6910, 6225, 6590, 65895, 5625, 3575, 5150, 4840, 4325Å
Zoisite	5950 (orange), 5280 (green), 4550Å (blue).

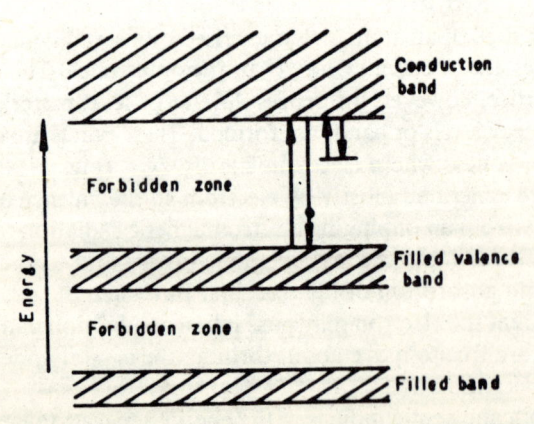

Fig. 3.11 Diagram of energy levels in a crystal lattice causing colour and luminescence.

gap" (Fig. 3.10). Transition across this gap results in strong absorption of energy which produces a *fundamental absorption* termed also as absorption edge or band edge.

Impurity atoms in an insulator can have electrons in such energy levels which are just below the conduction band and can push electrons to conduction band on receiving necessary *activation energy*. Just above the valence band there can also be *acceptor levels,* which accommodate electrons, excited from the valence band, which in turn accommodates the same number of 'holes'. Conduction may occur via these holes. Such impurities make up *extrinsic semiconductors.*

Some minerals may absorb radiation of wavelengths beyond the visible range, and part of the energy thus gained may be radiated again with wavelengths in the visible region and one can see some luminescence colours from the minerals. Scheelite ($CaWO_4$), willemite (Zn_2SiO_4), fluorite (CaF_2), U-bearing minerals, diamond (some varieties) etc. when irradiated by the invisible ultraviolet light (250 *nm*) in darkness, emit the luminescence. In these cases, the phenomenon is called *fluorescence,* since the emission dies as soon as the irradiation is stopped. The fluorescent property in these is due to the presence of small impurity elements called *activators* in the crystal lattice. The common activators are manganese (in silicate minerals), copper and silver (in sphalerite and some sulphides) etc. The scintillations of light from thallium activated KI crystal are the basis for one of the common types of radiation counter (*vide* p. 141). The electrons in the activators are raised to higher energy bands by the absorption of the incident radiation and on falling back to their original ground state level, emission of visible light takes place. Prolonged and strong irradiation may, in some minerals, lift the electrons to the conduction band giving rise to *photoconductivity* of the mineral.

In cases, where the light emission continues even after the irradiation source is taken away, the phenomenon is called the *after-glow* or *phosphorescence.** The emission occurs due to the gradual release of energy stored in the lattice. Phosphorescence may last for a millionth of a second to several hours. Phosphors with long decay times are useful for making television screens, cathode-ray oscilloscopes etc.

The long decay-time can be explained by a model of energy levels of the activators. In this model the excited electrons from the atoms, ionised by irradiation, while falling back are trapped at intermediate energy levels, created by impurity centres, positive holes or such defects. Here the falling electron may be held up till the statistical thermal vibrations allow it to escape. Often infra-red radiation can accelerate this escape and allow the electron to return to the ground state with consequent emission of light.

Lattice defects in a crystal can also cause absorption of light giving rise

---

*If the after-glow is short in time ($t < n.10^{-2}$s) and does not depend on temp., it is named fluorescence. If the duration of the afterglow ($t > n.10^{-2}$s) increases under cooling and decreases on heating, the phenomenon is called phosphorescence.

to a colour. Such defects are termed F (from German, *farbe*, meaning colour) centres. These centres are formed in the following situations.

(i) Presence of impurity ions or colloidal particles in the lattice; Ex., Cr^{3+} in ruby.

(ii) Excess of a cation violating chemical stoichiometry of the compound. Ex., K in sylvite.

(iii) Defermation of lattice causing cationic or anionic vacancies.

Normally, F centres are thought to be a positively charged vacancy with an electron moving around it.

The luminescence of some minerals is connected with the transition of boundaries of the *'molecular' centre*. The uranium bearing hypogene minerals show bright green luminescence due to the presence of complex $(UO_2)^{2+}$ ion. Yellow luminescence in barite, fluorite etc. are due to the presence of S_2^- and O_2^- ions (Fig. 3.12a).

The electronic absorption spectra of a yellow fluroite (from Amba Dongar, India) at room temperature (300K) and at liquid nitrogen temperature (77K) are shown in Fig. 3.12b. The molecular oxygen ion occurring as substitutional impurity show eight peaks, and flanked on both sides by two satellite peaks. The separation of the peaks is of the order of $1120 cm^{-1}$ (Mitra, 1981).

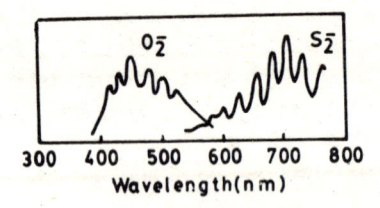

Fig. 3.12(a) Luminescence spectra of O_2^- and S_2^- ions in minerals.

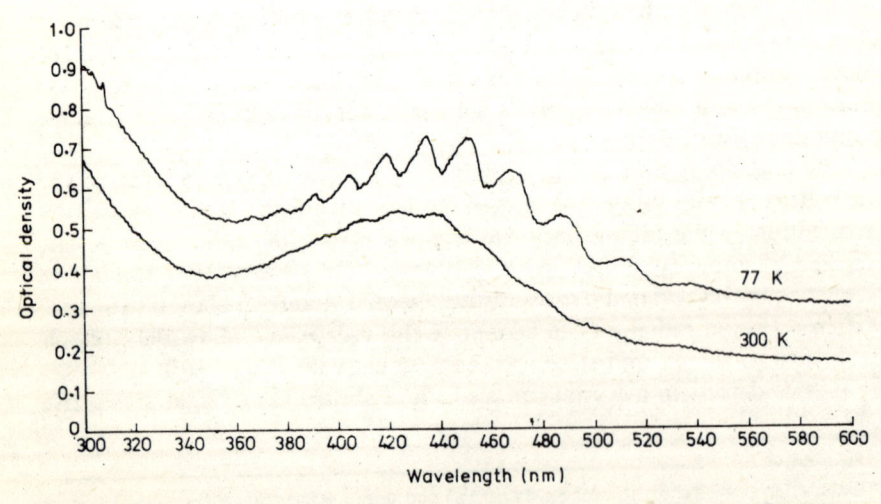

Fig. 3.12(b) Electronic absorption spectra of yellow fluorite at 300K and 77K (Mitra, 1981).

Molecular oxygen ions (O_2^- and O_3^-) are quite soluble in alkali halides and generate similar specta. The other isoelectric series Se_2, SeS, S_2 give similar spectral patterns but they have different positions and separation of about 580 cm^{-1}. (Fig. 3.12a).

3.15 UV FLUORESCENCE OF MINERALS

The opaque minerals are nonfluorescent. Some light transmitting minerals fluorescence under UV exposure. Fluorescent minerals respond bettern in longer wavelengths of the UV range.

Some common fluorescent minerals with the flourescence colours under UV are listed below (var. = variety):

Andalusite	:	weak green to yellow green
Apatite	:	lilac pink (yellow var.), blue (blue var.). greenish yellow (violet var.)
Benitoite	:	light blue
Beryl	:	mostly non-fluorescent, except emerald
Calcite	:	some are fluorescent, not all; may give any of the colours: white, red, orange, yellow, green, light-blue
Chrysoberyl	:	only alexandrite variety gives red fluorescent colour
Corundum	:	Strong red, blue sapphire is usually non fluorescent
Diamond	:	some fluoresce, not all
Emerald, Ruby	:	red
Felspar	:	orthoclase gives weak orange fluorescence colour, amazonite variety gives yellow-green colour, plagioclases are non fluorescent
Fluorite	:	blue green (only some var.)
Kyanite	:	dull red
Opal	:	pale green or blue (white var.) greenish (fire opal)
Scapolite	:	yellow to purple
Scheelite	:	pale to deep blue
Smithsonite	:	brilliant organge red (greenish yellow var.)
Spinel	:	red to orange (reddish var.) green (bluish var.)
Spodumene	:	orange to pitch (usually kunzite var.)
Topaz	:	weak yellow or green
Turquoise	:	weak greenish yellow to bright blue
Zircon	:	yellow (uncommon var.)

Common rock forming minerals are mostly non-fluorescent under UV. Some of these are: quartz, plagioclase felspars, tourmaline, garnets, olivines, pyroxenes, axinite, epidote, gypsum, idocrase, cordierite, carbonates, rhodonite, serpentine, sodalite, sphalerite, sphene, staurolite, zoisite etc.

References and Selected Reading

Bargeron, C.B., Avinor, M. and Drickamer, H.A. (1971) The effect of pressure on the spin-state of iron (II) in manganese (IV) sulfide. *Inorganic chemistry, 10,* 1138-39.

Burns, R.G. (1970) Mineralogical Applications of Crystal Field Theory, Cambridge University Press, England, p. 224.

Burns, R.G. (1985) In: Chemical Bonding and Spectroscopy in Mineral Chemistry, edited by F.J. Berry and D.J. Vaughan, Chapman and Hall, London, 63-101.

Cotton, J.A. and Wilkinson, G. (1972) Advanced Inorganic Chemistry, Interscience, New York.

Goodenough, J.B. (1963) Magnetism and the chemical Bond. Wiley Interscience, New York.

_____ (1969) Description of outer d-electrons in thiospinels. *Journal of Physics and Chemistry of Solids, 30,* 261-80.

_____ (1972) Energy bands in TX_2 compounds with pyrite, marcasite and arsenopyrite structures. *Journal of solid state chemistry, 5,* 144-52.

James, T.L. (1971) Photoelectron spectroscopy. *Journal of chemical education, 48,* 712-18.

Khan, M.A. (1976) Energy bands in pyrite-type crystals, *Journal of Physics,* series *C9,* 81-94.

Low, W. and Weger, M. (960) Paramagnetic resonance and optical spectra of divalent iron in cubic fields. *Physical Review, 118,* 1130-36.

Loeffer, B.M. and Burns, R.G. (1976) Shedding light on the colour of gems and minerals. *American Scientist, 64(6),* 634-647.

Marfunin, A.S., Platonov, A.N. and Federov, V.E. (1968) Optical spectra of Fe^{2+} in sphalerite. *Soviet Physics,* solid state, *9,* 2847-48.

Mitra, S. (1981) Nature and genesis of colour centres in yellow and colourless fluorites from Amba Dongar, Gujarat, India. *Nues Jahrbuch fur Mineralogie, Abh. 141,* 3, 290-308.

Murrell, J.N. Kettle, S.F.A. and Tedder, J.M. (1970) Valence Theory, Wiley, New York.

Nassau, K (1975) The Origin of colour in minerals and gems, pts. I, II, III. *Lapidary Journal, 29,* 920-28 (Aug.), 1060-70 (Sept.), 1250-1258 (Oct.).

Plantonov, A.N. and Marfunin, A.S. (1960) Optical absorption spectra of sphalerite. *Geochemistry International, 5,* 245-59.

Smith, G. and Strens, R.G.J. (1976) Intervalence transfer absorption in some silicate, oxide and phosphate minerals in The Physics and Chemistry of Minerals and Rocks. ed. R.G.J. Strens, pp. 583-612, John Wiley and Sons Ltd., New York.

Tossell, J.A., Vaughan, D.J. and Johnson, K.H. (1974) The electron structure of rutile, wustite and hematite from molecular orbital calculations. *American Mineralogist, 39,* 319-34.

Vaughan, D.J., Burns, R.G. and Burns, V.M. (1971) Geochemistry and bonding of thiospinel minerals. *Geochim et Cosmochim Acta, 35,* 365-81.

Wood, D.L. and Nassau, K. (1968) The characterisation of beryl and emerald by visible and infrared absorption spectroscopy. *American Mineralogist, 53,* 777-800.

CHAPTER IV

Reflection optics

4.1 DIELECTRICS AND OPAQUES
A part of the light that falls on a polished surface is transmitted through it or is absorbed and the rest is reflected. Metallic and opaque substances do not generally transmit light and have a very high absorption capacity. Polished surfaces of these substances show strong reflectivities.

When transmission is major or perfect, the substance is called translucent or transparent respectively; this happens in the case of dielectrics. The dielectrical properties are discussed below.

4.2 DIELECTRICAL PROPERTIES OF MATTER
Hard substances are of three types, dielectrics or insulators, electrical conductors or metals and semiconductors. Dielectrics are generally transparent and have ionic bonding; while in conductors with metallic bonding, a cloud of electrons exist.

When light impinges on the electron cloud in a metallic conductor, much of the energy of the electric vector of light is dissipated as a conduction current is imparted to the mobile electrons. In good reflector metals this dissipation is dominant.

In a dielectric substance the optical behaviour can be explained in terms of its refractive index. In a strongly light-absorbing medium the explanation can be made using a complex refractive index, which involves the *absorption coefficient*, K, of the substance. K is involvd in the Lambert's Law of intensity relationship thus:

$$I_x = I_o e^{-kx}$$

where I_x = intensity of light after passing through a thickness of x of the substance.
I_o = initial intensity

4.3 REFLECTANCE
The ratio of the itensity of the reflected light to the incident light falling on a surface is called the reflectivity. In ore microscopy it is measured in terms of the percentage of reflection.

Theoretically, the reflectivity has the following relationship,

$$R\lambda \% = \frac{(n_\lambda - N_\lambda) + K_\lambda^2}{(n_\lambda + N_\lambda) + K_\lambda^2} \times 100 \qquad (4.1)$$

where λ is the wavelength of the incident light, n and N are the refractive indices of the substance and air respectively, and K is the absorption coefficient.

It is because of the variation in reflectance with wavelength that pyrrhotite appears slightly coloured in a polished section. In a transparent medium, an increase in refractive index generally increases the reflectivity.

The equation 4.1 is strictly valid for linearly polarised light under normal incidence. In air N equals 1 and for transparent minerals K equals 0. In this section a mineral will look *opaque* if its K, the absorption coefficient, is greater than 0.01.

Absorption coefficient (K) is a measure of opacity. In cubic minerals the absorption coefficient is of one value, while in uniaxial minerals there are two values; in biaxial minerals $K_\alpha < K_\beta < K_\gamma$. A slight increase in K will greatly increase the reflectance.

The relationship between refractive index (n), absorption coefficient (K) and reflectivity (R) is shown in Table 4.1, using some common minerals.

Table 4.1: Relation between n, K and $R\%$ of some minerals

		n	K	R air (%) (N = 1)
Transparent				
Fluorite		1.434	0.0	3.2
Sphalerite		2.38	0.0	16.7
Weakly absorbing				
Hematite	(O)	3.15	0.42	27.6
	(E)	2.87	0.32	23.9
Strongly absorbing (opaque)				
Galena		4.3	1.7	44.5
Silver		0.18	3.65	95.1

Standard wavelengths: Four wavelengths are used as standards for reflectance studies. These are 470, 546, 589 and 650 nm.

4.4 COLOURS OF OPAQUES

Some ore minerals show strong colours (viz. gold, bornite, covellite etc.), some are distinctly coloured (e.g. pyrite, chalcopyrite, ilmenite etc.) others may have distinctive shades of colour, which human eyes only through practice can identify and distinguish one from the other.

In ore microscopy the colour response of the eye is dependent on the colour and reflectance of the associated minerals viz., chalcopyrite appears greenish yellow in contrast to the adjacent gold, while it looks distinctly yellow against galena or sphalerite association.

Colour sensation is also dependent on the illumination used. Illumination and light optics change with microscopes (make, alignment etc.). It is, therefore, always advisable to adjust the eyes with any microscope by studying some minerals, very 'well-seen' by the investigator.

The causes of colours of opaque minerals are discussed in the following sections.

4.4.1 Band structures

A large number of opaque minerals fall in the category of semiconductors or metals, the rest being insulators. The latter are translucent to transparent in thin (<0.035 mm) sections; their optical behaviour has been discussed in Chapter II. The optical behaviour of the opaques can be explained by considering their electronic structures.

Metals: The electronic structure of a metal is essentially a continuous band of allowed energy levels. This band is filled from the ground state upto the Fermi energy level. All energy states higher than Fermi level are empty and therefore, can accommodate excited electrons (Fig. 4.1).

This electronic configuration results in absorption of all wavelengths from infrared through visible to the ultraviolet and beyond. Matter absorbing light of all colours should be black in colour but metals are not black because, as explained earlier (*vide* section 3.12.7), an excited electron immediately returns to its original state by re-emitting a quantum with the same wavelength as the absorbed one. The polished metallic ores, therefore, have high reflectivities. The electronic structure of a metal is shown in Fig. 4.1.

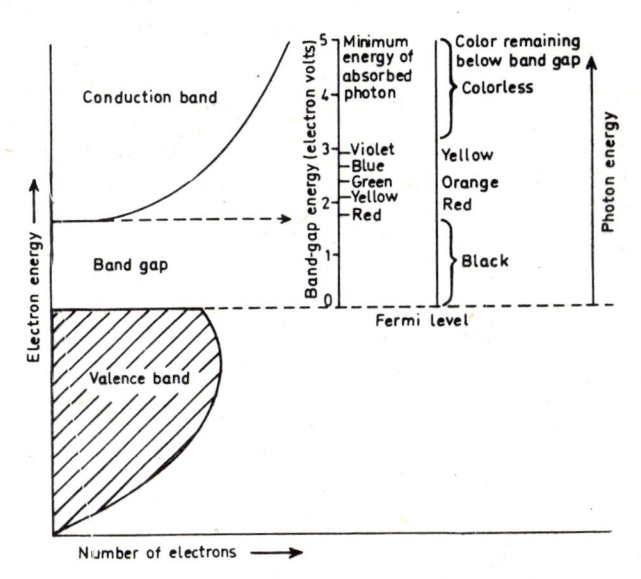

Fig. 4.1   Electronic structure of a metal

Semiconductors: The band structure of a semi-conductor is similar to that of a metal, except that a gap of forbidden energies exist between the filled valence band and the empty conduction band. Therefore, there is a minimum energy that a quantum of radiation must have in order to be absorbed. A certain quantum of energy must be introduced to promote an electron from the top of the valence band to the bottom of the conduction band.

The colour of a pure semiconductor is determined by the magnitude of the band gap. If it is in the infrared region, all visible wavelengths are absorbed and the mineral is black. When the gap energy lies in the ultraviolet region all visible wavelengths are transmitted and the mineral looks colourless. When the gap energy is in the visible region, some colours are transmitted and the mineral assumes any colour between red and yellow.

The band structure of semiconducting ore minerals are shown in Fig. 4.2. The other causes of colours observed in ore microscopy are discussed in the next sections.

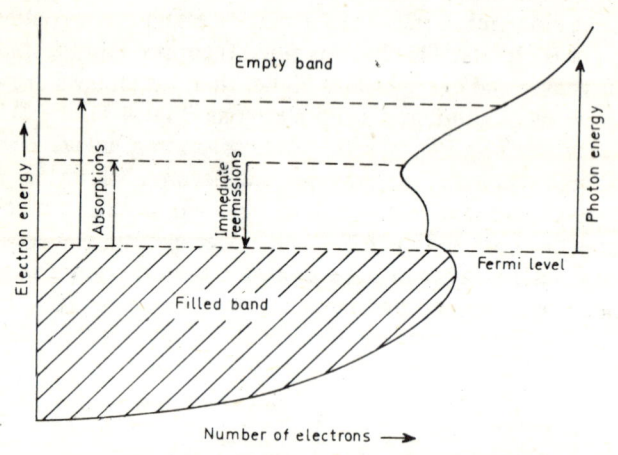

Fig. 4.2   Band structure of a semiconductor.

4.4.2 Colour related to interband transitions (in Brillouin Zone)

In specular reflection two major colours are usually observed (a) grey or white with shades of blue, pink etc. and (b) yellow.

Gray and white colours of specular reflection of coloured (blue, green, red) samples of minerals are due to the dispersion of reflectivity which is a function of refraction and absorption. The dispersion of refraction affects the dispersion of reflectivity. This explains the difference in colour of some minerals in reflected light when observed in air and in a different immersion medium. A typical example is covellite, which is dark blue in air, and red with different shades in immersion oil.

4.4.3 Absorption edge effect

If the absorption edge cuts off the spectrum near the UV region, its position at different wavelengths in the visible range causes bright and pure colours. That is how we see the red colour of cinnabar (absorption near 600 nm) and of some sulfosalts, yellow colour of greenockite (with energy gap 2.41 eV, corresponding to 514 nm) and orpiment (2.50 eV i.e. 496 nm, *vide* table 1.1.)

4.4.4 Internal reflection

Hematite which absorbs strongly at blue gives a red internal (transmitted) reflection and takes on a blue hue in ordinary light. Red ink absorbs the green of transmitted light. Therefore a thick layer of dry red ink will give a surface reflection for the frequencies of green light (*vide* section 4.7). Internal reflection colours of minerals have been presented in section 4.4.1.

4.5 COLOUR PERCEPTION

Sensation of colour is due to hue, saturation and brightness.

Hue is the characteristic colour of a range of wavelength, viz. 630-780 nm = red; 450-490 nm = blue etc. Pure colour has the maximum saturation of 100 per cent; white light is regarded as of no colour or zero saturation. Brightness affects the colour sensation.

4.6 CHROMACITY DIAGRAM

If we take three primary colours, red, green and blue, and label them as A, B and C, then any colour (X) may be made by mixing certain amounts of them. viz. a of A, b of B and c of C.

Hence,
$$X = aA + bB + cC.$$

The proportion of each colour is called the *tristimulus* value for that colour. The chromacity coordinate (x,y,z) are such that the sum of the three is taken as unity (Fig. 4.3). In this the X and Y values are plotted on orthogonal axes and the pure spectrum colours fall on an inverted U-shaped curve, the ends of which are joined by a straight line which constitutes the purple line. The boundary of the chromacity diagram represent pure colours of 100% saturation, while the central part is regarded as of 0% saturation.

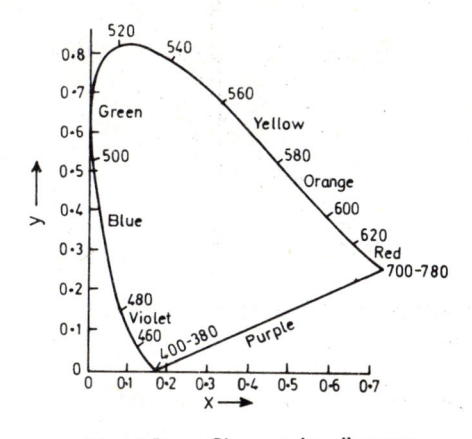

Fig. 4.3   Chromacity diagram

4.7 COLOURS OF STRONGLY REFLECTING SURFACES

Any strongly reflecting substance, such as a native metal like silver etc., has an index of refraction, which for some frequencies has a large imaginary part. Therefore when light falls normally from air ($n = 1$) to a surface of such material with $n = -in_i$, the intensity of reflected wave with reference to the incident wave can be represented as:

$$\frac{I_r}{I_i} = \frac{|1+in_i|^2}{|1-in_i|^2} = \frac{1+n_i^2}{1-n_i^2} = 1$$

Therefore, from a material with an index which is a purely imaginery number, there is 100 per cent reflection and it works as the ideal mirror.

But we know that a large imaginary part in the index means strong absorption. Therefore, this contradiction leads to the rule that if any substance is a good absorber at a particular frequency, the waves are strongly reflected at the surface and only a very small number of the waves can enter to be absorbed. Stated simply, a material which absorbs a light of frequency ω will reflect light of that frequency.

Crystals of deep coloured dyes have metallic lustres. Thickly dried purple ink on a glass plate gives a golden metallic reflection, while dried red ink will give a greenish metallic reflection. Red ink absorbs the greens out of transmitted light, so if the ink is thick and dry it will give a surface reflection for the frequencies of green light. An interesting experiment will illustrate this further. Coat a glass plate with red ink and let it dry. Flash a beam of light from a torch at the back (Fig. 4.4). The transmitted light will look red while the reflected one will be green. Hematite which absorbs strongly at blue gives a red internal (transmitted) reflection and takes on a bluish hue in ordinary light.

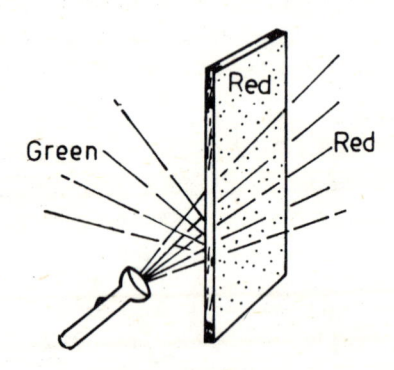

Fig. 4.4   White light reflected from a glass coated with red paint can appear green in colour.

4.8 MICROSCOPY FOR OPAQUE'S STUDY

Ore microscopy (called mineralography or mineragraphy) was first initiated by Berzelius in Sweden in 1806. This technique is now extensively used for the following studies:

Identification of opaque minerals; sequence of ore-mineral precipitation; degree of oxidation; incidence of periods of ore deformation; relative temperature of ore formation; textural analysis of minerals; ore dressing etc. Various economic applications of ore microscopy have been found.

Ore samples, chipped or cut first to a surface or mounted in bakelite or cold-setting plastic, are progressively polished using a polishing machine and a series of abrasives to smoothen out reflecting surfaces. The techniques of polishing are available in standard books on metallography. The polished samples are studied under an ore microscope.

4.8.1 Ore. Microscope

The polarising ore microscope (Fig. 4.5) is fitted with a special attachement that illumines the surface with vertically incident light. The light source is of high-intensity and is connected with a transformer to vary light-intensity. Light from the lamp is polarised by a polariser placed in the light-entrance of the microscope. The objectives and the light source in reflecting microscopes are of a special nature and significance.† These two are discussed below.

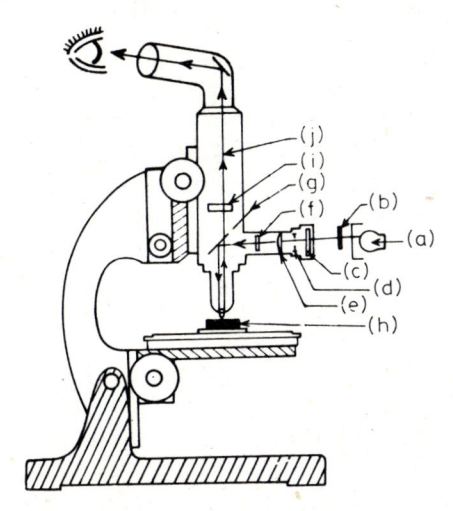

Fig. 4.5 A sectional view of an ore-microscope showing the lamp (a), filter (b), polariser (c), iris diaphragm (d), condensing lens (e), diffuser (f), reflector (g), sample (h), analyser (i) and line-path (j).

Objectives: On the objective N.A. (Numerical Aperture) is imprinted. It is related to the refractive index of the medium (n) and the angle (μ), which is ½ the angular aperture of the lens as:

$$N.A. = n \sin \mu$$

Table 4.2. Optical specifications of the objectives

Objectives	N.A. (Numerical Aperture)	A.A. (Angular Aperture)	Focal Length
5 X(oil)	0.09	7°	50
5 X(air)	0.09	10°	50
10 X (air)	0.20	23°	25
20 X(oil)	0.40	31°	12.5
50 X(oil)	0.85	69°	2.0

† Microscope objectives must be corrected for spherical and chromatic aberrations, since they must receive rays over as large an aperture as possible.

92 *Fundamentals of Optical, Spectroscopic and X-ray Mineralogy*

The objectives commonly used in ore microscopy are of five magnifications. They have the optical specifications and focal distances as shown in Table 4.2. As in petrological microscopes higher magnifications reduce the free working distance between the objective and the simple.

The objective for oil immersion requires a drop of immersion oil (R.I. ~ 1.515, at 589 nm at 20°C) which reduces the diffusion of light scattered from the mineral surface and consequently enhances colour contrast*. This enables observation of weak anisotropism and bireflectance, which are otherwise not visible in air medium. The oil is removed from the objective lens by employing solvents, usually alcohol.

Illumination: For illumination, normally tungsten filament lamps are operated by a variable rheostat. The temperature of the filament is around 3000K (2850K - 3300K). To eliminate the 'yellowness' of the light a pale blue filter is put before the lamp to make day-light. However, different colour-effects of minerals in different microscopic set-up are due to the temperature of incandesence of the filament.

The light path is guided usually by two lenses, two or three diaphragms and a polariser before it reaches the sample surface.

The polariser is placed between the lamp and the collector lens or between the diaphragms. Many workers find it convenient to put polars a few degrees (3-5 degrees) from true zero or parallel position to observe the pleochroism or polarisation colours.

The properties of minerals are studied on the basis of the following observations:
   a) Colour and colour change on rotation i.e., pleochroism
   b) Reflectivity and bireflectance
   c) Crystal form and habit; cleavage and partings
   d) Relative polishing hardness (followed by quantitative micro-hardness determination by indentation methods)
   e) Isotropic or anisotropic character between crossed polars
   f) Nature of internal reflections, when present
   g) Twinning
   h) Micro-chemical tests.

The principles of some of these are discussed in the following sections.

4.8.2 Reflecting and Interference Colour

The light, incident on the polished sample, is partly reflected and the rest is absorbed or transmitted through it. Transparent or translucent substances, which are dielectrics, will transmit much of the incident light.

Metallic substances have high absorption coefficients and are better at reflecting. Most ore minerals have high absorption and a little transmission and they show whitish colour in reflected light.

---
*Oil immersion objectives admit a wider cone of rays than dry (in air) objective of the same diameter. In the latter the cone of rays reflecting from the surface are bent outwards and thus the angular divergence is increased; while in an oil immersion objective, the angular divergence is minimised. For this reason, only with oil immersion objectives the highest numerical aperture ~ 1.4 can be attained.

Unlike the interference colours observed in a thin-section, the anisotropic colours observed in cross-nicols in ore-microscopes are due to rotation of the polarisation plane of the reflected beam of light.

Absorption and reflectance are selective. A highly transmitting substance may allow certain wavelengths of white light to pass through, while other wavelengths are absorbed and reflected. Ex., cuprite, a copper oxide, and the ruby coloured silver ores, proustite and pyrargyrite, are pale blue (to bluish green) in reflected light but blood-red in transmitted light (*vide* Table 3.2).

In white light, each wavelength undergoes a different rotation, the resolved parts are integrated in the direction of the analyser and the diagnostic interference colours are observed under crossed nicols.

The intensity of the light coming out of the analyser is greatly reduced and it varies with the resultant amplitude thus:

$$I \propto (amplitude)^2$$

4.9 BIREFLECTANCE AND REFLECTION PLEOCHROISM

The difference in maximum and minimum values of the reflectance is a measure of reflection pleochroism or bireflectance. In an anisotropic medium, the variation in absorption depends on orientation with respect to the polarisation plane of the incident light. Strong bireflectance is shown by covellite, molybdenite, stibnite, vallerite, graphite etc. Moderate bireflectance is shown by cubanite, pyrrhotite, marcasite, hematite, niccolite, cubanite, and weak bireflectance is seen in arsenopyrite, ilmenite, energite etc. In studying minerals having weak to moderate bireflectance it is advisable to observe the minerals in a neighbourhood which does not show any change in colour i.e. nonpleochroic.

4.9.1 Internal Reflection Colours

Some minerals may be translucent and allow the light to penetrate the surface and get reflected from cracks and flaws within the grains, giving a diffused internal reflection. This may be seen in polarised light and also under crossed nicols with strong illumination. Usually the body colour of the mineral is exhibited due to the preferential absorption of other wavelengths from the visible light. The following minerals show characteristic internal reflections:

Hematite	— red
Pyrargyrite-Proustite	— ruby-red
Cassiterite	— yellow or yellow-brown
Sphalerite	— yellow to brown (green to red)
Cinnabar	— blood red
Rutile	— yellow to steep reddish brown
Anatase	— blue
Azurite	— blue
Malachite	— green
Wolframite	— blood red
Chromite	— deep brown.

The colours become more pronounced when the minerals are seen in oil, with high-power objective. The grain boundaries are sites where the internal reflections are more commonly seen.

4.9.2 Anisotropism

For studying very weak anisotropism it is advisable to move the analyser through a small angle (5-10°) and increase the illumination intensity and reduce the aperture of the diaphragm.

4.10 THE HARDNESS

The hardnesses of opaques are qualified as:
1. Polishing hardness
2. Scratching hardness
3. Micro-indentation hardness.

Polishing hardness

Polishing hardness reflects the resistance offered by the mineral grains to abrasion during polishing. The relative variation in this hardness gives the *polishing relief,* in which the harder minerals stand out higher than the more deeply eroded softer minerals in association.

Some times the *Kalb light line* is used for studying this property, much like Becke line test, although they differ in origin. The light boundary line between two grains of minerals will move to the mineral of lower polishing relief when the tube is raised. Reversion will take place when the grain is of higher polishing hardness.

The relative hardnesses are mentioned as 'harder than' or 'as hard as' or 'less hard than' any associated common mineral like pyrite, chalcopyrite or galena.

Under a microscope, the relative polishing hardnesses can be determined using the procedure similar to Becke test in thin section. But the results are interpreted in the opposite way; viz., with medium power objective and illumination diaphragm partly closed, the light line at the boundary will move towards the grain of lower hardness when the microscope tube is raised.

Scratching hardness

Some minerals take smooth polish viz., arsenopyrite, niccolite, ilmenite, galena, magnetite, wolframite, covellite, bismuthinite. Some soft minerals commonly have scratch marks on them viz., molybdenite, graphite, gold, etc.

The width of the scratch marks, extending beyond the boundaries of individual minerals, is indicative of the relative scratching hardnesses of the affected minerals. The scratch will become deeper and wider in softer minerals.

Micro-indentation hardness

Talmage (1925), first used the scratch testing for relative hardness measurement. This has been later modified by diamond indentation testers. Simple diamond indenters attached to the barrel of the microscope have improved the measuring device. Micro-indentation hardness of several grains of the same mineral is measured and averaged.

Micro-indentation hardness is a vector quantity and is dependent on crystallographic directions (with respect to cleavage and border of the grain etc.). In many sulphides the microhardness is found to be inversely related to cell sizes.*

Micro-indentation hardness and reflectivity, plotted against one another (Fig. 4.6) have been used in delineating the regions of minerals classified as metals, sulphides, sulphosalts, Ni-Co-Fe sulphides, and oxides (Bowie 1967, Bowie and Taylor, 1958; Gray and Millman, 1962).

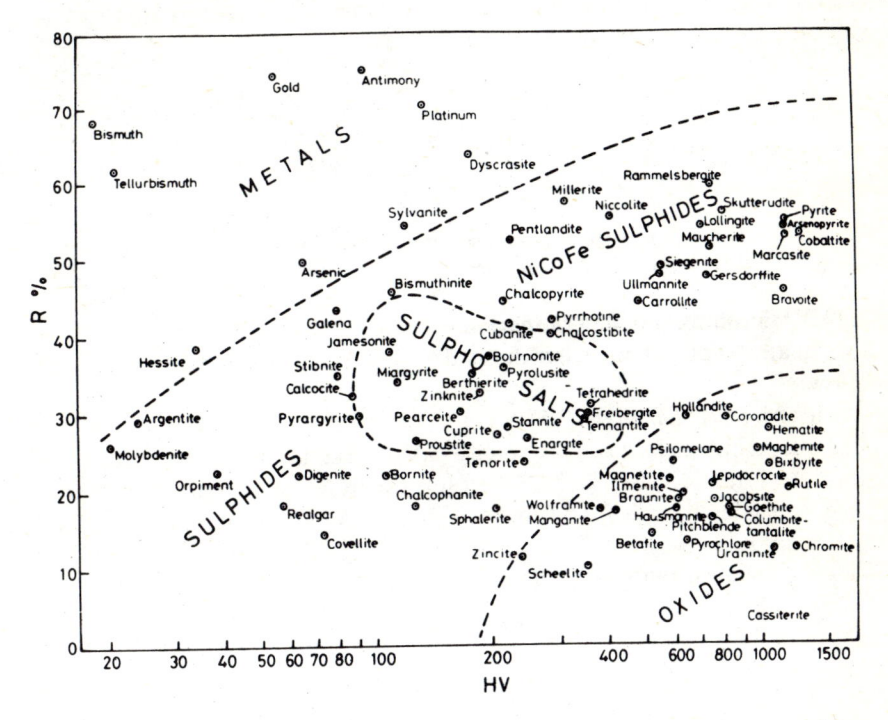

Fig. 4.6   Reflectivity (in R%) in white light and Vickers microhardness numbers (HV) for the common ore-minerals (after Bowie 1967).

Area of grains of less than 0.2 mm be used for reflectance measurement by closing the microscope diaphragm. The reflected light, however, is too small for sensitive registration by a galvanometer. A photomultiplier in circuit with selenium cell or cadmium sulphide cells greatly increases the sensitivity.

4.10.1 Set-up

The indenter is attached to the objective. To avoid any vibration effects the hardness tester should be set-up on a concrete or heavy bench. The identations are normally measured in ordinary light or preferably in a wavelength of 500 nm, using a N.A. objective of 0.85.

*This can be explained in terms of the occupancy of antibonding orbitals (e_g^*), which force the ligands away and cause increase in cell-dimensions.

4.10.2 Shapes of indentation

Micro-indentation forms are dependent on the structural characteristics of minerals. Young and Millman (1964) identified four distinct indentation types: (a) straight edge, (b) concave edge, (c) convex edge and (d) sigmoidal edge as illustrated in Fig. 4.7. Combinations of some of these types may occur. Minerals having distinct cleavages or fractures may develop fracturing and deformations around the indentations. These have been classified by Young and Millman (1964) as: (1) star radial fractures, (2) side radial fractures, (3) cleavage fractures, (4) parting fractures, (5) simple shell fractures, (6) cleavage shell fractures and (7) concentric shell fractures.

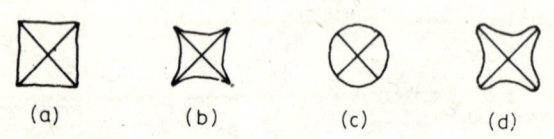

Fig. 4.7   The microhardness indentation forms (a) straight edge (b) concave edge, (c) convex edge and (d) sigmoidal edge (after Young and Millman, 1964).

4.10.3 Microhardness determination

Two major types of microhardness determination methods are discussed below:

1. *Vickers Microhardness* (VH)

A diamond indenter, which is a square based pyramid, is pressed on the sample under a load (L) of 25, 50, 100 grams, the impression is of a square and has a superficial area of $d^2/2\sin 68°$. The area of the impression is related to the hardness as:

Vickers hardness number (VHN) = load/area of impression

$$= \frac{2\sin 68° \times L}{d^2} \quad (\text{g } \mu m^{-2})$$

$$= \frac{1854.4 \times L}{d^2} \quad (\text{Kg mm}^{-2})$$

where 'd' is the length of the diagonal of the indentation square in micro metres. The diagonals of the indentation are measured and averaged.

2. *Knoops hardness (KH)*

When the indentation shape is of a rhombus, and the ratio of the diagonals is 1:7, Knoops hardness (KH) is determined from the length of the longer diagonal, 1, (in micron) as:

$$KH \ (KP/mm^2) = \frac{14230 P}{1^2}$$

where P is the load in grams.

4.10.4 Scale of equivalence

The equivalence of the Moh's scale with the measured Vickers (VH) and Knoops (KH), hardnesses is as follows:

Moh's scale	Vickers hardness VH	Knoops hardness KH
1	40	50
2	50	72
3	110	119
4	190	215
5	550	616
6	750	707
7	1100	∼ 850
8	1600	∼1200
9	>2000	2000
10	—	8000

4.10.5 Microhardness anisotropy

Microhardness is dependent on the bonding directions. Defects and dislocations also control microhardness. Hence anisotropy in microhardness is common even in cubic minerals. More often than not it is observed that VHN (100) >> VHN (110) > VHN (111).

But no relationship exist between optical and hardness anisotropies.

4.10.6 Microhardness variations

Microhardness changes with composition. In sphalerite, (Zn, Fe)S, microhardness increases with small substitution upto 2% by iron atoms, which fill vacant positions. But further substitution causes the expansion of cell structure and decrease in hardness.

For the disulphides, the cohesive energies (determined from thermochemical data) are almost equal but the metal-sulfur bond strengths decrease as $FeS_2 < CoS_2 < NiS_2$, causing hardnesses in that order.

Sample preparation methods may alter the hardness values. Cutting, grinding and polishing may increase the hardness by 5 to 30%, due to the 'work hardening' effect. This effect also increases reflectivity. Natural processes like tectonic movement may cause work hardening e.g. galena from Broken Hill (Stanton & Willey, 1971). Careful heating and hardness testing of such naturally deformed sulfides give valuable clues to their post depositional histories viz., whether a polycrystalline aggregate is a result of primary deposition or recrystallisation.

4.11 CRYSTAL FORM AND HABIT

Some minerals show euhedral forms viz., pyrite, arsenopyrite, magnetite, wolframite etc. Some are characteristically anhedral viz., chalcopyrite, bornite etc. Examples of some euhedral forms are described as:

Cubic	—	pyrite, galena
Octahedral	—	chromite, spinel, magnetite
Dodecahedral	—	pyrite, bravoite
Tabular	—	hematite, graphite, covellite, molybdenite
Rhombic	—	marcasite, arsenopyrite
Lath-shaped	—	hematite, ilmenite
Skeletal	—	magnetite, marcasite
Acicular or needle-like	—	stibnite, jamesonite, rutile.

4.12 CLEAVAGE AND PARTING

These are less common in polished specimens than in thin sections. These are commonly brought out by etching or natural weathering or by deformation.

When three sets of octahedral cleavages are present the polishing plane intersects all the planes and triangular pits may be formed as is characteristically found in galena. Similar pits may also be found in pentlandite, magnetite, gersdorffite and other minerals.

Prismatic cleavages give triangular or rectangular patterns while pinacoidal cleavages give parallel cracks.

4.13 TWINNING

Three types of twinning viz., growth, inversion and deformation, can be seen in ore minerals. In isotropic grains, twinning can be seen by etching, abrupt change in pattern of inclusions etc. In anisotropic minerals sometimes distinct composition planes are identifiable.

Lamellar deformation twins in hematite and chalcopyrite and growth twins in marcasite are common.

4.14 MEASUREMENT OF REFLECTIVITY

Reflectivity is measured as a percent of light (of a particular wavelength) reflected from a polished surface at a normal (or near to it) incidence †. It is measured as

$$\text{Reflectivity (R\%)} = \frac{\text{intensity of reflected light}}{\text{intensity of incident light}} \times 100$$

The equipment consists of a reflected light microscope with a stabilized light source, monochromatic filters or a continuous band interference filter, and a photomultiplier mounted on a microscope tube. The amount of photons impinging on the photomultiplier is indicated by the deflection in a galvanometer or by a digital output.

The amount of light reflected from a mineral specimen can be measured by replacing the eye-piece by a photon-detector like a selenium photo-electric cell or a photomultiplier tube. The current developed by photon impingement is amplified and led to a sensitive galvanometer. Calibration is done against a set of standard samples of known reflectivities at known wavelengths of light.

Quantitative determination of reflectivity of ore minerals was first attempted by Orcel (1927) using a potassium-silver photo-emission cell. E. Leitz Co. introduced a visual photometer for reflectivity measurement in 1930. Hallimond (1953) improved the visual photometer using three polarising discs. In 1964, Bowie and Henry developed a more sophisticated instrument using a simple selenium-barrier-layer cell. Recent measurement techniques have been discussed by Singh (1965), Bowie (1967), Mitra (1971), Galopin and Henry (1972).

† The relationship between electronic structure and reflectivity has been discussed by Burns and Vaughan (1970) and Vaughan (1973).

Photometers show sensitivities similar to the human eye i.e. maximum at ~ 550 nm and insensitive to red and blue end of the spectrum (Fig. 4.8.).

Now-a-days, photomultiplier tubes are more extensively used in many laboratories (*vide* Murchison, 1964). It has a greater sensitivity at the blue end of the spectrum and lower sensitivity at the red end (in contrast to the silicon cell).

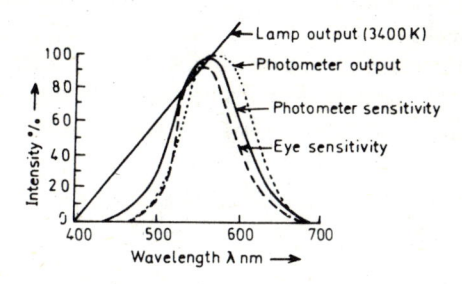

Fig. 4.8 Sensitivity of human eye, photometer and the outputs of the lamp and photometer (after R. Galopin and N.F.M. Henry, 1972).

Light source

For measuring reflectance against a standard for which absolute values are known by direct measurement, non-linearity in source and the sensor equally affect the reflectivity response of the sample and the standard, and hence cancel each other. But it is always advisable to check the linearity by using a series of neutral density filters of varying transmission. This can be achieved by employing a light source supplied by an electronically stabilized power unit or using a storage battery as a buffer.

Filters or monochromators

Monochromators are generally interference filters of continuous band type. The interference filters are of two kinds:
  (i) line type: this has ½ height of about 10 nm
  (ii) band type: this has ½ height of about 20 nm.

The other type called running filters, cover the whole range of the spectrum.

Set up

Measurements are made in air or under oil immersion. The following specifications of measurement were recommended by S.H.U. Bowie, Secretary of the Commission on Ore Microscopy (COM) in 1966 (*vide* NPL, England, Report, N. 2538, August 22, 1966):

       Mean angle of incidence .... 7°
       Maximum inclination to any ray of the principal ray ..... 2°
       Semi-angle of cone of collection of reflected light .... 2°
       Diameter of the circular patch of light on the sample   2 mm
       For this set-up the ambient temperature should be about 25°C.
  The recommended wavelengths for discrete measurements of R_o are: 470,

546, 589 and 650 nm. Of these 546 nm is most commonly used because the R_o values at 546 nm are close to the luminescence values for most minerals.

Standards

To eliminate the effect of microscopic condition affecting the reflectivity, a series of standards are chosen to cover the range of reflectivity of ore minerals. Standards are normally of isotropic material free from cleavages, inclusions and cracks.

The measurement can be made by using a single or a set of standards. But the accuracy of the procedure depends on the premise that the standard has the maximum polish, and the reflectivity of it is accurately known. The relationship used is

$$R_o = \frac{A \times C}{B}$$

where R_o = per cent reflection of the sample.
A = galvanometer deflection for the sample.
B = galvanometer deflection for the standard.
C = reflectivity of the standard.

Some standards, originally prepared by NPL, England, in 1966, have the following reflectivities.

Reflectivities of Standards

Wavelength (nm)	Black glass N. 2538.40	Carborundum N. 2538.27	Silicon N.2538.37
500	4.47	20.5	38.9
546	4.435	20.25	36.9
589	4.415	20.05	35.75

The COM's specification for another set of standards at 546 nm in air R% are:

black glass ∼ 4.5%
silicon carbide (SiC) ∼ 20%
tungsten-titanium carbide (WTiC) ∼ 50%

The standards which can be purchased from Zeiss, W. Germany, give reflectivities in air for ordinary light as:

Black glass, NG-1 .... 4%
Black silicon carbide 20%
Tungsten carbide 47%

The standard silicon-carbide prepared by NPL, England, for the wavelengths give the following reflectivities.

Wavelength (nm)	R%
400	21.8
420	21.5
440	21.2
460	21.0
480	20.8
500	20.6
546.1	20.4
589.3	20.2
620	20.0
700	19.8

Measurements

Normally a black light-tight cap is mounted on the objective to determine the back ground. This reading which is due to primary glare may also be offset by adjusting the zero-setting control till the galvanometer reads zero. The wavelength of incident light is changed and the counts for the back ground and the sample are noted. The counts from standards (viz., glass, carborundum and silicon) are obtained several times. After measurement of each of the standards the polished ore sample is measured and the counts of the sample are collected together and averaged. Using the relationship of wavelength used and the counts obtained for a known reflectivity, the reflectivities of the samples are determined against each of the standards. Average of the values obtained against each standard are made to obtain precision.

The results are expressed as per cent of the incident light reflected. Refined values are obtained by using a series of wavelengths, normally corresponding to some sensitive colours of the visible region.

Reflectivity ranges of some sulphide minerals are given in Table 4.3 along with their reflection pleochroism.

Table 4.3 Reflectivities and Colours of Some Sulphide Minerals

Reflectance (R % at 546 nm)	Colour range	Mineral
7 — 24	Bluish white—deep blue	Covellite
19 — 39	White—greyish white	Molybdenite
34 — 40	Brownish yellow—pinkish brown	Pyrrhotite
35 — 40	Yellow—pinkish brown	Cubanite
38 — 45	Yellowish white—whitish grey	Bismuthinite
50 — 56	Light yellow—yellow	Millerite

4.15 REFLECTION SPECTRAL DISPERSION

The optical constants (K and n) vary as function of the wavelength of the incident light.

102 *Fundamentals of Optical, Spectroscopic and X-ray Mineralogy*

The spectral dispersion curves of opaque surfaces, as shown in Fig. 4.9, show that blue substances have R% towards blue end (i.e. 400 *nm*) of the spectrum while red surfaces tend towards red (i.e. 700 *nm*). The white substances show reflectance independent of the wavelength of light.

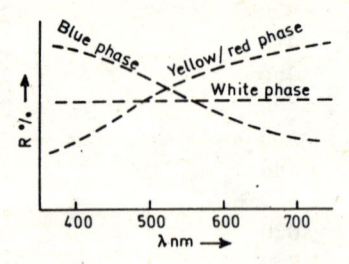

Fig. 4.9   Spectral dispersion curves of ore minerals.

But anisotropic substances show variation of R (and *n, k*) with the orientation. In the case of uniaxial minerals the dispersion curves of R_ω and R_ϵ (i.e., reflectivities in ordinary and extra-ordinary vibration directions) can be plotted against the wavelengths. Ex. covellite has the spectral reflectance curve as shown in Fig. 4.10. The curves meet at red wavelength i.e., 700 *nm*. The differing shapes of the curves as a function of wavelength (dispersion) illustrate the property of reflection pleochroism.

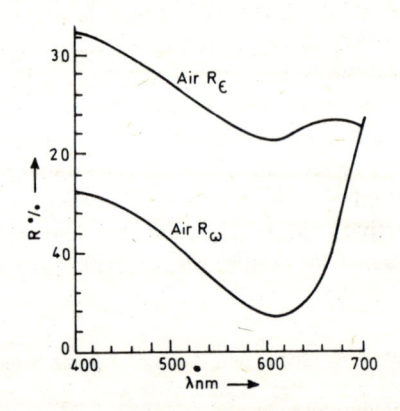

Fig. 4.10   Spectral reflectance curves for covellite in air (Ref. IMA/COM Quantitative Data File, 1977), accounting for its strong pleochroism.

4.16 STUDIES IN REFLECTED LIGHT

In plane polarised light the polished opaque samples are studied in air or oil for the properties such as colour and pleochroism, form and habit, cleavage, reflectivity, microhardness (polishing and indentation). With crossed polars the properties viz. bireflectance colour, dispersion of reflectance at different wave lengths, internal reflection, rotation of the vibration of the incident light etc. are studied.

Microchemical tests and etching using reactive solutions are done to identify the distinguishing chemistry and textural features. For opaque mineral studies under ore microscope the readers are advised to consult the standard books by Galopin & Henry, Uytenboogardt, Romdohr and other authors.

4.17 SOME ILLUSTRATIONS

At the conclusion of this chapter a few illustrations depicting how opaqe minerals look like under microscopes are given in Figs. 4.11 (a-e).

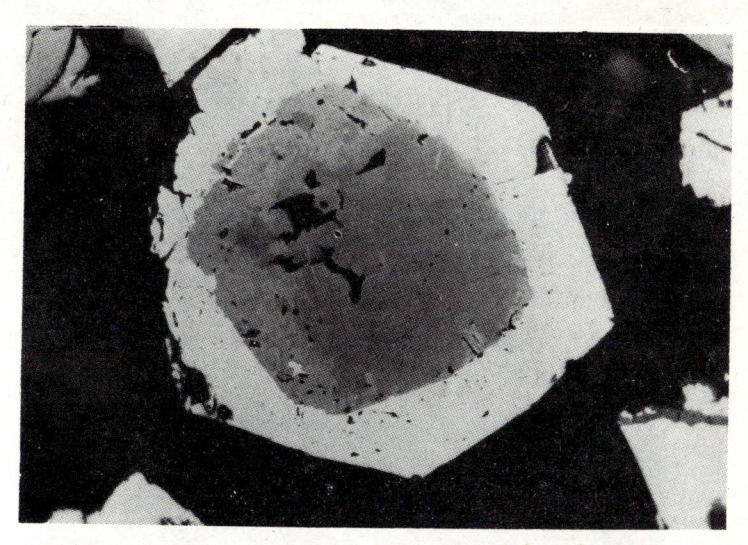

Fig. 4.11(a) An euhedral chromite (dark grey) replaced peripherally by ferroan chromite (patchy, light grey) followed by a wide rim of magnetite (whitish) grey). The original crystal faces of chromite are retained despite stream action. Sample: Chromite from Indus river, Pakistan (in oil, polarised light, wide side = 2.5 mm).

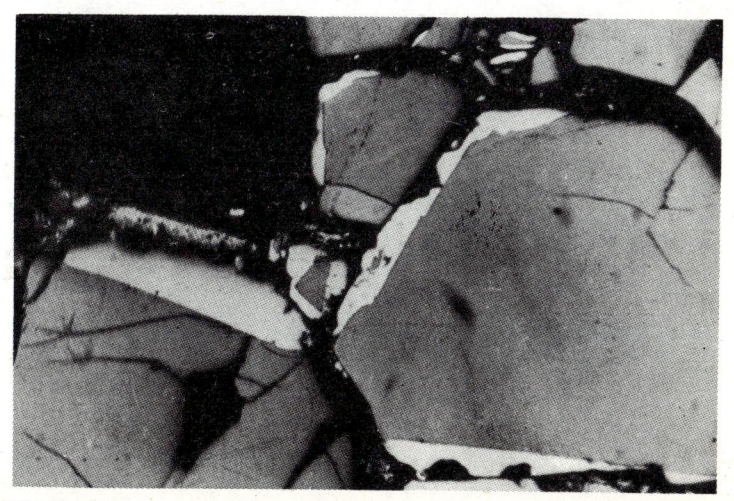

Fig. 4.11(b) Straight zone replacement of chromite (grey) by magnetite (white). Sample: Placer chromite in Aliakmon river, Greece (in oil, polarised light, wide side = 2.5 mm).

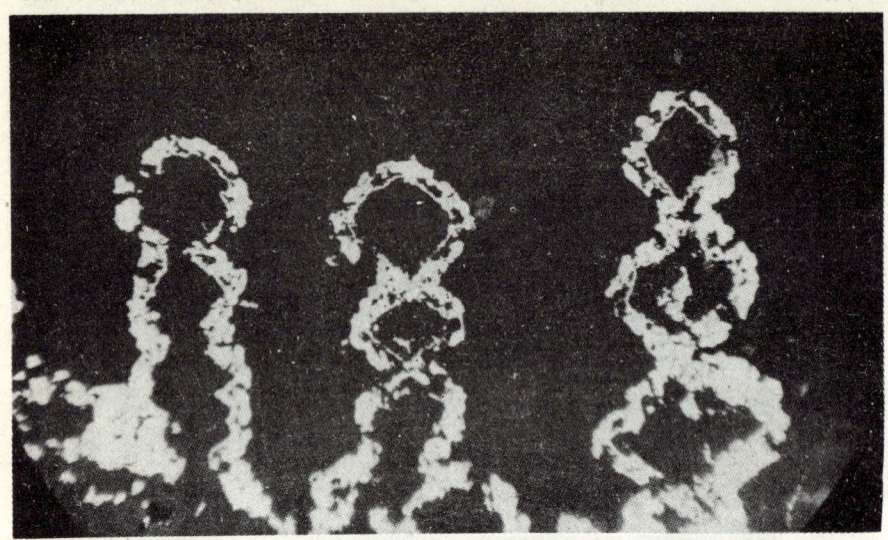

104 *Fundamentals of Optical, Spectroscopic and X-ray Mineralogy*

Fig. 4.11(c) Growth of native silver (white) around carbonate crystals (black) outlines, Sample: Sudbury, Canada, (in oil, polarised light, wide side = 2.5 mm).

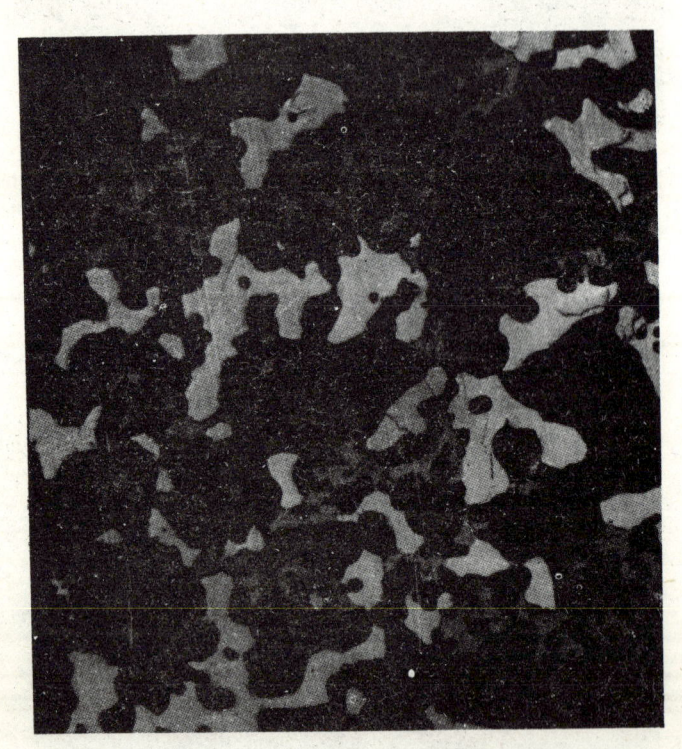

Fig. 4.11(d) Myrmekitic (eutectic) intergrowth of silicate minerals mainly orthopyroxenes (black) and iron minerals viz. troilite (dark grey), kamacite and taenite (light grey). The latter occurs with kamacite grains and has higher polishing hardness and is seen as of higher relief on kamacite.
Sample: Iron meteorite from Reckling Peak, Antarctica, RKPA 79015 (in oil, polarised light, shorter side = 2 mm).

Fig. 4.11(e) Reflection microscopic study of lunar limenite (*vide* Fig. no. 2.20(a)) coarse grained, anorthitic plagioclase (88-76 An) and subcalcic augite to pigeonite. The skeletal reflecting grains are limenite; a single grain of armalcolite (Fe Mg Ti_4O_{10}, darker than ilmenite; anisotropic, pale grey to dark-blue grey) is seen at the top left corner, enclosed in augite (in polarised light in air, longer length = 1.5 mm).

References and Selected Reading

Bowie, S.H.U. (1967) The photometric measurement of reflectivity. *Mineralogical Magazine, 31*, 476-486.

Bowie, S.H.U. and Taylor, K. (1958) A system of ore mineral identification. *Mining Magazine, 99*, 265.

Burns, R.G. and Vaughan, D.J. (1970) Interpretation of reflectivity behaviour of ore minerals. *American Mineralogist, 55*, 576-586.

Commission on Ore Microscopy (1977) IMA/COM Quantitative Data File, Edited by N.E.M. Henry, Applied Mineralogy Group, Mineralogical Society, London.

Cameron, E.N. (1961) Ore Microscopy, John Wiley & Sons, New York.

Galopin, R. and Henry, N.F.M. (1972) Microscopic Study of Opaque Minerals, McCrone Research Associates, London.

Gray, I.M. and Millman, A.P. (1962) Reflection characteristics of ore minerals. *Economic Geology, 57*, 325-349.

Hallimond, A.F. (1953) Manual of Polarising Microscope, York, Cooke, Troughton & Simms Ltd.

McLeod, C.R. and Chamberlain, J.A. (1968) Reflecting and Vickers microhardness of ore minerals, charts and tables. *Geological Survey of Canada Paper, 68-64*, Ottawa.

Mitra, S. (1971) A technique of accurate measurement of spectral reflectance of metallic minerals and an investigation of Sukinda chromites. *Quarterly Journal of the Geological Mining & Metallurgical Society of India, 44* (2), 89-96.

Mitra, S. (1972) Metamorphic 'rim's in chromites from Sukinda, Orissa, India. *Neues Jahrbuch fur Mineralogie, 8,* 360-375.

Murchison, D.G. (1964) Reflectance technique in coal petrology and their possible application in ore mineralogy. *Transactions of the Institute of Mining and Metallurgy, 73,* 479.

Oelsner, O. (1961) Atlas of the Most Important Ore Mineral Parageneses under the Microscope, Pergamon Press, Oxford.

Orcel, J. (1927) Sur L' emploi de La pile photo-electrique pour la measure de pouvoir reflecteurs des mineraux opaques, *Comptes Rendus Hebdomadaires Des Seances Del Academie Des Sciences, 148,* 1141.

Piller, H. (1966) Colour measurements in ore microscopy, *Mineralium Deposita, 1,* 175-192.

Ramdohr, P. (1976) The Ore Minerals and Their Intergrowths, Pergamon, Oxford.

Schouten, C. (1962) Determinative Tables for Ore Microscopy, Elservier, Amsterdem.

Singh, D.S. (1965) Measurement of Spectral Reflectivity with the Reichert microphotometer. *Transactions of the Institute of Mining and Metallurgy,* London, *74,* 907-16.

Stanton R.L. & Willey, H.G. (1971) Recrystallisation softening and hardening in sphalerite and galena. *Economic Geology, 66,* 1232-38.

Talmage, S.B. (1925) Quantitative standards for hardness of ore minerals, *Economic Geology, 20,* 535-553.

Uytenbogaardt, W. and Burke, E.A.J. (1971) Tables for the Microscopic Identification of Ore Minerals, Elsevier, Amsterdam.

Vaughan, D.J. (1973) Spectral reflectance properties of minerals and their interpretation. Mineralogy and Materials News Bulletin for Quantitative Microscopic Methods, nos. 3 and 4, Cambridge, England.

Ward, J.C. (1970) The structure and properties of some iron sulphides, *Review of Pure and Applied Chemistry, 20,* 175-206.

Westbrook, J.H. & Jorgensen, P.J. (1968) Effects of water desorption on indentation microhardness anisotropy in minerals. *American Mineralogist, 53,* 1899-909.

Winchell, H. (1945) The Knoop microhardness and deformation characteristics of ore minerals. *Transactions of the Institute of Mining & Metallurgy, 73,* 437-66.

Young, B.B. and Millman, A.P. (1964) Microhardness and deformation characteristics. *Transactions of the Institute of Mining & Metallurgy, 73,* 437-466.

CHAPTER V

Reflection Spectroscopy

5.1 INTRODUCTION

Radiations reflected by a material have two components: (a) a specular component, which consists of first surface reflections and is described by Fresnel's laws for absorbing dielectrics and (b) a diffuse component, which is composed of light that has entered at least one grain and has been scattered back towards the observer. The diffuse component contains the most compositional information.

Diffuse reflectance spectrophotometry is applied both in laboratory and in the field (for remote sensing mineralogy). This method allows a rapid acquisition of data about energy (or wavelength), width and intensity of the characteristic absorption bands of a material surface. The observed spectral bands can often be ascribed by a fair accuracy to transition elements or to anionic groups like OH^-, CO_3^{2-} etc., which mostly account for the absorption bands observed within the visible to near infrared spectral range.

The absorption bands that appear in reflection spectra are primarily due to (a) electronic transitions and charge transfers by d-shell electrons in transition metal ions, (b) overtones of molecular vibrations, (c) photoconduction and (d) photoelectronic emission.

The wavelength positions of absorption band centres depend on (a) the types of the ions present and (b) on the dimensions and symmetry of the sites in which the ions are situated. To a large extent, these two factors are responsible for bringing out the spectral signatures of mineralogy.

Reflectance spectroscopy, therefore, is in contrast to X-ray fluorescence or gamma-ray spectroscopy, which give only the chemical information in remote sensing.

Field (mostly telescopic and satellite) reflection spectra are interpreted by using the combination of theoretical and laboratory techniques. The theories that are mostly used are: (a) ligand field theory (inter-electronic transitions), (b) molecular orbital theory (charge-transfer transitions), (c) band theory (photo-conductivity and photoemission) and (d) theories of molecular vibrations.

In interpreting the spectral signatures obtained over a land area by a satellite, the atmospheric absorption (due to the presence of CO_2, H_2O etc.) is also taken into account.

This method of mineralogical remote sensing using powerful telescopes has been able to map successfully the petrological units on the Moon, Mercury, Mars, asteroids and other planetary bodies.* Using ground-based telescopes (occasionally set on satellites), the pyroxene composition, Ti-content, soil maturity and mare basalt types have been determined for units across the lunar surface. The abundance of ferric oxides and water content in Martian soils have been determined, the presence of water-ice on Galilean satellites and rings of Saturn has also been detected.

5.2 LABORATORY STUDY

The equipment used in the laboratory consists of a double-beam spectrophotometer, usually Cary-14 or-17D, fitted with an appliance for diffuse reflectance, which is provided with internally $BaSO_4$ lined integrating sphere, suitable for sensing any radiation reflected within a hemispheric space. The measurement of the mineral powder (40-50μm) is taken in relation to the $BaSO_4$ reference surface whose diffuse reflectance is about 100 per cent in the 400 to 1100 nm range. (Hedelman and Mitchell, 1968).

In a linear scale where the wavelength is expressed in energy (i.e., wave number in cm^{-1}), the absorption coefficients are represented by Gaussian or Lorentzian shaped functions. Most of the *d-d* transition bands approximate Gaussian while the intense CT bands near the UV region approximate Lorentzian curves.

Example: **Silicates (pyroxenes)**

As stated earlier, reflection bands are due to absorptions effected by electronic transitions, charge transfer, molecular vibrations etc. The principles are the same as in transmission optics. The diffuse reflectance spectrum of a mineral powder is an average of the spectra of individual crystals.

The reflectance spectra of several lunar pyroxenes are shown in Fig. 5.1. The position of reflectance minima near 0.9 μm and 1.9 μm compares well with Fig. 3.6(b), where the absorption maxima correspond to the reflectance minima. But, of course, the absorption bands in the absorption spectrum are stronger than the corresponding reflection bands. This is because the latter contain the specular components.

Spectra of Clinopyroxenes

In clinopyroxene, M_1 site is more distorted than the orthopyroxene M_2 site. In addition to these two Fe^{2+} crystal field bands in stoichiometric calcic clinopyroxenes, another absorption band around 8000Å, which originates from $Fe^{2+} \rightarrow Fe^{3+}$ charge transfer, is also present.

The spectra of pigeonites $(Mg, Fe, Ca)_2 Si_2O_6$ are very similar to bronzites. Fe^{2+} ions are predominantly in M_2 position, which is 7-coordinated, and the principal absorption bands occur at about 9600Å and 19000Å.

In diopsides and hedenbergites from skarn deposits, which have Ca^{2+} ions completely filling the M_2 sites in $Ca (Mg, Fe^{2+}) Si_2O_6$, contributions

---

* However, vegetation and cloud on the Earth, windblown dust on Mars and agglutinitic glass on the Moon and Mercury cause masking effects. To a lesser extent, reflection geometry and polarisation effects complicate the interpretation.

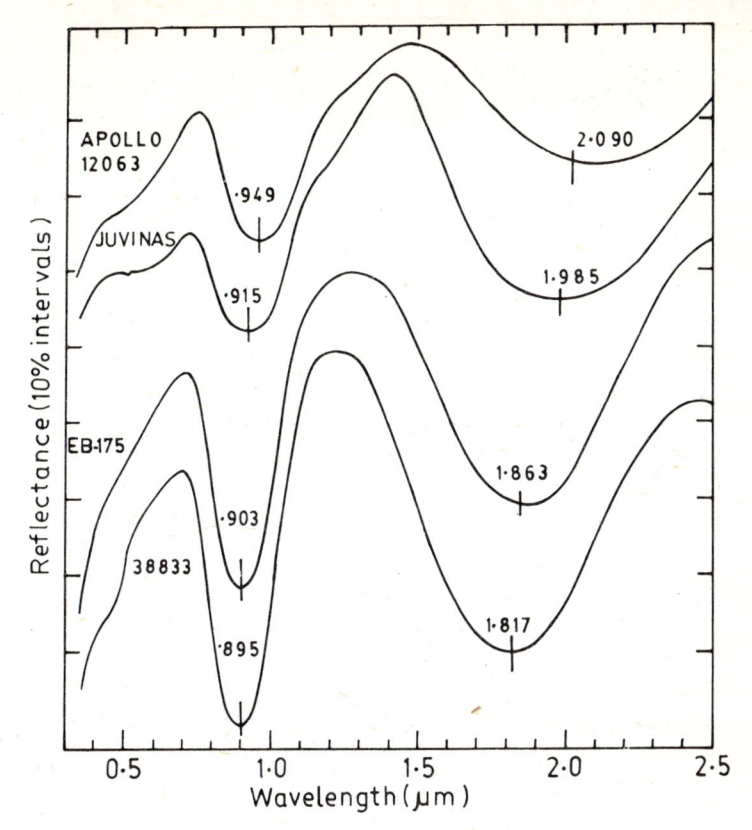

Fig. 5.1   Reflection spectra of lunar pyroxenes.

from Fe^{2+} in the M_1 sites dominate the spectra with crystal field bands at 10,000-10,400Å and 12,000 cm^{-1} (Burns, 1982).

In comparison, augites are deficient in Ca, which helps Fe^{2+} ions in occupying the M_2 sites. This results in the abserved 9,800 − 10,300Å and 20,200 - 20,300Å bands. Fe^{2+} ions at M_1 site produce a band at 12,000Å and broadens the band at 10,000Å. The Fe^{3+} ions present in augites cause $Fe^{2+} \rightarrow Fe^{3+}$ intervalence transitions which are observed as a peak around 8000Å.

Significantly, a plot of bands near one micron versus two micron bands found in pyroxene samples give a trend demonstrating the relationship between the spectral characters and the composition and structural types of pyroxenes. This is illustrated in Fig. 5.2.

Soil or rock powders are essentially a mixture of several minerals and the reflection spectrum is an weighted average of the spectra of individual minerals. Minerals with optical depth-to-grain size ratios near unity will contribute stronger absorption bands to the reflection spectrum than the transparent or opaque constituents.

Amongst the common rock-forming minerals, the iron-bearing pyroxenes typically have the deepest and best-defined absorption bands. The two characteristic bands are due to Fe^{2+} in highly distorted octahedral sites.

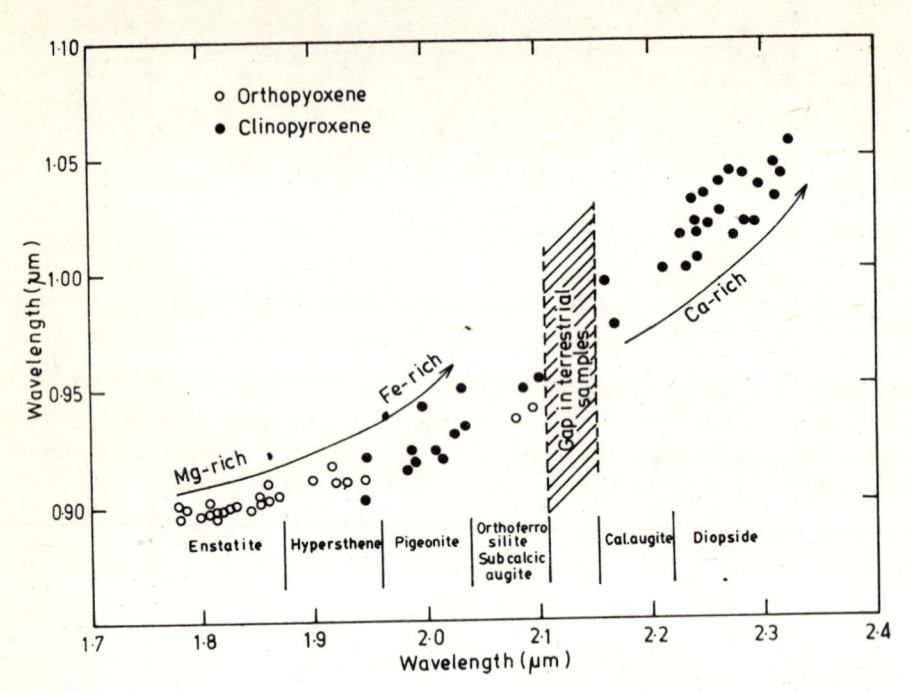

110 *Fundamentals of Optical, Spectroscopic and X-ray Mineralogy*

Fig. 5.2 Wavelengths of centres of absorption bands near 1μm and 2μm in the diffuse reflectance spectra of pyroxenes. Orthopyroxene bands shift to longer wavelengths with increasing amounts of Fe, while clinopyroxene bands shift to longer wavelengths with increasing calcium content.

With compositional change of a mineral, the wavelength positions of bands change. The wavelength positions of absorption bands in diffuse reflectance spectra of pyroxenes are diagnostic of their composition and structure. In orthopyroxenes and low-Ca-pigeonites, Fe^{2+} occurs mainly at M_2 sites. As Ca^{2+} content increases, M_2 sites get gradually filled by Ca^{2+} and M_1 accommodates more of Fe^{2+}. Therefore, the absorption band centres (due to Fe^{2+}) of orthopyroxene and low-calcium clino-pyroxenes change as a function of $Fe/(Mg+Fe+Ca)$. With increase in Ca in clino-pyroxenes the centres of the main absorption bands change as a function of $Ca/(Mg+Fe+Ca)$.

A plot of the wavelength centres of absorption bands near 1μm against the 2μm in the diffuse reflectance spectra of pyroxenes is given in Fig. 5.2. It shows that orthopyroxene bands shift to longer wavelength with increasing amounts of iron, while clino-pyroxene bands shift to longer wavelengths with increasing calcium content (Adams, 1974). Common orthopyroxenes show bands in the region of 0.90μm and 1.80μm, hyperstenes have bands at 0.91μm and 1.90μm; pigeonite occur next, the rare orthoferrosilites may occur near 0.93μm and 2.05μm. Subcalcic augites are next along the trend. Then there is a break for terrestrial pyroxenes; Ca-augite and diopside occur next in the upper right corner with Ca increasing toward longer wavelength.

Such criteria for other minerals viz., olivines, felspars, amphiboles, carbonates, transition metal oxides, clay minerals, glass etc. can be established by the use of a reflectance spectrometer.

Opaques

Diffuse reflectance spectra of some opaque oxides and minerals are given in Fig. 5.3. (Adams, 1975).

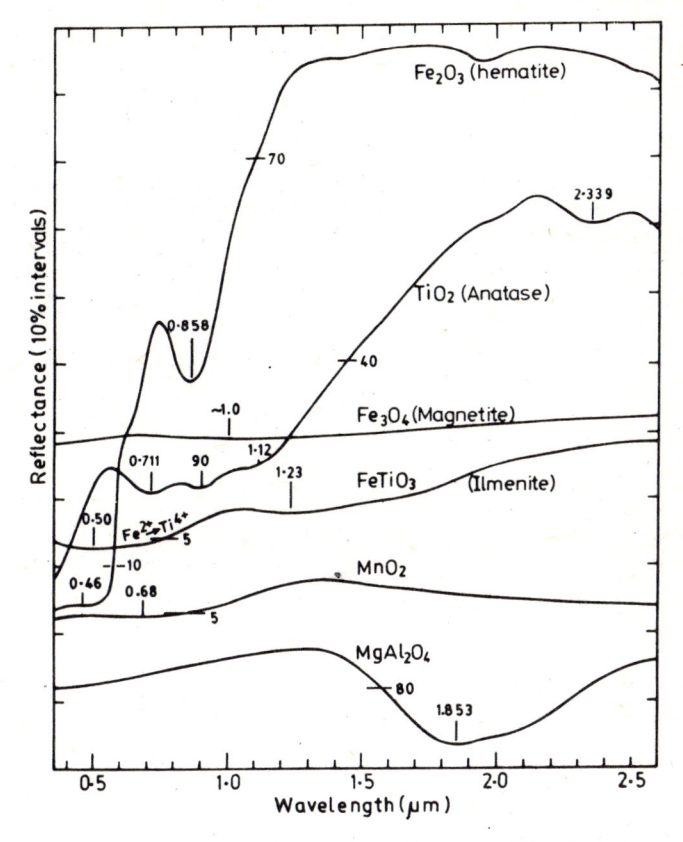

Fig. 5.3   Reflection spectra of some oxide minerals.

Ilmenite, *FeTiO₃*, shows a broad band near 0.96μm. This is due to $Fe^{2+} \rightarrow Ti^{4+}$ CT.

Anatase, *TiO₂*, gives Fe^{2+} bands at 0.90, 1.12, 2.34 μm. and a CT band at 0.71 μm.

Magnetite, *Fe₃O₄*, spectrum is complex and looks flattish with hardly any discernable peak. The spectrum of magnetite has been extensively studied by Huguenin (1973 a, b) who explains that β-spin electrons migrate over the octahedral site sublattice causing a continuum of optical absorption.

Hematite, *Fe₂O₃*, spectrum shows an intense blue to uv charge transfer absorption feature, developed due to $O^{2-} \rightarrow Fe^{3+}$ transition. Superimposed on the CT shoulders are a set of Fe^{3+} transitions bands. These bands arise from d manifold interelectronic transition, which CFT predicts to be forbidden, both by parity and symmetry; because Fe^{3+} has five electons, each occuppying one of the five $3d$ orbitals with parallel spin.

In the range where Fe^{3+} transitions occur, the O^{2-} and Fe^{3+} electronic states are in part symmetry-and parity-coupled. This coupling is greatest at the CT band centres and decreases towards its band edge. This causes a progressive increase in intensity of the transition bands toward shorter wavelengths.

5.3. MIXTURES AND MODAL ANALYSIS

Diffuse reflectance spectra of mineral mixture offer valuable clues to the proportion of the constituents. This is done by studying the positions and depths of the occurring bands. This effect is illustrated in Fig. 5.4, which shows the variation of band depths with different mixtures of plagioclase, pyroxene and magnetite (Adams, 1974).

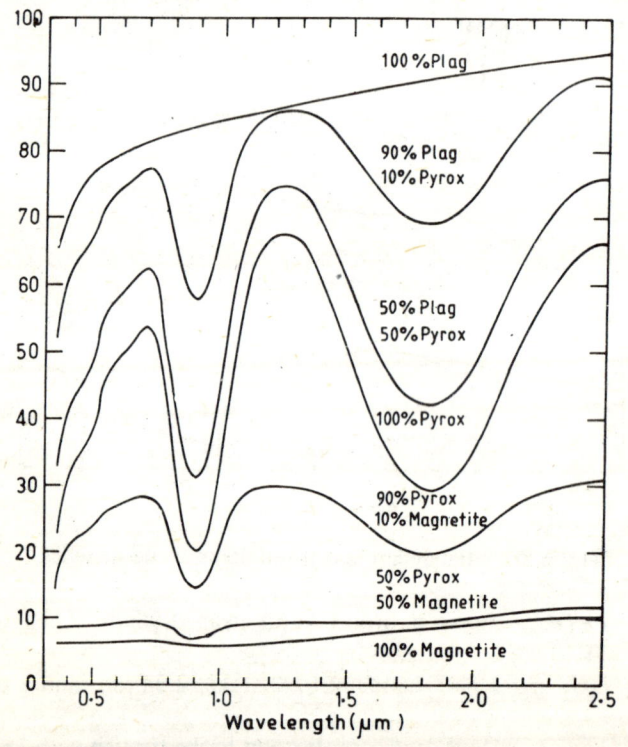

Fig. 5.4 Variations in the depths of 1 μm and 2 μm bands in mineral mixtures.

Technique for determining modal abundances from reflectance spectra can therefore be developed by using the spectral characters of a series of mixtures of a series of common minerals (Gaffey, 1975).

Analysis of reflectance spectra gave a fairly accurate petrological mapping and Ti-content variations on lunar surface before the lunar samples were available.

References and Selected Reading

Adams, J.B. (1974) Visible and near-infrared diffuse reflecting spectra of pyroxenes as applied to remote sensing of solid objects in the solar system. *Journal of Geophysical Research, 79,* 4829.

Adams, J.B. (1975) Interpretation of visible and near infra-red diffuse reflectance spectra of pyroxenes and other rock forming minerals, in: Infrared and Raman Spectroscopy of Lunar and Terrestrial Minerals, edited by C. Karr, Jr. Academic Press, New York, pp. 91-116.

Burns, R.G. (1982) Lectures on Geochemistry and Spectra of Transition Elements in Minerals, delivered at Department of Geological Sciences, Jadavpur University in May-June, 1982, (unpublished).

Fredericks, P.M., Tattersall, A, and Donaldson, R. (1987) Near Infra-red reflectance analysis of iron ores. *Applied Spectroscopy, 41* (6), 1039-1042.

Gaffey, M. (1975) Qualification of spectral reflectance characteristics for mineralogical and petrological information, paper presented at Annual Meetting of the AAS/DPS, Columbia, Maryland, 17-21 February, 1975.

Hall, W. (1981) International Symposium on Near Infra red Reflections Analysis, Tarytown, New York.

Hedelman, S. and Mitchell, W.N. (1968) Some new diffuse and specular reflectance accessories for the cary models 14 and 15 spectrophotometers. *in* Modern Aspects of Reflectance Spectroscopy (*ed.* W. Wendlandt), Plenum Press, New York.

Huguenin, R.L. (1973a) Photostimulated oxidation of magnetite, 1. Kinetics and alteration phase identification, *Journal of Geophysical Research, 78,* 8481-8493.

_____ (1973 b) Photostimulated oxidation of magnetite, 2. Mechanism, *Journal of Geophysical Research, 78,* 8495-8506.

Pieters C.M., Head, J.W., Adam, J.B. et al (1980) *Journal of Geophysical Research, B85*, 3913.

Strens, R.G.J. and Wood, B.J. (1979) Diffuse reflectance spectra and optical properties of some iron and titanium oxides and hydroxides, *Mineralogical Magazine, 43,* 347-54.

Tandon, S.P. and Gupta, J.P. (1970) Measurement of forbidden energy gap of semiconductors by diffuse reflectance technique, *Physica Status Solidi, 38,* 363-367.

Wendlandt, W.W. and Hecht, H.G. (1966) Reflectance Spectroscopy, Interscience, New York.

CHAPTER VI

Infrared Spectroscopy: An Outline

6.1 INTRODUCTION

The range of the electromagnetic spectrum extending from 0.8 to 200μm is referred to as infra-red (IR). Infra-red method is sensitive to short-range ordering or nearest-neighbour relation while X-ray analysis is responsive to periodic arrangement of atoms, i.e., long-range order.

Infra-red spectroscopy has been widely used in the characterisation of a large number of organic and inorganic compounds. Absorption of infra-red energy by a mineral is associated with the vibrational and rotational motion of molecules within it.

Commonly the infra-red analysis employs the group vibrational concept to ascertain the presence or absence of various functional groups in the molecule. Certain chemical groups have characteristic absorption bands that are consistent among minerals containing these groups. The spectra originate primarily from vibrational stretching and bending modes within molecules. This method offers a fingerprint for identification of molecular structures.

Empirical correlations of vibrating groups with specific observed absorption bands offer the possibility of chemical identification and possibly quantitative analysis through intensity measurements. The band positions are noted in μm with corresponding wave number in cm^{-1} (Fig. 6.1a), *vide* Appendix III.

In the IR study of inorganic compounds and minerals the spectra usually run through the range of 2.5 to 50 μm (i.e. 4000 to 200 cm^{-1}).[†] On the other hand, UV-visible spectra generally cover 0.2 to 0.8 μm. The intermediate range, i.e., 0.8 to 2.5 μ, termed as near IR (NIR), can be obtained in most of the modern UV-visible spectrometers.

6.2 THEORY

In 1913 Niels Bohr first proved that atoms and molecules cannot have any arbitrary energy and can exist only in discrete energy states. Consequently, transitions between energy states result in emission or absorption of characteristic units of energy (quanta), which are observed variously as emission lines from excited molecules or as absorption bands in the IR, visible, and UV regions. The subject of spectroscopy is based on this concept.

---

† The three regions in IR are: near infrared (12,500 – 4000cm^{-1}), medium infrared (4000 – 650cm^{-1}) and far infrared (650 – 100 cm^{-1}).

Development of quantum mechanics offered more accurate method of describing atomic phenomena than was possible by classical mechanics.

A molecule can have the following types of motions: (a) translation of the whole molecule, which can be regarded as translation of the centre of mass, (b) rotation of the molecule as a framework around its centre of mass, (c) vibrations of the individual atoms within the framework, and (d) spins of the electrons and atomic nuclei.

Situations of (a) and (d) are not of direct relevance in analytical IR studies.

The total energy of a molecule is the sum of its translational, rotational, vibrational and electronics energies. That is

$$E_{total} = E_{trans} + E_{rot} \ E_{vib} + E_{elect}$$

Transitional energy has little effect on molecular spectra. The vibrations of CO_2 and OH are shown below (Fig. 6.1b).

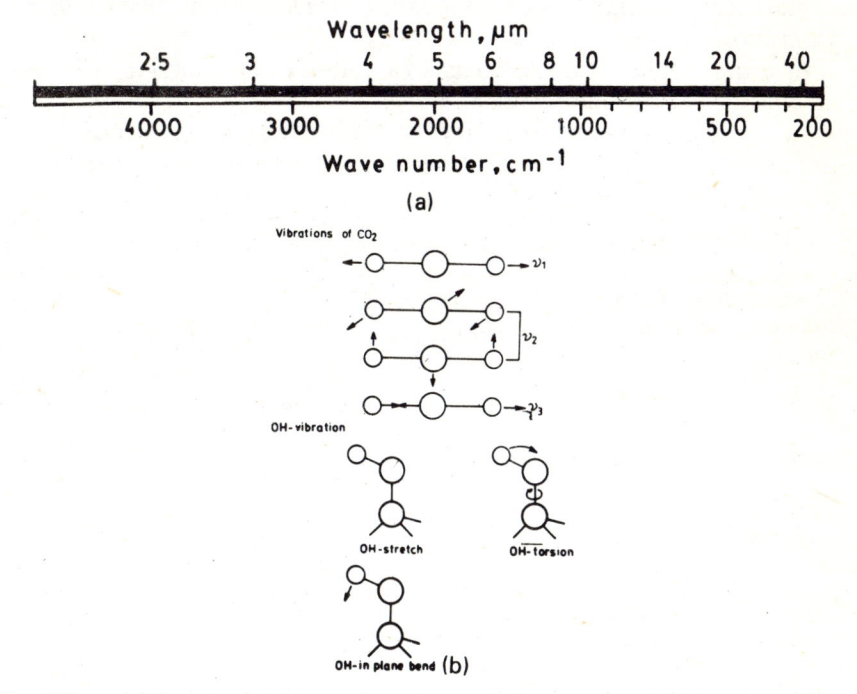

Fig. 6.1 (a) A scale of correspondence between the wavelength and wavenumber in the IR Region. (b) Vibrational modes of CO_2 and OH-group.

Spurious bands:

Laboratories involved in IR studies should have a listing of spurious bands due to the presence of some compounds in that laboratory conditions. A list of such bands have been prepared [P.J. Launer, *Perkin-Elmer Instrumentation News*, 13(3), 10, 1962], some of which are given in Table 6.1.

6.3. IR SPECTROMETERS

All IR spectrometers, whether simple or sophisticated, have a source, optical system, detector and an amplifier. The IR spectrometer types can be classified as below:

```
                          IR spectrometers
        ┌─────────────────────┼─────────────────────┐
    Sequential              Spatial            Multiplex devices
       │                       │                       │
     Single              Multichannel               Single
    detector              detector                 detector
       │                       │                       │
   ┌───┴────┐             ┌───┴────┐             ┌───┴────┐
Dispersive  Non-      Dispersive  Non-       Dispersive  Non-
           dispersive            dispersive              dispersive
   │           │            │         │            │          │
┌──┼──┐    ┌───┼───┐        │         │            │          │
Scanning Multislit Grill  Tunable Rotating Opto-  Hadamard  Fourier
                          laser   filter   acoustic transform transform
```

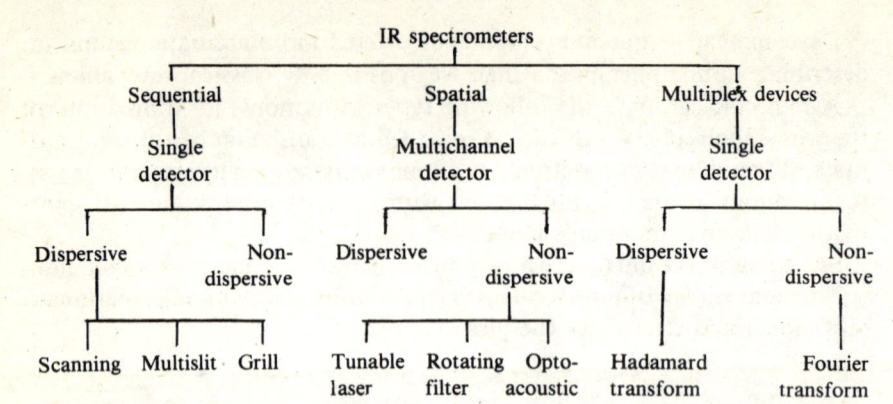

In *sequential devices* the information is collected sequentially in time. A single detector is used to collect information as each spectral element is scanned in turn.

In *spatial devices* multiple detectors are used and a photographic recording as in spectrographs is done.

In *multiplex devices* a single viewer (detector) receives many signals and are transmitted by a single channel and can be decoded for interpreting the individual spectral element.

Table 6.1 Some Common Spurious Absorption Bands

Approximate frequency (cm^{-1})	Wavelength (μm)	Compound or Group	Origin
3700	2.70	H_2O	Any source
3650	2.74	H_2	Any source
3450	2.9	H_2O	Hydrogen-bonding in water, usually in KBr discs
2350	4.26	CO_2	Atmospheric absorption
2000–1400	5–7	H_2O	Atmosphere
1640	6.1	H_2	Water of crystallisation
1430	7.0	CO_3^{2}	Contaminant in halide window
1360	7.38	NO_3^{-}	Contaminant in halide window
1270	7.9	$SiCH_3$	Silicone oil or grease
1110	9.0	?	Impurities on KBr discs
1000–1110	9–10	SiOSi	Glass, siliceous
667	14.98	CO_2	Atmosphere

Common infra-red spectrophotometers cover wavelength region from 4000 to 650 cm^{-1}. The spectrum is a plot of sample absorbance or percent transmission versus wavelengths. A grating or several prisms are used for changing wavelengths and for monochromatisation.

Sodium chloride prisms are used for the 4000 to 600 cm^{-1} region while cesium bromide prisms are used for 600 to 250 cm^{-1} range. However, the wavelength scale usually changes with use, and therefore, frequent calibration is necessary.

Mineral samples can be studied as mulls in Nujol. The mull is prepared by grinding the sample to a fine powder and mixing it with a mulling agent to make a paste. A thin layer of the paste is studied between NaCl or other optical plates. But this *Nujol method* produces spectra which are poorly defined because of the reflection and refraction of incident radiation by crystal particles.

In *KBr disc method,* the mineral powder is mixed intimately with KBr and is pulverised and pressed into a clear disc which is mounted and examined directly.

A double-beam technique is usually used with one beam impinging on a reference and the other on the sample.

Reflectance spectra of a mineral rock powder can be obtained by using commercially available reflectance attachments. Cary spectrophotometers are fitted with specially constructed bidirectional reflectance attachments in both the sample and reference beams.

The beam impinges from above on the horizontally placed cups containing reference and the sample. The minimum thickness of the sample layer required is ~ 3.5. mm. The reference is usually freshly prepared MgO powder. Because of experimentel difficulties diffuse reflectance has not been used much in the I.R. region, but an interferometer spectrometer has become available that is specially designed for this type of work (*vide* R.R. Willey, *Applied Spectroscopy, 30,* 593, 1976).

Transmission spectra

Transmission spectra are obtained from powder films made up of only large particles ($> 20\mu m$) and absorption spectra are obtained from small particles ($<5\mu m$). Transmission bond acquires prominence at the wavelength where the R.I. of the mineral approaches one.

Spectral positions of the transmission bands are characteristic for different minerals and could be used for analytical work if the particle size is controlled. The peak positions shift with average particle size variation.

Bands below cm^{-1} are often caused by *lattice vibrations,* i.e., the translational and torsional motions of molecules in the lattice. These vibrations often form combination bands with intramolecular vibrations and cause pronounced frequency shifts in the higher frequency regions of the spectrum. Complications arise when the unit cell of the mineral contains more than one chemically equivalent molecule e.g., CO_3^{-2} in calcite. In such cases, the vibrations in individual molecules couple with each other and give rise to frequency shifts and band splitting.

As a result of complications, interpretation of solid samples is more difficult as compared to liquid or gaseous states.

The *site symmetry* of a molecule in a crystal determines the selection rules for transition. Change in site symmetry, as in polymorphs, causes change in band positions. Example: calcite with CO_3^{-2} ion in D_3 symmetry, has $v_1 = 1087$ cm^{-1}, $v_2 = 879$ cm^{-1}, $v_3 = 1432$ cm^{-1} and $v_4 = 710$ cm^{-1}; while aragonite having CO_3^{-2} ion in C_s symmetry has infrared active bands, while v_3 and v_4 each split into two bands.

6.4 FINGER PRINTING

Normally the spectra of unknown samples are compared with bands of known samples.

Strong absorption bands due to vibration and rotation of atoms within the silicate minerals occur in the region of 9 to 12 µm. As the Si:O ratio progresses from 0.25 in the isolated silica tetrahedron (e.g. SiO_4 in olivine) to 0.50 (e.g. SiO_2 in quartz), the region of strong absorption shifts to shorter wavelengths and the wavelength ranges tend to narrow down.

An increase in the number of coordinated atomic bonds will increase the number of the modes of vibration or the rotations possible within wider range of wavelengths.

The spectral peaks of quartz are due to: Si-O stretching at 9.10-12.5µm (1100-800 cm^{-1}), Si-Si stretching at 12.5-16.67µm (800-600 cm^{-1}) and Si-O-Si bending and distortion at 21.74-23.3µm (460-430 cm^{-1}).

Above 2500 cm^{-1} nearly all fundamental vibrations involve a hydrogen stretching mode. The O-H stretching vibration occurs around 3500 cm^{-1}. Hydrogen bonding lowers the frequency and broadens the band. Absorption of Si-H occur at 2300 cm^{-1}.

Absorption spectra of minerals containing the hydroxyl group are characterised by an absorption band at the shorter wavelengths, usually greater than 2800 cm^{-1}. The absorption is due to the stretching vibrations within the hydroxyl group. The minerals like serpentine, goethite, lepidocrocite etc. contain hydroxyl group and show the characteristic absorption band in their infrared spectra.

6.5 CHARACTERISATION

6.5.1 H_2O bands

The presence of H_2O in a mineral can be detected by the two characteristic absorption bands in the 3600 to 3200 cm^{-1} region and in the 1650 cm^{-1} region. If water is present as lattice water, an additional band occurs in the 600 to 300 cm^{-1} region. When H_2O is coordinated to a metal ion another band appears in the 880 to 650 cm^{-1} region. These peaks are due to combinations and overtones of the fundamental vibration modes of the water molecule.

Combination and overtone modes of H_2O occur near 1.9µm and 1.4µm respectively. The bands near 1.9µm arise from a combination of the fundamental bending (v_2) and asymmetric stretching (v_3) modes, and are polarised in the H-H direction of H_2O molecules (assuming C_{2v} symmetry).

Information about the orientation of molecular plane is obtained from the fundamental modes in which v_1 (symmetrical stretch) and v_2 (fundamental bending mode) are polarised along the molecular two-fold rotation axis, and v_3 (asymmetric stretching) is polarised along the H-H direction. The combination mode from ($v_2 + v_3$) appear at about 1.9µm.

The presence of these two bands is a characteristic of the water of hydration or of lattice-trapped water. The presence of both the 1.4 and 1.9 µm bands in a mineral spectrum indicates the presence of H_2O molecule, while 1.4 µm band without the 1.9 μm band indicates the presence of structural O H and *not* H$_2$O.

Study of channel constituents of cordierite by Goldman *et al* (1977) revealed two types of H$_2$O oriented 90° apart with their H-H directions parallel to [001] and [010]. Similar IR results were obtained in the study of beryl by Wood and Nassau (1967).

The superposition of lattice vibrations on the more intense bands may give rise to side-bands of lower intensity. In general, the resolution of the bands improves at low temperatures, especially in hydrogen-bonded systems.

6.5.2 Hydroxyl or OH⁻ groups

Absorption spectra of minerals containing the hydroxyl group are characterised by absorption bands near 3600 cm⁻¹.
Example: The 3677, 3662 and 3645 cm⁻¹ bands in talc are due to the presence of OH groups in three different sites in the crystal lattice.

The $2\nu_{OH}$ overtone gives rise to a band at about 1.4μm NIR region. In some minerals like mica, where Si-Al substitution causes different environments for the OH⁻ groups present, the 1.4μm band gets broadened.

6.5.3 H$_2$O and OH⁻ spectra in silicate minerals: Examples

H$_2$O and OH⁻ spectra in silicate minerals are illustrated with the examples of some common minerals viz., gypsum, apophyllite, mesolite and tourmaline. The first three are hydrated crystals while the last contains hydroxyl group.

Spectra were taken at 300 and 80K, using a cold-finger type cryostat. The light beam of a Cary-14 spectrophotometer passed through two silicon windows with the specimen placed in-between and the interspace evacuated. The samples were made into thin plates and the thickness (*t*), was measured with an indicating micrometer reading accurately to ± 1μm. From the optical density and thickness values, the absorption coefficients were calculated. The spectra were taken with or without a prism polariser, depending on the crystal sample, in the range 1 to 2.5 μm in the transmission mode and 1 to 1.8μm for the diffuse reflectance mode. The diffuse reflectance spectra were taken only for the purpose of comparison of results and hence at room temperature (300K) only. The absorption peaks and assignment are presented in Table 6.2.

(a) *Gypsum*, CaSO$_4$, 2H$_2$O; monoclinic. This has a layered structure with CaSO$_4$ layers separated along the b-axis by sheets of H$_2$O. There is hydrogen bonding with O-H .....O distances as 2.811 and 2.823Å.

The bands at 1.75, 1.93 and 2.21 μm resemble other hydrated salts. viz., hydrated fluosilicates (De and Datta, 1978). The diffuse reflectance spectrum at the room temperature shows these characteristic peaks although of lowered intensity and with some loss of fine structure Fig. 6.2(a).

(b) *Apophyllite*, KFCa$_4$ (Si$_8$O$_{20}$), 8H$_2$O; tetragonal: There are silicon-oxygen tetrahedra forming sheets between alternate layers of which K, F and Ca ions and water molecules occur. The last ones are assumed to be hydrogen-bonded

120 *Fundamentals of Optical, Spectroscopic and X-ray Mineralogy*

TABLE 6.2 Principal peaks in the NIR absorption spectra of single crystals at 80 K (in μm).

Apophyllite	Mesolite	Gypsum	Fluosilicate	Calc. value	Assignment
0.994		0.978	0.989	0.954	$2v_1 + v_3$
		0.998			
1.168 (0.8)		1.187	1.181 (1.8)	1.162	$v_1 + v_2 + v_3$
1.345 (0.4)		1.378 (0.8)	1.352 (0.8)	1.350	$2v_2 + v_3 + lm$
1.447 (6)	1.44 (4.2)	1.445 (5)	1.42 (7.2)	1.419	$v_1 + v_3$
1.525 (0.2)	1.46 (4.4)	1.495 (1.2)	1.46 (10)	1.486	$2v_2 + v_3$
	1.485 (4.3)	1.538 (5.2)	1.48 (8)		
1.715 (1.3)	1.737 (0.8)	1.753 (2.0)	1.718 (1.8)	1.72	$v_2 + v_3 + lm$
1.745 (1.0)	1.75 (0.6)	1.77 (2.1)	1.771 (1.8)	—	—
1.94 (10)	1.938 (14)	1.953 (10)	1.934 (10)	1.952	$v_2 + v_3$
	2.03	2.065	2.07 (0.4)	2.078	$3v_2$
	2.17 (2)	2.22 (2)	2.21 (2.6)	—	

Figures in parentheses indicate absorption coefficient values (in cm^{-1})
$v_1 = 3480$, $v_2 = 1604$, $v_3 = 3520$ cm^{-1}, lm = lattice mode.

to oxygens of the silicate network. The characteristic H$_2$O bands appearing in the sample are shown in Fig. 6.2(b).

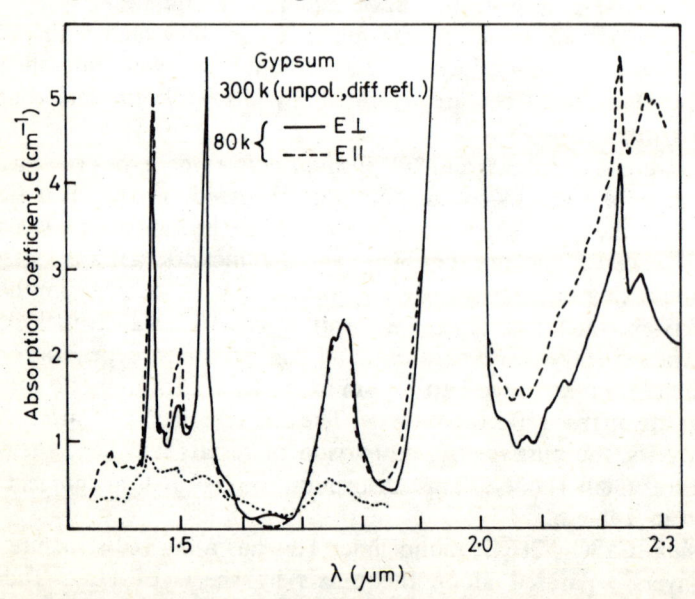

Fig. 6.2(a) Near infrared spectra of gypsum at 300K (unpolarised, diffuse reflectance for the powder) and 80K (polarised).

(c) *Mesolite*, Na$_2$Ca$_2$(Al$_2$Si$_3$O$_{10}$)$_3$.8H$_2$O; monoclinic: Hydrogen bonds have lengths 2.86 and 2.99Å between oxygen and water. As the sample is not very thick the peaks lying at 0.985 and 1.18μm are not clearly discernable and the one around 1.45μm shows clear resolution into three sharp peaks at 80K. Intense peak at 1.93 μm as well as the medium sharp peak at 2.17μm are

Infrared Spectroscopy: An Outline 121

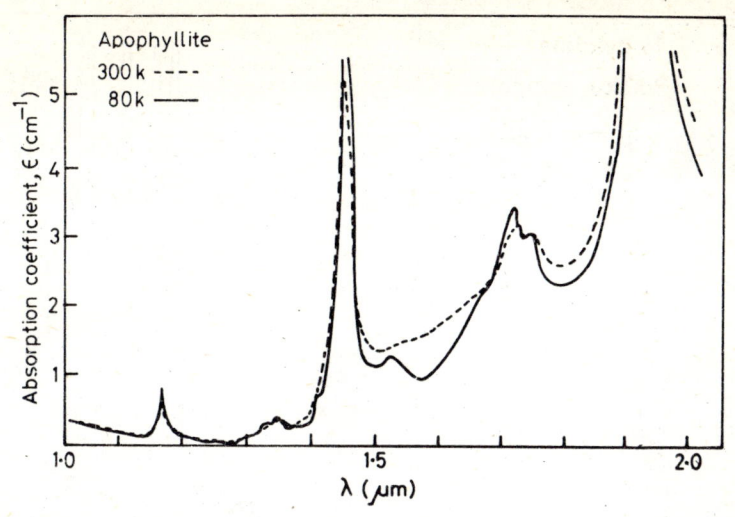

Fig. 6.2(b) Near infrared spectra of apophyllite (unpolarised) at room temperature (300K) and at liquid air (80K).

present (Fig. 6.2(c)). The band at 1.98μm which is very conspicuous at E//C almost vanishes at E⊥c.

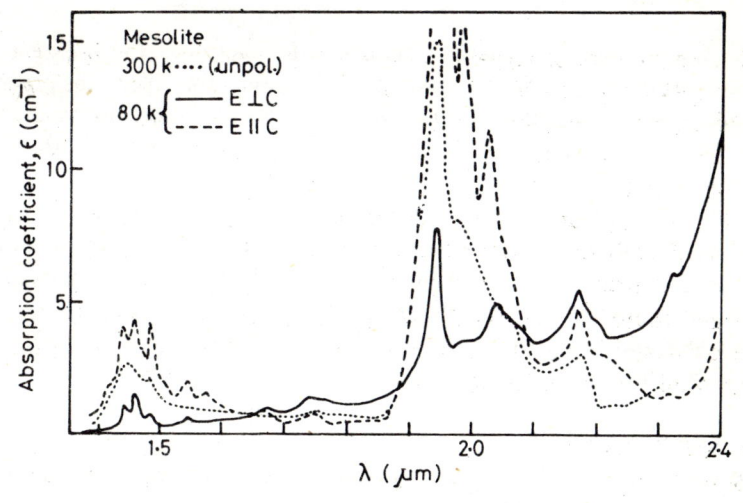

Fig. 6.2(c) Near infra red spectra of mesolite at 300K (unpolarised) and at 80°K in polarised E⊥C and E‖C) light.

(d) Tourmaline, colourless, achroite, $XY_3Al_6(BO_3)_3 Si_6O_{18}(OH)_4$, where X = Na, Ca, Y = Al, Fe^{3+}, Li, Mg; hexagonal. There are borate triangles, 6-membered silicate rings and Mg and Al octahedrally coordinated with oxygens. The near infrared spectrum of tourmaline (Fig. 6.1 d) shows three fairly sharp peaks even at room temperature around 1.45 μm as in the hydrated crystals but the peak frequencies and shapes are quite different with

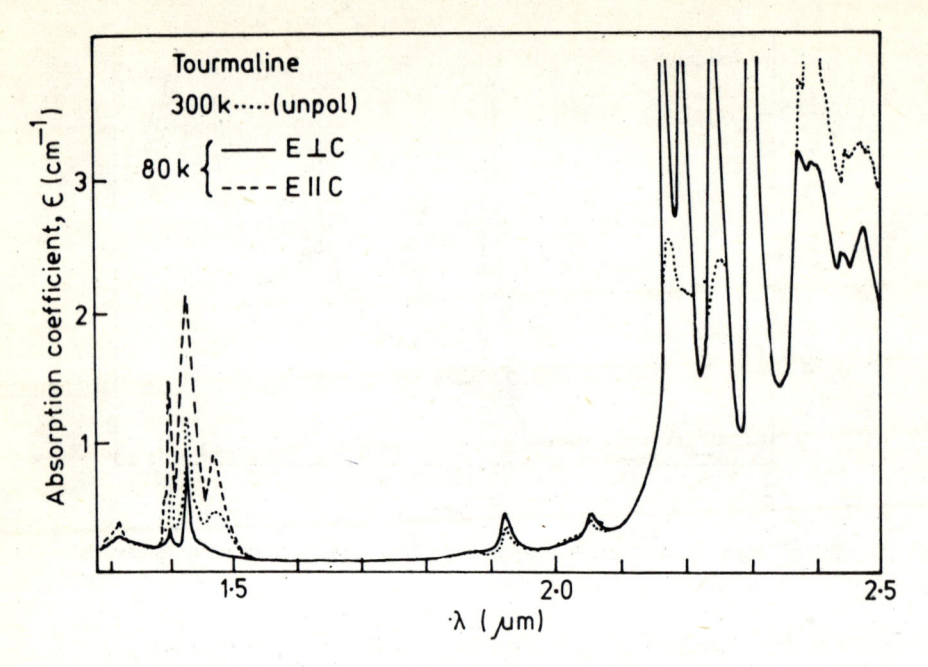

122 Fundamentals of Optical, Spectroscopic and X-ray Mineralogy

Fig. 6.2(d) Near infrared spectra of tourmaline at 300K (unpolarised) and 80K (polarised).

much lower absorption (ε) values (Table 6.2). Furthermore, the intense water band at 1.93 μm is reduced to a very low peak, presumably due to the presence of channel water (Goldman et al., 1977). In addition, there are several intense and characteristic peaks in the range 2.2-2.5 μm, which are absent in hydrated crystals.

In this range, polarisation studies on tourmaline are very intersting. In the E//c (where c is the hexagonal axis of the crystal) polarisation there is a fairly strong peak at 1.47μm which disappears in the E⊥c polarisation. This is due to the fact that the OH-groups are oriented in a direction parallel to the c-axis and hence the first overtone of its fundamental vibration frequency vanishes in the perpendicular polarisation.

6.6 C-N-O-H BANDS
While studying minerals with elements H, O, N, C etc and with bonds (as common in organic molecules), a knowledge of the following band positions may be helpful.

(a) > 2500 cm^{-1}: nearly all fundamental vibrations involve a hydrogen stretching mode. The O-H stretching vibration occurs around 3600 cm^{-1}. Hydrogen bonding lowers the frequency and broadens the band. C-H stretch occurs in the region 2850 to 3000 cm^{-1} for an aliphatic compound and in the region 3000 to 3100 cm^{-1} for an aromatic compound.

(b) 2500-2000 cm^{-1}: stretching vibration for triply-bonded molecules, e.g. C≡N give a strong peak in 2200-2300cm^{-1} region.

(c) 2000-1600cm^{-1}: stretching vibrations for doubly-bonded molecules and bending vibrations of O-H, C-H and N-H groups. Stretching vibrations from C=C, C=N etc. occur in this region.

(d) < 1600 cm^{-1}: single bond region. Generally, coupling occurs in individual single bonds which have similar force constants and connect similar masses e.g. C-O, C-C and C-N stretches often couple. The absorption bands in this region for a given functional group occur at different frequencies depending upon the skeleton of the molecule. It is because each vibration often involves oscillation of a large number of atoms of the molecular skeleton.

Inorganic anion groups have strong, usually simple, absorption spectra; some of these are presented in Table. 6.3 and Fig. 6.3.

TABLE 6.3 Wavelengths and frequencies of the principal anion absorptions

Group	Absorption peaks	
	Wavelength (μm)	Wave number (cm^{-1})
CO_3^{2-}	6.90-7.09	1450-1410
	11.36-12.50	880-800
NO_2^-	7.14-7.70	1400-1300
	11.90-12.50	840-800
NO_3^-	7.09-7.46	1410-1340
	11.63-12.50	860-800
PO_4^{3-}	9.09-10.50	1100-950
SO_4^{2-}	8.85-9.26	1130-1080
	14.71-16.40	680-610
All silicates	9.09-11.10	1100-900

There are six fundamental vibrations of CO_3^{2-} ion, two of which are doubly degenerate: v_1, the symmetric C-O stretching mode; v_2, the extraplanar bending mode; v_3, the antisymmetric stretching mode (doubly degenerate); and v_4, the in-plane bending mode (doubly degenerate).

The strongest combination and overtone vibrations in carbonates occur near 2.53μm ($v_1 + 2v_3$), 2.33μm ($3v_3$) 2.13 - 2.14 μm ($v_1 + 2v_3 + v_4$ or $3v_1 + 2v_4$), 1.98 μm ($2v_1 + 2v_3$), and 1.86 μm ($v_1 + 3v_3$).

In 1 to 4μm wavelength region hardly any vibrational absorption occurs for other minerals. The SiO_2 fundamental occur near 10μm (Si-O stretching modes) and 20 μm (O-Si-O bending modes). The fundamental vibrations of the SO_4^{-2} ion occur near 22.17μm, 16.31μm, 10.19μm, and 9.06 μm.

6.7 CO$_3$-BANDS

Calcites with bands at 6.1, 11.2 and 13.9 can be distinguished from dolomites having 5.9, 10.9 and 13.5μm bands.

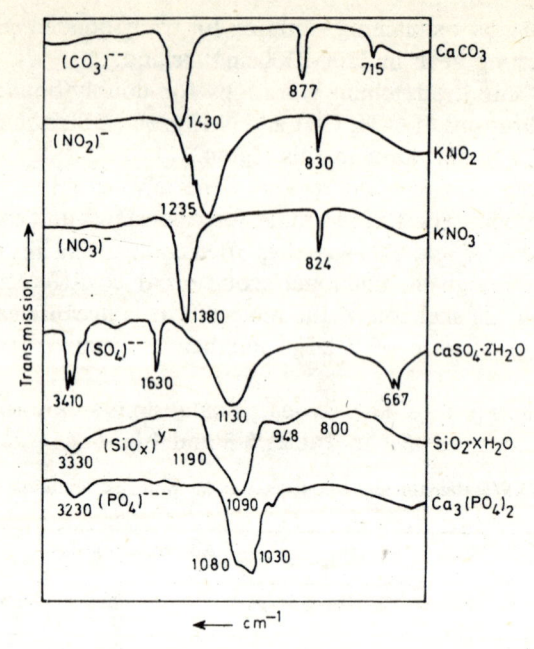

124 *Fundamentals of Optical, Spectroscopic and X-ray Mineralogy*

Fig. 6.3 The infrared absorption bands of some common inorganic and anionic groups.

The absorption frequencies of carbonate ions in different carbonates groups of minerals are presented in Table 6.4.

Table 6.4 Absorption frequencies of carbonate ions in different minerals

Mineral	Wavelength (in μm) (Wave number in cm^{-1} in brackets)		
A. Calcite group			
Calcite. CaCO$_3$	6.97 μm (1435)	11.45 μm (873)	14.04 μm (712)
Rhodochrosite MnCO$_3$	6.98 μm (1433)	11.53 μm (867)	13.76 μm (727)
Siderite FeCO$_3$	7.03 μm (1422)	11.55 μm (866)	13.57 μm (737)
Magnesite MgCO$_3$	6.90 μm (1450)	11.27 μm (887)	13.37 μm (748)
B. Dolomite group			
Ankerite Ca,Fe(CO$_3$)$_2$	6.90 μm (1450)	11.27 μm (877)	13.77 μm (726)
Dolomite Ca,Mg(CO$_3$)$_2$	6.97 μm (1435)	11.35 μm (881)	13.70 μm (730)
C. Aragonite group			
Cerussite PbCO$_3$	6.94 μm (1440)	11.89 μm (841)	14.77 μm (677)

Witherite	6.82µm	11.61µm	14.43µm
Ba CO$_3$	(1445)	(860)	(693)
Strontianite	6.80µm	11.61µm	14.14µm
Sr CO$_3$	(1470)	(860)	(707)
Aragonite	6.90µm	11.61µm	14.04µm
Ca CO$_3$	(1450)	(860)	(712)

Compilation of infra-red data of large number of minerals are available (*vide* references and the cross-references therein), which may be used for comparison with the unknown samples.

References and Selected Reading

De, D. and Datta, S.K. (1978) Near IR spectra of some hexahydrated 3d metal fluosilicates, *Journal of Chemical Physics, 68,* 1865-70.

Farmer, V.C. (1974) ed. The Infrared Spectra of Minerals, Minerlogical Society Monograph 4, 305-330.

Gadsclem, S.A. (1978) Infra-red Spectra of Minerals and Selected Inorganic Compounds. Butterworth, London.

Goldman, D.S., Rossman, G.R. and Dollase, W.A. (1977) Channel constituents in cordierite *American Mineralogist, 62,* 1144-1157.

Ham, F.S. and Slack, G.A. (1971) Infra-red absorption and luminescence spectra of Fe^{2+} in cubic ZnS, role of Jahn-Teller coupling, *Physical Review, 48,* 777-95.

Hunt G.R. and Salisbury, J.W. (1970 a) Visible and near-IR spectra of minerals and rocks: I. Silicate minerals, *Modern Geology, 1,* 283-300.

Hunt, G.R. and Salisbury, J.W. (1970 b) Visible and near-IR spectra of Minerals and Rocks II. Carbonates, *Modern Geology, 2.* 23-30.

Hunt, G.R. Salisbury, J.W. and Lenheff, C.J. (1971 a) Visible and near IR spectra of minerals and rocks-III. Oxides and hydroxides. *Modern Geology, 2,* 195-205.

_____ (1971 b) Visible and near IR spectra of minerals-IV. Sulphides and sulphates. *Modern Geology, 3,* 1-14.

Lyon, R.J.P. (1962) Minerals in the Infra red. Menlo Park, California, Stanford Research Institute, 76 pp.

Moenke, H. (1962) Mineral Spektren, Deutsche Akademische Weisenschaften, Berlin, Akademie-Verlag 42 pp.

Nakamoto, K. (1978) Infra-red and Raman Spectra of Inorganic and Coordination Compounds, John Wiley, New York.

Wood, D.L. and Nassau, K. (1968) The characterisation of beryl and emerald by visible and infrared absorption spectroscopy. *American Mineralogist, 53,* 777-800.

CHAPTER VII

X-ray Optics

The first materials used for testing the electromagnetic theory of x-rays were minerals. No wonder x-ray diffraction would be an important method for mineral studies.

7.1 NATURE AND PRODUCTION OF X-RAYS

X-rays are caused by inner electron transitions in an atom. This electromagnetic radiation normally, has the wavelength (λ) in the range from 0.1 to 50 Å*. X-rays consist of photon energies (E) in the range from 120 Kev to 0.25 Kev. It is derived from the relationship

$$E = h\upsilon = \frac{hc}{\lambda} = \frac{\text{Planks constant} \times \text{velocity of light}}{\lambda}$$

$$= \frac{6.63 \cdot 10^{-27} \text{ erg sec.} \times 3 \cdot 10^{10} \text{ cm sec}^{-1}}{\lambda}$$

$$= \frac{19.89 \cdot 10^{-7} \text{ erg cm}}{\lambda} = \frac{19.89 \times 10^{-9} \text{ erg}}{\lambda} \text{ Angstrom Units (AU)}$$

$$\lambda_{min} \text{(AU)} = \frac{hc/e}{V} = \frac{6.63 \cdot 10^{-27} \times 3 \times 10^{10}}{\frac{V}{300} \times 4.80 \times 10^{-10} \times 10^{-8}}$$

$$= \frac{12,400}{V} \qquad (7.1)$$

where υ is the frequency and V is in volts.

X-rays are produced when a beam of electrons of sufficient energy strike any matter. These can also be produced by irradiating matter with primary x-rays or γ-rays or particles with energy sufficient to knock out one or more electrons from an inner atomic level like the k-shell. These secondary x-rays are of the nature of fluorescence.

7.2. X-RAY SPECTRA

X-ray spectra are of two types: continuous spectrum and characteristic spectrum.

*Bek-radiation, has λ = 115.7Å, while the white radiation from million volts x-ray generators and synchrotrons have $\lambda \sim 0.1$Å.

Continuous spectrum (Bremmstrahlung)

When electrons of sufficient energy (but not high enough to knock out electron from an inner shell of an atom) strike any matter, x-rays with a continuous spectrum of energies are produced. These energies can range from that of the incident electrons to down beyond the lower limit of x-ray photon energies.

The maximum photon energy and the corresponding minimum wavelength depend only on the energy of the incident electrons but are independent of the material emitting the x-rays. It should, however, be mentioned that when x-rays are produced by irradiating matter with photons of sufficiently high energy, continuous radiation or white radiation is not necessarily produced.

In continuous spectrum the applied voltage (V) and the minimum wavelength λ bear the relationship, $V = \frac{12,400}{\lambda}$. With increase in applied voltage the minimum wavelength progressively shifts to lower values and the intensities at all wavelengths increase i.e., the peaks of the curves move to the left.

Characteristic spectrum

Characteristic or line spectra are produced when the incident electrons possess sufficient energy to remove electrons from the inner shells of an atom. The x-ray photons that result, when outer electrons fall into the vacancy, have an energy that is characteristic of a particular element. These characteristic spectral lines are superimposed on the continuous spectrum (Fig. 7.1).

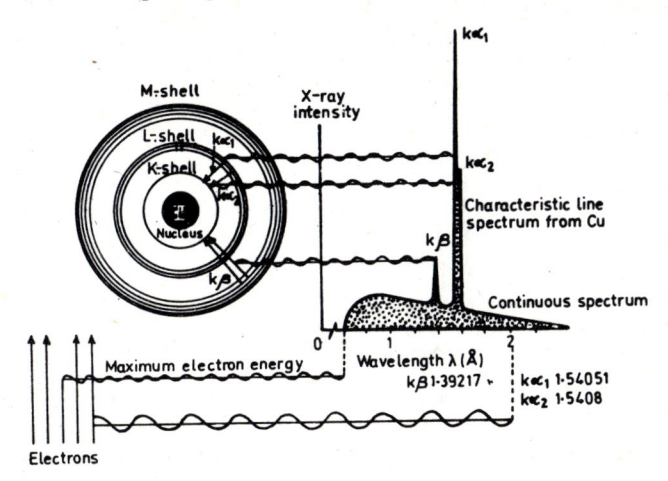

Fig. 7.1   Continuous x-ray line spectrum of a copper target. The generation of x-ray line spectrum of copper is due to the transfer of electrons into the K shell, and the generation of the continuous spectrum is due to complete or partial electron collisions.

When an electron is expelled, the atom is said to be excited and the vacancy is rapidly occupied by an electron falling from an outer shell. The energy of an electron in its initial state in an outer shell (E_i) is higher than when it is in its final state in an inner shell (E_f). As a result, an x-ray photon is emitted with an energy equal to the difference ($E_i - E_f$). If the energies E_i

and E_f are expressed in electron volts, then from equation (7.1) the wavelength of the emitted x-ray will be given by

$$E_i - E_f = \frac{12{,}400}{\lambda_{min}} = \frac{198.9 \times 10^{-10}}{\lambda (\text{Å})} \text{ ergs}$$

The energies of electrons in the atoms of a particular element are fixed; so the different $(E_i - E_f)$ can only take a limited number of fixed values which are characteristic for each element.

In Fig. 7.2, the x-ray spectrum for tungsten is shown for some applied voltages. When the voltage is greater than 40,000 volts, k electrons are removed from the tungsten target and the characteristic k-radiation appears superimposed on the continuous radiation.

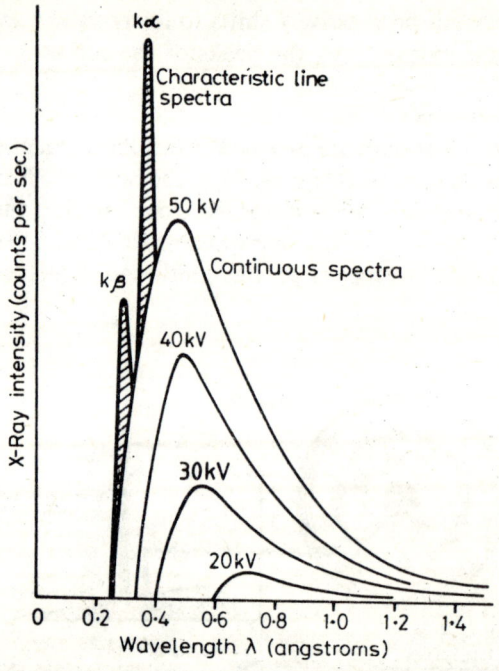

Fig. 7.2 The characteristics spectrum of tungsten superimposed on the continuous spectrum at a voltage higher than 50 KV. The intensity distributions with wavelength at different voltages (20, 30, 40, 50 KV) are also shown.

When the voltage applied is high enough to excite its harder characteristic radiation, the lines will be found superimposed on the curve of the continuous radiation as shown in Fig. 7.3 for molybdenum. The characteristic radiation from an x-ray tube is essentially unpolarised.*

The basic difference between a continuous or white radiation and characteristic radiation is that while the former is affected by the applied

---

* Like ordinary light, x-rays also get polarised after being scattered or diffracted by matter. The amount of polarisation depends on the angle through which it is scattered or diffracted.

X-ray Optics 129

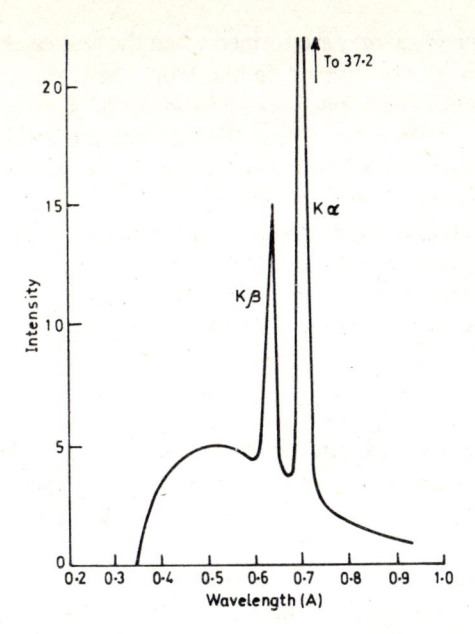

Fig. 7.3 The characteristic K_β spectrum of molybdenum on the curve of continuous radiation

voltage, the positions of the characteristic radiation peaks remain independent of the applied voltage. The peak positions are characteristic of the material irradiated by electrons and emitting the characteristic radiation.

Characteristic spectra are also produced when a matter is irradiated with x-rays. The analysis of these secondary x-rays forms the basis of x-ray fluorescence spectrography.

Incident electrons can expel electrons from the inner shells of atoms provided that their kinetic energy exceeds certain values, which are physical characteristics of the individual elements. The accelerating potentials required to give electrons these energies are referred to as *excitation potentials.*

X-ray excitation differs from electron excitation in that no continuous spectrum is produced; in other respects however, the principles are the same. X-ray photons of sufficient energy can remove an electron, known as *photoelectron,* from the inner shells of an atom. X-ray photons must possess an energy corresponding to the relevant excitation potential to remove a photoelectron.

The characteristic spectrum is divided into K,L,M ... series, depending on whether the spectrum originated from a vacant space in the K.L.M... shell, respectively. Electron transitions into these spaces are governed by selection rules, and some transitions are not possible, that is forbidden.

For any element, the energy differences $(E_i - E_f)$ are greatest when the K-shell is concerned, and the K-spectrum X-rays thus have the highest energies and the shortest wavelengths. The L-spectrum lines have lower energies, and so on.

Hard (high energy) x-rays are formed when the ejected electron comes from the K-shell ($n = 1$): an electron falling from the L-shell ($n = 2$) to K-shell gives rise to the K_α - line, one failing from the M-shell ($n = 3$) gives the K_β-line, and so on. *Soft x-rays* (long wavelength) are formed when the electron is ejected from the L-shell, and the line L_α, L_β etc. are formed as electrons drop from the M- and N- shells, etc.

In the K- spectrum, the doublet α_1 and α_2 lines are generally unresolved in commercial spectrographs and β is the only other line of appreciable intensity, so that this spectrum consists effectively of two main lines, the K_α and the K_β (*vide* Figs. 7.2 and 7.3). The latter line has about $\frac{1}{6}$ th the intensity of the K_α line, which is the one normally used analytically. $K_{\alpha 2}$ has about $\frac{1}{2}$ the intensity of $K_{\alpha 1}$.

7.3 X-RAY GENERATION

The shapes of characteristic radiation change with applied voltage following the equation

$$I = Ai\,(V - V_k)^{1.5}$$

Where V is the applied voltage, V_k is the critical excitation potential of the target material, i is the tube current and A is a constant. The product of the tube current (in milliamperes) with the applied voltage (in KV) measures the tube load. Suitable operating voltages for different targets are as follows:

Copper	: 20 to 40 KV
Chromium	: 25 KV
Iron	: 35 - 40 KV
Molybdenum	: 50 - 55 KV

A commercial copper-target at 1000 watts is run at 40 KV and 20 mA or at 35 KV and 25 mA.

7.3.1 Filters

The target of an x-ray diffraction tube emits under electron bombardment a spectrum of x-rays, which includes the characteristic emission lines of the target material and a continuum of 'white' radiation.

In most x-ray diffraction experiments it is desirable to use a very narrow range of wavelengths, usually the K_α emission lines of the target, and to have as little intensity as possible at other wavelengths.

The K_α radiation line is actually a close doublet, with α_1 having twice the intensity of α_2. Ordinarily the K_α doublet satisfactorily meets the requirements for diffraction studies but in special cases rigid monochromatisation may be made by using *crystal monochromators*.

One should use a crystal monochromator to obtain a strictly monochromatic beam. In Laue method, however, a polychromatic beam is actually required.

For normal investigations filters may be used. The filters are so chosen that the K-absorption edge of its element is just in the short wavelength side of the K_α line of the target material (fig. 7.4).

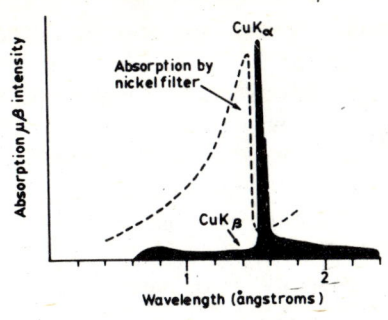

Fig. 7.4 The absorption of CuK_β by a nickel filter and allowing CuK_α to pass.

The thickness of the filter should be such that it would cut down the fairly intense K_β line to negligible intensity and yet not allow the intensity of K_α radiation to fall significantly. Fall in intensity of K_α would demand an increase in exposure time which is not desirable.

Table 7.1 lists the suitable thicknesses of various metal foils that may be used as β-filters for common radiations or targets.

Table 7.1 Filters for K_β absorption of different targets

Target	Wavelength of Kα (doublet) in Angstrom	Filter	Absorption of K_β (%)	Thickness (in mm)
Mo	0.711	Zr	96.3	0.080
Cu	1.542	Ni	97.9	0.015
Ni	1.659	Co	98.4	0.018
Co	1.790	Fe	98.9	0.012
Fe	1.937	Mn	98.7	0.011
Cr	2.291	V	99.0	0.016

By such metal foil filters the β lines are completely eliminated but long wavelength parts of white radiation may still linger. These may be effectively removed by a balanced filter. e.g. Ni-Co for CuKα radiation.

7.3.2 Mass absorption

Certain amount of x-ray is absorbed by matter while passing through it. This mass absorption property of x-ray, defined as k/ρ, importantly distinguishes it from the visible light. To ordinary light sooty carbon or graphite is strongly absorbing while diamond is transparent; but to x-rays both have the same mass absorption co-efficients.

7.4 OPTICAL VS. X-RAY ABSORPTION

X-ray absorption follows identical laws as in optical absorption for imperfect transmission as in the case of translucent and opaque minerals. The law is expressed by

$$I = I_o e^{-kx} \quad \ldots\ldots(\text{vide eq. 4.1})$$

where I is the intensity of transmitted waves, I_o is the intensity of incident beam, e is the Napierian base, k is the linear absorption coefficient of the material (at a particular wavelength it is constant), and x is the thickness of the absorbing layer in centimetres.

In contrast to optical absorption, x-rays manifest an important property known as *mass absorption*. The mass absorption of x-rays has applications in many fields like x-ray fluorescence analysis etc. The mass absorption coefficient is defined by k/ϱ, where ϱ is the density of the material.

The above equation therefore can be transformed by dividing and multiplying k by ϱ, when it becomes.

$$I = I_o e^{-(K/\varrho)\varrho x}$$

Thus the mass absorption coefficient distinguishes x-rays from visible light.

The mass absorption coefficient for an element in different phases remains the same although it might show different optical behaviour; e.g., mercury in liquid, solid and gaseous state has the same mass absorption coefficient (k/ϱ). Similarly, carbon has the same value of it in its graphite or diamond form. This property is taken care of during quantitative x-ray fluorescence analysis.

7.5 CHOICE OF A TARGET

The target elements most useful for diffraction purposes are those whose K_α radiation falls within the wavelength range of 0.56 to 2.29Å. Such targets are Ag, Mo, Cu, Ni, Co, Fe and Cr.

If an element, present in the specimen, gets itself excited by the primary x-ray beam, it emits fluorescent x-ray radiation. This results in a higher background and therefore poorer clarity in diffraction pattern.

Unwanted fluorescence is avoided by choosing a suitable target. For example, if the specimen has a high iron content, a copper target would not be suitable and cobalt or iron radiation should be used.

With increase in wavelength the dispersion increases. Longer wavelengths are more useful for determining complicated patterns with large d-spacings. In this respect, *copper target* is the most useful. Cobalt, iron and chromium targets are usually used when higher dispersion is required and fluorescence is avoided. But the intensities are lower than those obtained from copper. In terms of intensity and peak to back-ground ratio the copper target is the best.

The use of a short x-ray wavelength may sometimes be desirable in order to record a greater number of x-ray reflections. With *Mo K_α radiation* (λ = 0.71Å) (Fig. 7.3), although many more reflections could theoretically be recorded in accordance with Bragg's Law, in the record many are lost because of a rapid fall of diffracted intensity with increasing Bragg angle. In minerals where the unit-cell dimensions are small, a short wave-length radiation (like Mo K_α) is specially suitable.

Complex minerals like clay minerals have very large unit-cells and for that the long wavelengths of a *chromium target* can be advantageously employed, whereby the many closely spaced lines of large interplanar spacings are better dispersed and resolution is greatly improved.

Longer wavelengths (from *cobalt and iron targets*) may be used in order to increase resolution between close reflections and also in obtaining better reflections at high Bragg angles. The power rating of cobalt and iron x-ray is usually lower than that of copper tube. Therefore, they have lower intensities and longer exposure time compared to copper tube.

7.6 X-RAY DIFFRACTION

Diffraction by crystalline materials occurs due to (i) scattering and (ii) interference.

In a crystal the atoms are arranged with periodicity in three dimensions, and the repeat distances are of the order of the x-ray wave-lengths. X-rays incident on a crystal are scattered. Scattered rays from neighbouring atoms cause diffraction by interference. The rays scattered by a crystal are concentrated at a series of points at regular intervals where the maxima occur.

The diffraction maxima occur in a direction along which the path difference between the rays scattered by two neighbouring atoms is $n\lambda$, where n is an integer and λ is the wavelength. In a crystal, the scattering atoms are arranged in three-dimensional arrays with periodicities $\bar{a}$, $\bar{b}$, $\bar{c}$ in three noncoplanar directions. Therefore, diffraction maxima would be observed along those directions for which the above condition of path difference is satisfied simultaneously for neighbouring pairs of scatterers by all the three vectors $\bar{a}$, $\bar{b}$, $\bar{c}$. Three integers, h, k and l, are associated with each maximum of diffracted intensity.

The three integers, h, k, l, label a particular family of planes, and are known as *indices*. The interplanar spacing between the members of the family of parallel planes is designated as d (hkl).

Different atoms have different numbers of electrons, consequently their relative scattering varies. The diffraction pattern, depicting the positions and intensities of diffracted beams, serves for speedy identification as well as complete elucidation of the structure i.e., the size and shape of the unit cell and distribution of atoms in the unit cell. By measurement and analysis of the intensities of diffraction spots, the position of individual atoms in the cell can be accurately located. In locating the positions of atoms from the diffraction intensities, the important factor which is taken into consideration is known as the *structure factor*, discussed in section 7.24.3.

From the powder diffraction study the following information can be obtained.
1) The interplanar (d−) spacings of the lattice planes
2) The intensities of the reflections
3) The Miller indices of reflection planes
4) The unit cell dimensions and the lattice type.

7.7 BRAGG LAW

An impinging beam of x-ray gets scattered by the electrons of atoms and a diffracted beam is formed when certain geometrical conditions are satisfied (following Bragg or Laue equations).

Bragg Law, first formulated in 1912, states the condition for diffraction of an incident beam of monochromatic x-rays. It states that if the path difference for the waves, reflected by successive sheets of atoms, is a whole number of wavelengths, the wave trains will combine to produce a strong reflected beam. In more formal geometric terms, if the spacing between the reflecting planes of the atom is d and the half angle of deviation between the beam's incidence and reflection by the plane is θ, the path difference for waves reflected by successive planes is $2d \sin \theta$ (Fig. 7.5).

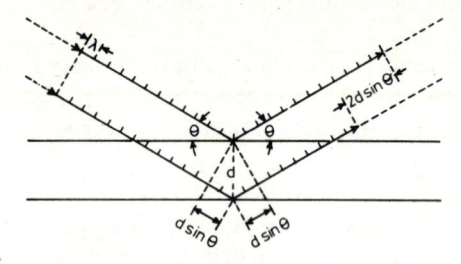

Fig. 7.5 Diffraction of x-rays from lattice planes at angles satisfying Bragg's equation.

Constructive interference between the radiations reflected from the successive planes occurs when the path difference is an integral number, n, of wavelength, λ. Therefore,

$$n\lambda = 2d \sin \theta.$$

This is known as Bragg's Law. Obviously, only for certain values of θ will the reflections from all parallel planes add up in phase giving a strong diffracted beam. At other angles, there is no reflected beam because of interference. The reflection can only occur for wavelengths $\lambda \leq 2d$. Because of this visible light (having $\lambda > 2d$) can not be used.

Example: An x-ray radiation of 1.54Å incident on a cubic crystal of lattice constant 4Å, will have the first order ($n = 1$) reflection from planes parallel to pinacoids (100) at $\theta = \sin^{-1} (\lambda/2d) = \sin^{-1} (1.54/8) = 11°$.

X-ray studies are done using single crystals or powdered samples. Single crystal methods are discussed in sections 7.8 and 7.19 to 7.21; while powder methods are treated in sections 7.9 to 7.18.

7.8 LAUE PATTERN

When the x-ray beam is polychromatic, spots can be obtained from a stationary single crystal (Fig. 7.6). This is how the first Laue pattern was obtained. Laue diffraction method will be discussed in section 7.19.1.

The beam can also be recorded as spots on a photographic film or plate by rotating the crystal so that greater number of planes can reflect. The diffracted beam can be recorded by a Geiger counter or a proportional counter or a scintillation counter. In such cases, the monochromatic beam is used.

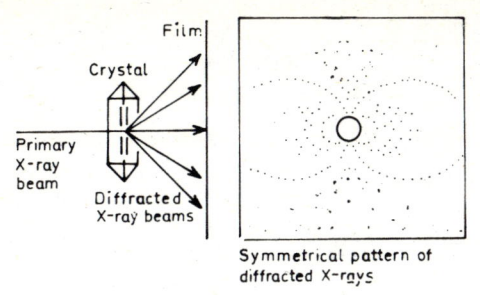

Fig. 7.6 The symmetric diffraction pattern of x-rays from a stationary single crystal.

7.9 POWDER METHOD

In powder method, the required grain size should be small. The larger the grain size, the fewer the number of grains in a given specimen, and so all grain orientations are less likely to be represented. The result is spotty, instead of evenly blackened circular, areas, and greater difficulty in positional measurement and intensity estimation is encountered. A grain size of < 325 mesh (i.e., less than 45 microns) is suitable.

In order to ensure that a sufficient number of planes are present in the specimen in all orientations, it is the usual practice to have the specimen rotating about the camera axis by means of a motor drive.

The powder is mixed with a gum (canada balsam, Duco cement, collodion or tragacanth) and rolled into a rod or put over the tip of a glass needle. The powder rod is then mounted on the sample holder and centred by a centering device. By repeated trial with the device, the axis of the rod is so perfectly aligned with the axis of the camera that the sample looks stationary when the sample holder is rotated.

7.9.1 The Debye – Scherrer Camera

It is a cylindrical camera having diameter preferably of 11.46 cm or 5.73 cm (Fig. 7.7). The apparently strange diameters are used because they have the effect of making one millimetre on the film equal to ½° and 1° angles (2θ) respectively. On the inner side wall of the cylinder, forming a cassette, is inserted the strip film which has emulsions on both side. The camera has a sample holder with a centering device and can be rotated about the axis of the camera, and two holes are inserted for collimators.

A sample holder is placed at the centre of the camera and is rotated by a motor. By a spindle the specimen axis is made to coincide with the axis of the camera. Two tubes called *collimators* are inserted through the opposite ends of the diametrical axis. X-rays enter through one collimator, called the *receiving tube,* and get trapped on the fluorescent screen on the other collimator, which is called the *beam trap.* Debye-Scherrer cameras are of several types, the most commonly used is that designed by Buerger.

136 *Fundamentals of Optical, Spectroscopic and X-ray Mineralogy*

X-ray diffraction cones generated from the mineral grains intersect the film as pairs of arcs, the distances between which are proportional to their respective 4θ values (Fig. 7.7).

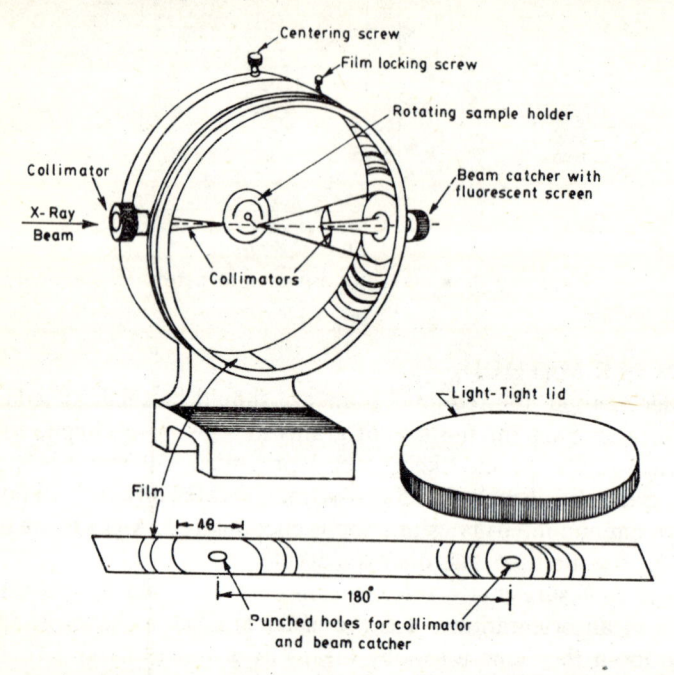

Fig. 7.7 Internal view of the Debye-scherrer powder camera and the diffraction arcs on the film.

As stated earlier, in camera with diameter of 11.46 cm, the 2 mm distance on the film will measure as 1°; while in the camera of radius 5.73 cm, the distance of 1 mm will be 1°. Thus from the direct measurement of the film the angular values of the cone i.e., 4θ values can be determined. For accurate measurement, however, camera radii are to be calibrated because of constructional faults and factors like film shrinkage etc. The arrangement records the diffraction pattern on the film through 360° and is also used in precise measurement of lattice constants.

We know from the principle of light, the intensity of illumination is inversely proportional to the square of the distance. Therefore, the larger camera would need a longer time of exposure but, of course, the line spacings would be better separated and measurement accuracy would improve.

Knowing the target used the wavelength of x-ray is known and consequently *d* values can be determined using 2θ values in accordance with the Bragg equation. But, for greater convenience, tables are available for different wavelengths, from which the *d* values can be directly determined from known values of 2θ. The charts (Parrish and Mack, 1963) and table (National Bureau of Standards, 1950) effectively give solutions for Bragg equations for different x-ray wavelengths at close intervals of θ.

7.9.2 Film distortion correction

When the film is processed the distances, P, between pairs of arcs can be measured and from these the corresponding Bragg angles can be determined and d values are obtained from the Bragg Law (Fig. 7.8). The relation between P and θ would be simple

$$\frac{P}{2\pi r} = \frac{4\theta}{360}$$

If the radius, r, of the camera is accurately known, θ can be determined from P values. The method described in the following section overcomes the problem of distortion of the film on processing, assuming that the distortion due to shrinkage etc. has taken place uniformly over its length.

7.9.3. Film measurement

The film after developing and fixing in a dark room is allowed to dry in air. It is then placed on the measuring glass screen, illuminated by a fluorescent lamp from below. The 2θ measurements on the film are done using scales (with Vernier) with magnifying glass attachment. The procedure is illustrated with the powder photograph of a chromite sample (S_{17Q4}) below.

Holding the film with low-angle reflections in the left, all positions of lines on a scale are recorded. Measurements are made directly with a mm. rule of the distance between a pair of powder arcs on both sides of the hole (Fig. 7.8). Theoretically, all sums of $P_1 + P_2$ (where P_1 and P_2 stand for the readings of the same band on the left and right side, respectively, of the hole) would be equal. Any deviation greater than 0.005 will lead to the detection of an error of measurement (as in the next page in the line 5 of chromite sample no. S_{17Q4}). For a pair of powder arcs at $2\theta < 90°$,

$$\frac{P_2 - P_1}{2B - 2A} = \frac{4\theta}{360°} \text{ and } \theta = (P_2 - P_1) \cdot \frac{90°}{2B - 2A}$$

and for reflections with $2\theta > 90°$

$$90° - \theta = (P_2 - P_1) \cdot \frac{90°}{2B - 2A}$$

A = distance of the nearer hole
B = distance of the far hole

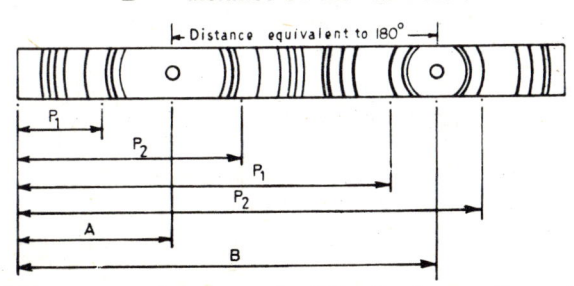

Fig. 7.8 Measurement of x-ray powder diffraction lines on films using Straumanis method and calculation of 2θ

From this a factor, determining the degree of difference between the averages of $P_1 + P_2$ ($2\theta < 90°$) and $P_1 + P_2$ ($2\theta > 90°$), is calculated. This

138 Fundamentals of Optical, Spectroscopic and X-ray Mineralogy

factor multiplied by the (P_2-P_1) values gives a series of 2θ values corresponding to the lines, and N values are obtained. The procedure is illustrated with the powder photograph of a chromite sample (S_{17Q4} from Sukinda, Orissa) as below.

Sample No. S_{17Q4} (chromite in CoK_α with Fe-filter)

	Line No	P_1	P_2	P_2+P_1	P_1-P_2	(5)x (factor)* = 2θ (degrees)	Value of N
	(1)	(2)	(3)	(4)	(5)		
$2\theta < 90°$	1	9.040	4.720	13.760	4.320	21.583	3
	2	10.445	3.315	13.760	7.130	35.621	8
	3	11.085	2.680	13.765	8.405	41.991	11
	4	12.010	1.750	13.760	10.260	51.259	16
	5	13.720	0.065	13.785	13.655	68.220	27
$2\theta > 90°$	9	31.655	18.135	49.790	3.520	67.546	59
	10	30.915	18.885	49.800	12.030	60.102	64
	11	29.015	20.785	49.800	8.230	41.117	75
	12	27.845	21.945	49.790	5.900	29.476	80
	13	26.900	22.905	49.805	3.995	19.959	83

*Difference between average of $P_1 + P_2$ ($2\theta < 90°$) = 13.766 and P_1+P_2 ($2\theta > 90°$) = 49.797 is 36.031; $\dfrac{180°}{36.031}$ = 4.996° (factor); the difference, 36.031 = 2B−2A

The measurement of the film, therefore, yields the angular distances, which are transformed into interplanar (d) spacings, following the relationship $d = \lambda/2 \sin\theta$.

7.9.4 Calculations for d-spacing and cell edge

For a cubic system, $d_{hkl} = \dfrac{a}{\sqrt{h^2 + k^2 + l^2}}$

where a = side length of the unit cell.
Substituting this in the Bragg equation and squaring, we have

$$\sin^2\theta = \dfrac{\lambda^2}{4a^2} N \text{ where } N = h^2 + k^2 + l^2$$

Assuming that the lowest $\sin^2\theta$ value = 1 . $\dfrac{\lambda^2}{4a^2}$,

We divide the remaining $\sin^2\theta$ values by this value of $\dfrac{\lambda^2}{4a^2}$

A series of approximately whole numbers, which are also the sums of squares of whole numbers, will be obtained. When the series of numbers contains some which are obviously not whole numbers, the process is repeated assuming that the first $\sin^2\theta$ value has N = 2,3,4 etc. until a series of whole numbers is obtained.

Since the value of $\sin^2\theta$ is found to be most in error for low values of θ, the value of the factor using one of the higher values of N determined is used in the reversed value of $\lambda^2/4a^2$ to determine N for lines with even higher N values.

The values of N so obtained give the values of h, k and l i.e., the indices of the reflection by looking up a table of N and corresponding h, k, l values. Knowing λ, the factor $\lambda^2/4a^2$ gives the value of a, the cell edge of a unit cell.

To make the value of a more accurate, calculations are made from the N values of the higher angle lines ($\theta > 60°$) and the values are plotted against the extrapolation function of $\frac{1}{2}\left(\frac{\cos^2\theta}{\sin\theta} + \frac{\cos^2\theta}{\theta}\right)$

The value of a extrapolated to $\theta = 90°$ is then determined, since as $\theta \to 90°$, the error in determination of $a \to 0$.

An output of computer calculation of the cell-edges of an orthorhombic enstatite sample is presented in Appendix IV (from Mitra, 1973).

7.10 LINE INTENSITIES

Human eye is a remarkably good photometer and can visually estimate with fair accuracy the relative order of blackness of a series of spots. Experience has shown that eye estimation is often more accurate than microdensitometers. Thousands of crystal structure determinations have been made by visual estimation of intensity with the help of standard calibrated strips. Visual estimation, however, cannot give intensity distribution in a line for which a very fine beam microdensitometer is necessary.

Visually the intensities are chracterised qualitatively as very very strong (vvs), very strong (vs) strong (s), medium strong (ms), medium (m), weak medium (wm), weak (w), very weak (vw), very very weak (vvw).

The reflection intensities can also be recorded by ionization or Geiger-Muller counter (spectrometer).

7.11 USE OF STANDARD

A substance with known d spacings can be used for calibration purpose. In order to make the conditions of the standard and specimen as similar as possible, the standard can be used internally, i.e., mixed in suitable proportion with the specimen powder. For this method an internal standard must be chosen that does not form circular arcs overlapping reflections of interest arising from the specimen.

The standard must be readily obtainable in good purity and of suitable crystallite size to give sharp diffraction lines. The common internal standards are silicon, CaF_2, calcite. etc.

Example: To determine the line positions of orthopyroxene a silicon internal standard is used (Fig. 7.9) From the known 2θ positions of the silicon peaks (311,333) the 2θ values of 250, 12,0,0, 10,3,1, 060, 14,5,0 planes can be accurately determined. The accuracy of the 2θ values is important for lattice parameters determination as well as for determining the En/Fs molecular ratios.

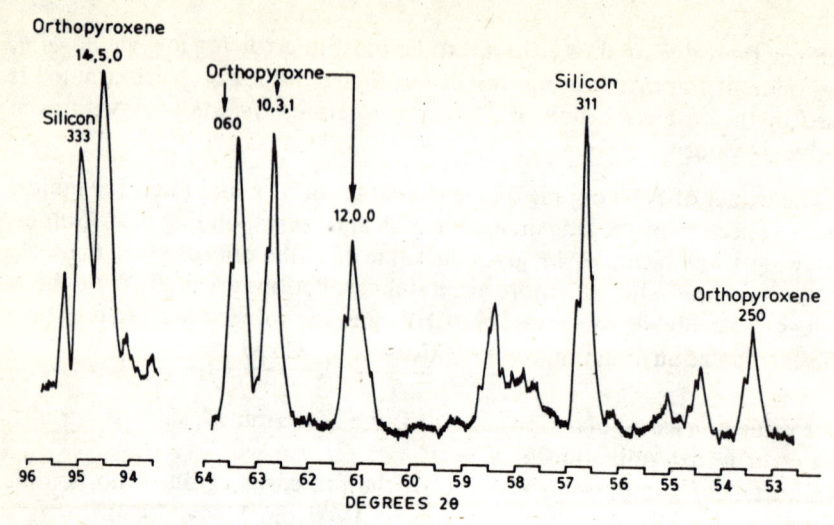

Fig. 7.9   Diffractometer record of orthopyroxene (with silicon internal standard).

7.12 X-RAY DIFFRACTOMETRY
Principles of operation

X-ray tubes, used for powder diffractometry, must be operated at a voltage several times greater than the critical excitation potential of the target element to obtain sufficient intensity. Usually 25 to 60 KV is used which produces a spectrum consisting of characteristic emission lines superimposed on a continuous band of radiation (*vide* fig. 7.2).†

It is necessary to optimise the experimental conditions to obtain maximum line intensity and peak-to-background ratio. Monochromatisation of X-rays is done by the use of a beta-filter, pulse height discriminator or monochromators.

Parafocussing geometry

The basic geometrical arrangement in a powder diffractometer equipped with a goniometer is shown in Fig. 7.10. A divergent beam (ribbon) of x-rays from the focus on the target of the X-ray tube impings on the flat surface of the powdered sample and gets diffracted at Bragg angles of differently oriented crystallites in the sample powder. This beam is converged in front of the receiving slit before it reaches the counter.

The parafocusing arrangement of the diffractometer is based on the theorem of geometry that the angle at the centre of a circle is twice the angle at the circumference standing on the same arc.

The diffracted beams are focused by setting the sample in the position tanget to the focusing circle. This position bisects the angle between the incident beam and the diffracted beam. The counter is moved at double the rate of the rotation of the sample holder. This keeps the holder tangent to

---

†Exposures are often reported in terms of tube currents in milliamps and time in hrs: thus, 2 hrs at 20 ma equals 40 milliampere etc.

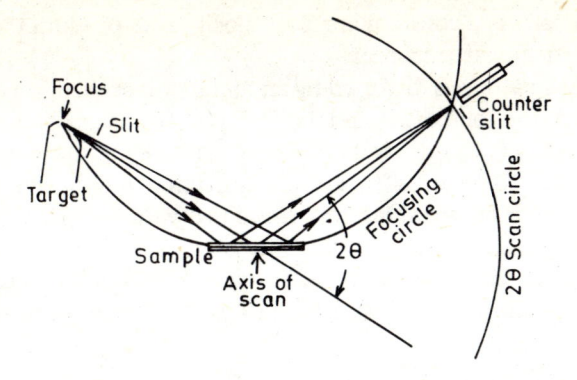

Fig. 7.10 Parafocussing arrangement in a powder diffractometer.

the focusing circle. The crystallites whose lattice planes are parallel to the surface of the specimen holder contribute to the reflection. By use of a narrow line focus the focusing error is kept small*.

The diffracted X-ray beams are controlled by the *receiving slit, receiving parallel slit* assembly and the *scatter slit*.

Through the scatter slit the X-ray beam gets into a detector system, which is filled with inert gas† (argon, xenon or krypton). Ionisation causes electric pulses. The pulses are amplified and counted by a device called *scaler*. This is linked to a *timer,* counting the number of pulses in a fixed time or the time for a fixed number of pulses. The connected *ratemeter* converts digital counts to a continuous electric current which actuates a *strip chart recorder*.

By gear control the counter, moving a double the speed (2θ) of the rotating angle (θ) of the sample holder, can be made to scan at different rates ranging from 1/16th degree per minute to 2 degrees per minute. A ½ degree mark is made on the lower side of the chart by the other pen in the recorder (*vide* Fig. 7.9).

Detectors: Geiger-Muller gas proportional and scintillation counters are most widely used detectors (besides the film) for powder diffractometry.

Gas proportional counters and scintillation counters with Na I(Tl) scintillators (*vide* p. 81) have greatly improved the counting efficiency. The scintillation counter has nearly 100 per cent quantum counting efficiency for the x-ray wavelengths used in diffractometry. The efficiency of G.M. counter for short wavelength x-ray diffraction is appreciably poor.

Recording

The most common method for powder diffractometry is to drive the counter

---

*This is done by viewing the 10mm × 1mm line focus at angles of 6° to 3° known as *take off angle;* the divergence *parallel slit* assembly delimits the divergence of the line focus parallel to its length. The *divergence slit* before the camera allows the beam to diverage in a controlled way in the direction perpendicular to the goniometer axis. At higher angles wider divergent slits are used.

†mixed with a suitable organic vapour, for quenching action.

tube automatically at a constant angular velocity and to record the ratemeter output on a strip chart recorder.

The recorded pattern is distorted by an amount dependent on the product of scan speed of the goniometer and the T C (time constant) of the ratemeter. When the product is large, say 2° (2 θ) min^{-1} and 8 sec. time constant, the peaks will be shifted towards the scanning direction and the line profiles get distorted. A moderate scanning rate combined with a small time constant gives accurately placed sharp peaks.

Normally good results are obtained using a ½° (2θ) scan speed and 2 second time constant or with ¼° (2θ) and 4 second time constant.

7.12.1 X-ray Diffractometer

Common x-ray diffractometers (with vertical goniometer, PW 1050 or horizontal goniometer, PW 1380 of Philips company) employ a flat-specimen geometry.

The sample powder, spread as a flat sheet (using some gum or acetone) on a glass slide or in an aluminimum specimen holder, is allowed to rotate with respect to the impinging x-ray beam. Instead of the film, the diffracted x-ray photons are recorded by a proportional counter or a scintillation counter which are connected to an electronic counting system. The latter is synchronised with a strip chart recorder, a ratemeter, a timer, a goniometer power supply, and a pulse-height analyser. These are often connected to a digital printer for a printed output.

Procedure

The sample powder (∼325 mesh) is put in the sample holder or on the glass slide and the lid of the camera is sealed tightly. The x-ray is switched on and the window is opened to the x-ray.

For a reasonable diffractogram, a narrow slit of 0.1 mm. is used with a scanning speed of 1° per minute and time constant of 2 seconds. For studying under high resolution any individual peak, a slow speed of 1/8° per minute with a narrow slit of 0.05 mm. can be employed with good effect. It should be remembered that the narrower the slit the greater is the resolution but the lower is the intensity. For studying a low intensity peak 0.1 mm. or 0.2 mm. slits yield better resolution.

The goniometer scanning speed should be synchronised with the chart recorder speed i.e., a slow goniometer speed should go with slow chart speed. To achieve 2 cm per degree 2θ a slow scan of ½° or ¼° per minute is used. The following combinations of goniometer and chart speed give a 2 cm per degree 2θ.

Goniometer speed 2θ/min.	Chart (recorder) speed mm/hour.
1°	1200
1/2°	600
1/4°	300
1/8°	150

A speed of 1° per minute is fairly good for high-precision lattice parameter determination.

When the angle of start is marked on the chart, 2θ values of the peaks are calculated by counting the ½° kinks on the 2θ marked at the bottom of the chart (Fig. 7.9).

Automated powder diffractometry
Microprocessor-based control and measuring system offers fully automated powder diffractometry. This type is found in the new generation PW 1700 series of Philips. A recent model is the PW 1710 (Fig. 7.11) which contains software for both qualitative and quantitative analysis. In this automated X-ray machine fast routine analysis is possible because of its inbuilt special intensity sampling technique and the four main functions: (a) X-ray diffractometer control (control of goniometer, control of sample changer and spinner, high voltage setting of detectors), (b) Measurement and further processing of x-ray intensities (pulse height selection and counting of x-ray photons, calculation and printing of d-values on diffractograms, evaluation of peak intensities and peak areas), (c) Input/output interface (input of commands, printout and display of data), (d) Execution of commands and user-made programmes (performing step-by-step commands entered via the front panel or remote key board, or automatic execution of complete measurement programmes upto 255 assembled by the user and stored in the system memory).

7.12.2 Powder photographs vs. diffractometer record
Powder diffractometry has a definite advantage over the photographic method. The powder film method is time consuming as it involves a long exposure time and time for developing, fixing, washing and drying. Besides time saving, the diffractometer recording offers the following advantages:
i) 2θ values can be directly read from the chart; with internal standard, the accuracy in 2θ determination can be improved.
ii) In film the relative intensities of the lines can be evaluated, while on a chart the relative intensities can be determined with more precision.

For a visual comparison the x-ray diffraction records of quartz done by the powder film method and diffractometer are presented in Fig. 7.12 (a,b,). Note that each film line corresponds to a diffraction peak shown. The Miller indices of the reflection planes are indicated on the diffraction lines. In the same operating condition when the scan speed is slowed down, the peaks of interest become more separated and the angles can be more accurately determined (Fig. 7.12c.).

7.12.3 Line intensities and spectrometer charts
Human eye is very sensitive to detect the relative blackening of the spots or lines. Its sensitivity response is often preferred by many over that of microphotometers. If the size of the crystallites in the sample is of colloidal range (10^{-5} to 10^{-7} cm), the lines will be broadened and the peak areas rather than heights give a more accurate measure of intensity.

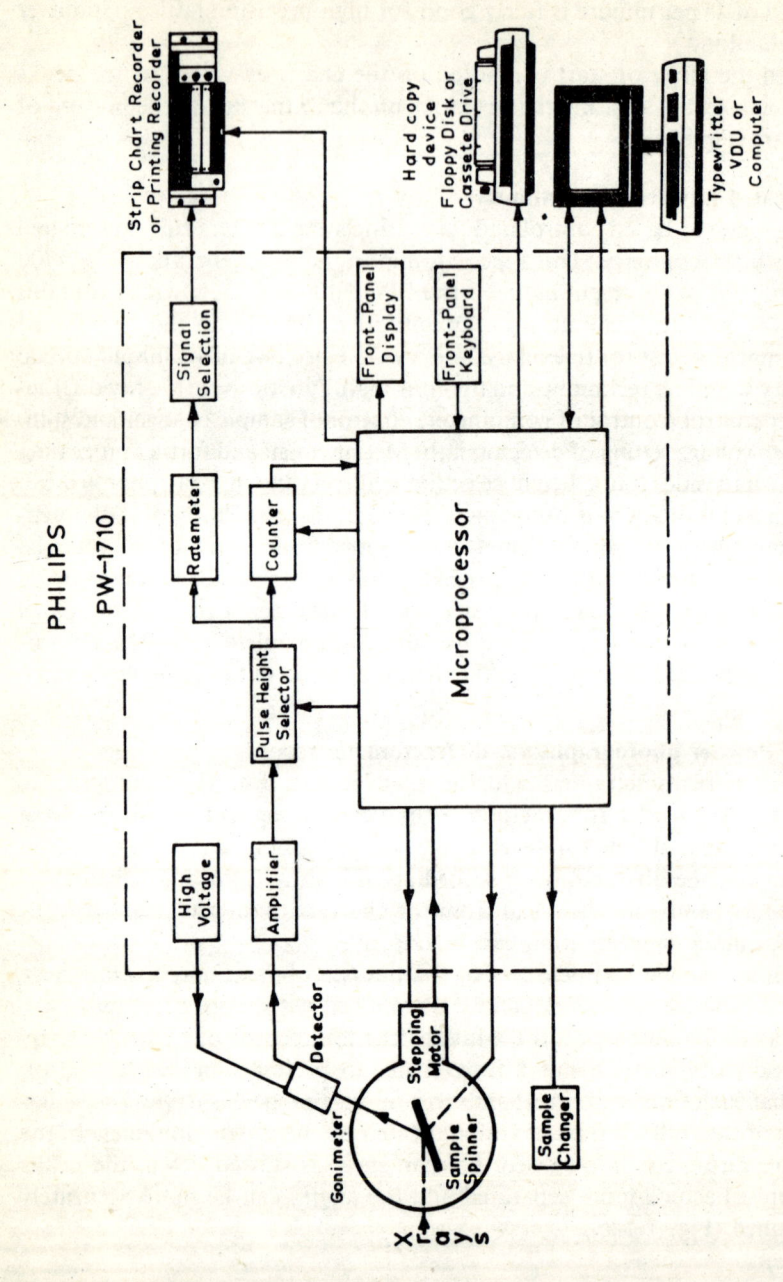

Fig. 7.11 The assembly system of an automated powder diffractometer (Philips PW 1710).

X-ray Optics 145

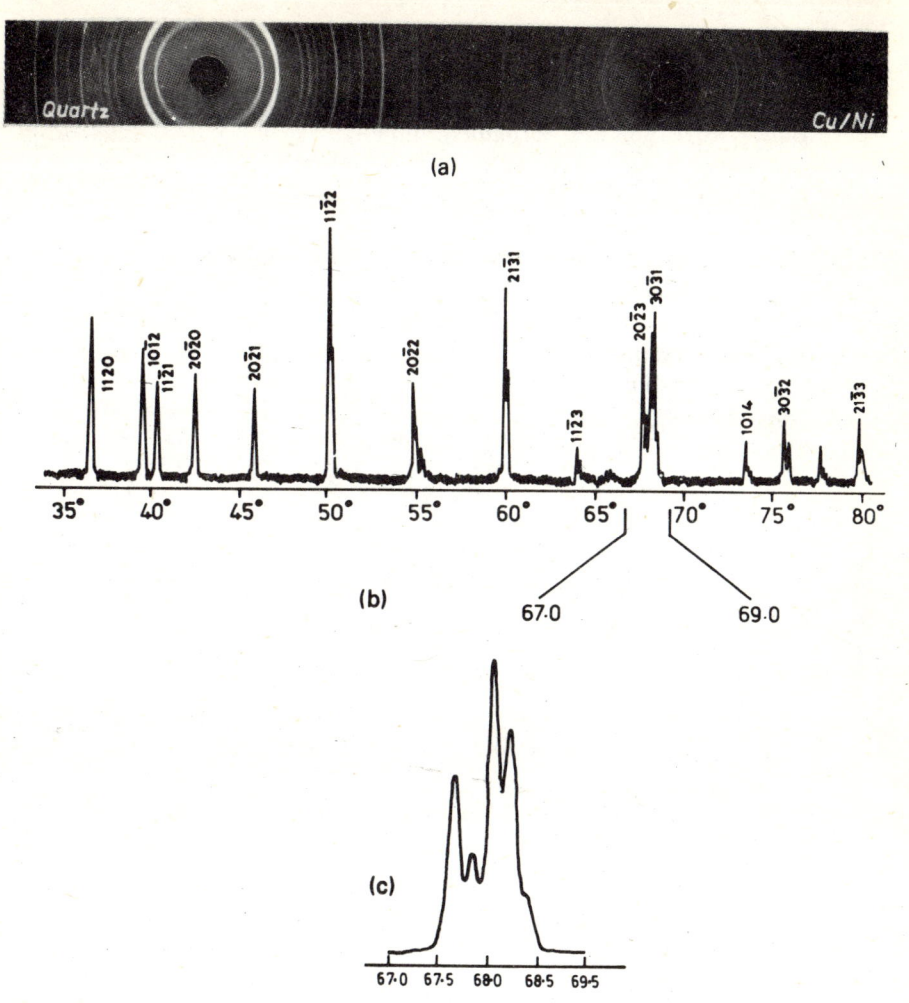

Fig. 7.12 Debye-Scherrer powder photographs and diffraction chart record of quartz powder in Cu K$_\alpha$ (Ni filter) at 40 KV/15 mA.
(a) A print of the film exposed in a small size (5.73 cm diam.) camera.
(b) The record on the diffraction chart in a run between 35-80° 2θ angles (with proportional counter 1600V, slits 1°, scan speed 1°/min.)
(c) The diffraction record of peaks between 67° and 69° with scan speed of ½° per minute.

Determination of the Crystal System

For the determination of the crystal system the following relationships of d, and h,k,l values are employed:

we know for a cubic system, $\dfrac{1}{d^2} = \dfrac{h^2 + k^2 + l^2}{a^2}$ and $d^2 = \lambda^2/(4 \sin^2\theta)$

$$\sin^2\theta = \dfrac{\lambda^2(h^2+k^2+l^2)}{4a^2}$$

146 *Fundamentals of Optical, Spectroscopic and X-ray Mineralogy*

For a particular crystal with specific *a* and known λ, the value $\dfrac{\lambda^2}{4a^2}$ becomes a constant (say, A). Now, putting $h^2 + k^2 + l^2 = N$ (which will have values other than the 'forbidden' numbers like 7, 15, 23, 28, 32 etc.)

$$\mathrm{Sin}^2\theta = A(N)$$

From the measured Bragg angles, θ's of the crystal reflections, a set of $\mathrm{Sin}^2\theta$ values would be obtained, which in a cubic crystal will be the integral multiples of constant A.

When the values do not tally, the crystal is not cubic in symmetry. Equations for other crystal systems then would be tested, keeping in mind the following relationships:

Tetragonal System:

$$\mathrm{Sin}^2\theta = \frac{\lambda^2}{4a^2}(h^2 + k^2) + \frac{\lambda^2}{4c^2}l^2$$

i.e. $\mathrm{Sin}^2\theta = AN + BM$

where $A = \lambda^2/4a^2 =$ const.
$B = \lambda^2/4c^2 =$ const.
$N = (h^2 + k^2)$
and $M = l^2$

The permissible values of *N* are:

hk	10	11	20	21	22	30	31	
N	1	2	4	5	8	9	10	etc.

while, the permissible values of *M* are

	1	1	2	3	4	
M	1	1	4	9	16	etc.

Orthorhombic system:

$$\mathrm{Sin}^2\theta = \frac{\lambda^2}{4a^2}h^2 + \frac{\lambda^2}{4b^2}k^2 + \frac{l^2}{4c^2}l^2$$

i.e. $\mathrm{Sin}^2\theta = AN + BM + CP$

where
$A = \lambda^2/4a^2$ (const)
$B = \lambda^2/4b^2$ (const)
$C = \lambda^2/4c^2$ (const)
$N = h^2$
$M = k^2$
$P = l^2$

The permissible values for N, M and P are 1, 4, 9, 16 etc.
Hexagonal system:

$$\sin^2\theta = \frac{\lambda^2}{3a^2}(h^2 + hk + k^2) + \frac{\lambda^2}{4c^2} l^2$$

i.e. $\sin^2\theta = AN + BM$

where
$A = \lambda^2/3a^2$ (const.)
$B = \lambda^2/4c^2$ (const.)
$N = (h^2 + hk + k^2)$
and $M = l^2$

The permissible values for N are:

hk	10	11	20	21	30	22	31	
N	1	3	4	7	9	12	13	etc.

while these values for M are 1,4,9,16 etc.

The equations in $\sin^2\theta$ for monoclinic and triclinic systems are very complex and are not treated here.

It is, however, evident from these that the θ values help in measuring the size and shape of the unit cell. Stating otherwise, the geometry of space-lattice determines the angle in which the rays are diffracted.

7.13 DETERMINATION OF THE CRYSTAL SYSTEM: INDEXING POWDER LINES

Indexing of the X-ray powder lines for minerals having orthogonal (i.e. cubic, tetragonal, hexagonal, trigonal and orthorhombic) symmetry can be done relatively easily when the unit cell is first determined.

For non orthogonal (i.e. monoclinic and triclinic) systems indexing can only be done by employing the basic principles of reciprocal lattice (*vide* section 7.20).

Indexing *orthogonal* system: A plane, $h\,k,\,l$, intercepts the crystallographic axes at a/h, b/k, c/l distances. The perpendicular distance, d, between the planes is related to those parameters as

$$d = 1/\sqrt{(h^2/a^2 + k^2/b^2 + l^2/c^2)} \tag{1}$$

In (a) *cubic* system $d = a/\sqrt{(h^2 + k^2 + l^2)}$

Since $a = b = c$, and since $\lambda = 2d\sin\theta$

$$\sin^2\theta_{(hkl)} = \frac{\lambda^2}{a^2}(h^2 + k^2 + l^2) \tag{2}$$

$$= \frac{\lambda^2}{4a^2} N.$$

When the series of numbers, N, corresponding to different $\sin^2\theta$ values, contains some which are obviously not whole numbers, the process is repreated assuming that the first $\sin^2\theta$ value has N = 2, 3, 4 etc. until a series of whole number is obtained.

Since the value of $\sin^2\theta$ is likely to be met in error for low values of θ, the value of the factor using one of the higher values of N determined, are employed. The value of $\lambda^2/4a^2$ is ascertained from known λ, and an assumed value of a, and multiplied by possible values of N. The results are then compared with the values of $\sin^2\theta$ obtained from the powder photographs. The first trial may not match and a probable value for hkl may be indicated. This value is again fed back into equation (1) and a new value of a is suggested which is again tried till all the lines are indexed.

Another method is to reverse the value of $\lambda^2/4a^2$ to determine N. The values of N so obtained give the values of h, k, and l by looking up a table of N and corresponding hkl values. Knowing λ, the factor $\lambda^2/4a^2$ gives the value of a.

To make the value of a more accurate, Nelson-Riley extrapolation function, discussed in section 7.16.1, is employed.

The value of a, extrapolated to $\theta \to 90°$, is determined, since as $\theta \to 90°$, the error in determination of $a \to 0$. A computer can do these calculations after the lines are accurately measured for θ values.

The formulae used in other systems are:

b) *Tetragonal* system

$$\sin^2\theta_{hkl} = \frac{\lambda^2}{4a^2}(h^2 + k^2) + \frac{\lambda^2}{4c^2}l^2$$

c) *Hexagonal* and *trigonal* system

$$\sin^2\theta_{hkl} = \frac{\lambda^2}{3a^2}(h^2 + hk + k^2) + \frac{\lambda^2}{4c^2}l^2$$

d) *Orthorhombic* system

$$\sin^2\theta_{hkl} = \frac{\lambda^2}{4a^2}h^2 + \frac{\lambda^2}{4b^2}k^2 + \frac{\lambda}{4c^2}l^2$$

Published tables give (i) values of d, for the θ's obtained with known wavelengths, (ii) the direct values of $\sin^2\theta$, (iii) the values of $(h^2 + k^2 + l^2)$, $(h^2 + k^2)$ and $(h^2 + hk + k^2)$. These tables make the calculations much easier.

7.14 GRAPHICAL METHOD OF INDEXING

A graphical method for indexing the lines when the unit cell is known, has been devised and can most conveniently be used in the case of cubic crystals. Based on the relation $d = a/\sqrt{(h^2 + k^2 + l^2)}$, lines giving d for various values of a for some simple Miller indices, are drawn on a graph of a in *nm*, against d in *nm* (Fig. 7.13a). On a strip of paper the d-values of the powder diffraction lines are plotted on the same scale as d of the diagram. The paper strip is then moved parallel to the d-axis until a few of the radial or divergent lines of the graph coincide simultaneously with the lines on the paper. The horizontal line marking the edge of the paper meets the a-axis to indicate the cell-parameter of the mineral. Conversely, when the cell-parameter is known the strip can be directly placed at that position on the vertical *a*-axis, with the strip length parallel to the *d*-axis, the intersection points of the radial lines would suggest the indices of the respective diffraction lines marked on the paper.

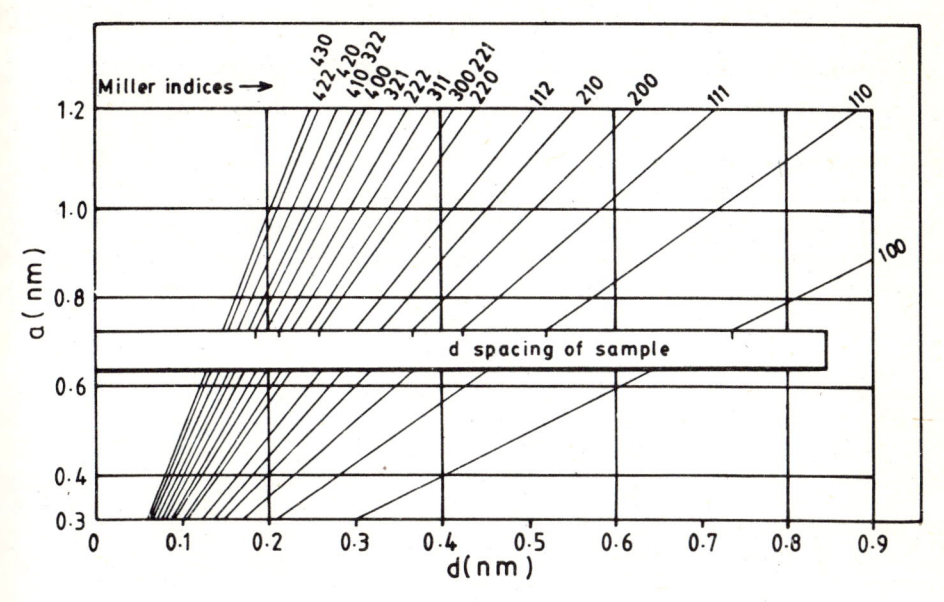

Fig. 7.13a Graphical method of indexing.

7.15 FIBRE DIAGRAMS
If the mineral is not sufficiently pulverised (i.e. >> 325 mesh) or the powder contains needle like fragments (as in the case of asbestos) or when the polycrystalline material is drawn in the form of a wire as in the case of metals, the arcs on the film get spotted. The spots represent individual spots due to diffraction from certain planes. The spots are non-random and make layer lines. From the distances between such layer lines the characteristic atomic distances in a lattice can be determined.

7.16 DETERMINATION OF LATTICE PARAMETERS
When the indices of the lines are known, the lattice parameters can be determined. Determination of the parameters of non-cubic minerals need measurement of a large number of *hkl* reflections. Calculation with a greater number of reflection minimizes the errors in the calculated parameter. J.B. Nelson and D.P. Riley (Proc. Phys. Soc., London, *57*, 160–177, 1945) listed systematic errors in determining d_{hkl} as (1) the *absorption error*, due to powder rod's absorption, (2) the *eccentricity error*, due to angular rotation of the rod axis, and (3) inaccurate knowledge of the camera diameter etc. All these errors are minimised at $\theta = 90°$.

7.16.1 The Nelson-Riley Method:

The systematic errors in 2θ resulting from absorption are obviated by applying a function determined by Nelson and Riley. It was first used in a crystal of cubic symmetry. Having calculated the a_0 values for each line from the powder photograph, using the equation

$$a_o = d_{hkl} \sqrt{h^2 + k^2 + l^2}$$

and plotting the values against what they named as extrapolation function, which is $\frac{1}{2} \left(\frac{Cos^2\theta}{Sin\,\theta} + \frac{Cos^2\theta}{\theta} \right)$, a linear relation is obtained for $\theta < 30°$. An extrapolation of the line to $\theta = 90°$ would give a specific a_0 value which would be free from absorption error (Fig. 7.13b). A table containing the extrapolation functions at $0.1°$ intervals of θ was prepared by them and is available in standard books on X-ray crystallography.

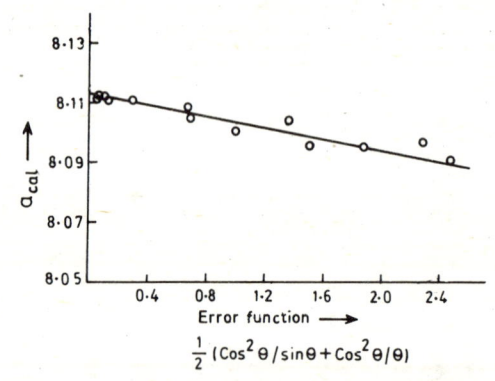

Fig. 7.13b Nelson-Riley extrapolation function in the correction of error due to absorption.

7.16.2 Goniometric determination of Cell-parameter ratio

Lattice constants of single crystals can be determined using rotation photographs. Rotation about different crystal axes helps in determining the lattice constants more accurately.

Measurement of the interfacial angles of a single crystal give the axial ratios which often tally very closely to the cell parameter ratios determined by X-ray. For example, the interfacial angles, measured by using contact or reflection goniometer, of a topaz crystal (orthorhombic) give the axial ratio a: b: c = 0.528: 1: 0.954 whereas by X-ray measurement the dimensions obtained are a = 4.65 Å, b = 8.80 Å and c = 8.40 Å. The axial ratio therefore come as a: b: c = 0.528 : 1 : 0.955 which is almost the same as obtained by goniometric measurements.

7.17 SEARCHING MINERAL POWDER DIFFRACTION FILE

The *Data Cards* provide the most efficient application of the Powder Diffraction File and they are accepted universally as standard reference for powder diffraction analysis. The Data Cards list "*d*" spacings and intensities of specific chemical compounds, reference source, radiation, name, formula, crystallographic data, optical characteristics, and the Powder Diffraction File number. Most patterns are indexed and the *hkl* values listed.

Mineral Powder Diffraction File, compiled by the Joint Committee on Powder Diffraction Standards (JCPDS), is of two volumes (a) Search manual and (b) Data book. The volume of 1980 contains 2300 species represented by 3000 patterns*.

The data book has been arranged aphabetically by mineral name and also has a chemical name section, Hanawalt numerical section and Fink numerical section.

The steps used for Hanawalt and Fink Search Manual for identification of minerals are as follows:

1) Tabulate the measured *d*-spacings and relative intensities of the experimental pattern in numerical order of *d*-spacings.
2) Start the search, using the strongest three (or more) reflections to locate the Hanawalt (or Fink) group in the Hanawalt (or Fink) search manual.
3) When the preliminary identification is done, the observed pattern of *d*-spacings and relative intensities are compared with all those obtained from the mineral data-card to achieve a final identification.
4) When this search with the strongest line fails, the next strongest reflection or the next smaller *d*-spacings is used to locate the position within the Hanawalt or Fink group. This procedure is repeated with weaker reflections till the matching is achieved from 1st to 8th *d*-spacings accompanied by their relative intensities. To help a research worker, in the absence of the mineral diffraction file, a guide to mineral identification has been arranged as an appendix to this part of the book (Appendix VB).

Microfiche

The Powder Diffraction File has been a successful means of compiling data on powder diffraction patterns and it is recognised as the standard reference source for powder diffraction analysis.

The original concept of making this data available on cards has remained unchanged. However, the file has grown considerably and future plans provide for greater expansion. As a result of this, the Powder Diffraction Microfiche File was developed as an alternative to the Data Card File.

The Microfiche File, like the Data Card File, is divided into two sections. These are organic and inorganic. While both sections consist of patterns from sets 1 to 27, the former is represented by 85 fiche and the latter by 220 fiche.

Search manuals

Hanawalt Method: This search manual is based on searching for and matching the most intense *d* spacings. Eight *d* spacings are given for each chemical

---

*Annually, a new set of data, consisting of compounds not previously included in the File, is compiled. To maintain an up-to-date File, users should acquire the current set. Each year, the File is expanded and updated.

compound and the three most intense d spacings permuted so that each compound is entered three times at different locations. Also, a systematic overlap of groups is provided to account for experimental error. Intensities are shown as subscripts and the chemical formula and Powder Diffraction File number are given for each entry.

Fink Method: This Search Manual is based on searching for one of the four intense d spacings and matching the remaining d spacings in descending numerical order. Eight d spacings are given for each chemical compound with each of the four most intense d spacings selected for entry. Intensities, although not imperative for searching with this manual, are shown as subscripts. The chemical formula and Powder Diffraction File number are given for each entry.

Alphabetical Index: The Alphabetical Index lists compounds alphabetically by their chemical name. The listing is rotated and the fragments of the chemical name entered alphabetically (KWIC) providing for added convenience when prior chemical knowledge is known. Where applicable, the chemical name is followed by the mineral name. Also, this manual lists minerals alphabetically by their mineralogical name. The three most intense d spacings, intensities, chemical formula and Powder Diffraction File number are given for each entry.

7.18 SELECTED POWDER DIFFRACTION DATA FOR MINERALS

Selected Powder Diffraction Data for Minerals are published in two volumes, the Data Book and Search Manual. The Data Book consists of copies of the Powder Diffraction *Data Cards* arranged in card number sequence, and includes an index by mineral names and groups.

The *Data Cards* listed therein, represent nearly 4,900 mineral species found in sets 1 to 23 of the Powder Diffraction File. Patterns have been selected, if obtained from natural material, from the synthetic counterpart of a mineral, or the end-member composition of a mineral solid-solution series. Powder diffraction data calculated from crystal structure determinations of minerals or their synthetic equivalents are also included.

The *Search Manual* provides for quick access to entries in the Data Book and includes an alphabetical Mineral Name section listing the three most intense d spacings and intensities, chemical formula, the Powder Diffraction File card number, an alphabetical-KWIC Chemical Name section and a Hanawalt Numerical Section.

7.18.1 Mineral File Workbook

A Mineral File Workbook was prepared by Professors D.K. Smith of the Pennsylvania State University and G.J. McCarthy of North Dakota State University.

This volume uses mineralogical examples from the Mineral Powder Diffraction File to illustrate the Alphabetical, Hanawalt and Fink search procedures for crystalline phase identification. Examples and problems illustrate solid solution effects, feldspar analysis, polytype characterisation, clay mineral identification and quantitative estimation. Additional problems in mineral, ore and rock identification permit the student to combine mineralogical knowledge with the search-match procedures. An example of computer-assisted mineral analysis of a complex multiphase assemblage is included.

7.18.2 The Data Card (ASTM/JCPDS)
The data cards have several segments (Fig. 7.14). The left corner segment gives the d-values and Intensity (I/I_1) values of the three strongest lines followed by the largest d-spacings recorded. The segments downwards give the details of experiments viz., radiation and filter used, the diameter of the camera, the method of determining the intensity and the references.

The next two segments give all crystallographic parameters and optical values in succession. Sample description and source come next.

At the right hand top the mineral name and its chemical formula in given. This is succeeded below by columns of d-values with respective intense and hkl values.

A JCPDS card for aluminium chromite is shown in Fig. 7.14 to cite an example.

7.19 SINGLE CRYSTAL METHODS

7.19.1 Stationary Crystal: Laue Method
In this method a single crystal is held stationary in a beam of polychromatic x-ray (or neutron) radiation of continuous spectrum, (from 0.2 to 2Å radiation) from a fine-focus x-ray tube. The x-ray from molybdenum or tungsten source (good Laue pattern may be obtained using copper as well) is collimated into a fine beam of about ½ mm and allowed to fall on a small single-crystal specimen of about 1 mm size.

The crystal diffracts the discrete wavelengths (λ) for which Bragg's Law is satisfied with certain d-spacings and the angle of incidence.

Geometrically speaking, x-ray diffraction takes place if and only if the sphere of reflection, corresponding to a given wavelength, intersects a reciprocal lattice point of the crystal under study (*vide* section 7.20). When the crystal is stationary, the reciprocal lattice points are also stationary. So if a monochromatic beam is used, only one or two reciprocal lattice points will be intersected by the corresponding sphere of reflection. When the radiation is white (continuous), there will be numerous wavelengths and innumerable spheres of reflection. Thus, each reciprocal lattice point will be intersected by one of these spheres of reflection and consequently will be recorded as a diffraction spot on the photographic plate or film or any other detection system.

The symmetry of the Laue spots, therefore, is the symmetry of the reciprocal lattice.

The Laue pattern consists of a series of spots and indicates the symmetry of the crystal (Fig. 7.15). If a crystal with four-fold symmetry is placed with its 4-fold axis aligned along the x-ray beam then the Laue pattern will show a four-fold symmetry, and so on.

Laue patterns for some common crystals are shown in Fig. 7.15.

154 *Fundamentals of Optical, Spectroscopic and X-ray Mineralogy*

3-873					
d	2.52	1.60	1.46	4.82	FeCr₂O₄
I/I₁	100	90	90	50	Iron Chromium Aluminum Oxide Chromite, aluminian

Rad. MoKα1 λ 0.7093	Filter	Dia	d Å	I/I₁	hkl	d Å	I/I₁	hkl
Cut off	I/I₁ Visual	I/1 cor.	4.82	50	111	.850	60	844
Ref. Clarke and Ally, Am. Mineral., 17 69 (1932)			2.95	60	022	.815	10	1020
			2.52	100	113	.805	40	951
Sys. Cubic S.G. Fd3m (277)			2.40	10	222	Indexed	by LGB	
a₀* b₀ c₀ A C			2.07	70	004			
α β γ Z 8 Dx 5.09			1.69	40	224			
(FeCr₂O₄)			1.60	90	115			
			1.46	90	044			
εα 2.08¹ nωβ 2.16² εγ Sign			1.40	10	135			
2V D 4.5–4.8 mp 1850°C Color Brown to			1.31	20	026			
Ref. Dana's Systems of Mineralogy. Brownish-black			1.26	50	622			
7th Edition			1.20	30	444			
			1.16	20	117			
			1.11	30	246			
*a₀ varies from 8.36 at 0 percent Al₂O₃ co 8.20 at 31			1.10	60	n.i.			
percent Al₂O₃.			1.04	30	008			
¹Nottingham, Pa., Cr₂O₃ 51.21 percent.			0.979	20	660			
²North Carolina.			.960	40	751			
Contains minor Al.			.931	30	840			
Spinel-linnaeite group			.873	30	931			

Fig. 7.14 A JCPDS data card on chromite.

X-ray Optics 155

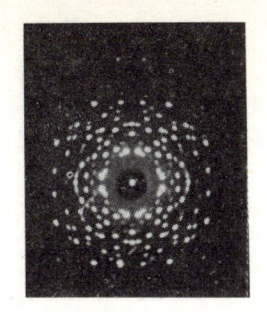

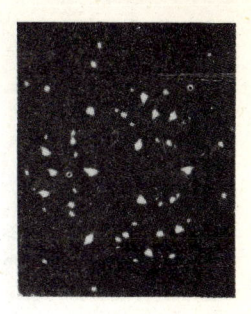

Fig. 7.15 Laue Photographs (a) *Cubic* minerals: (i) Halite, NaCl ⊥ (100) (ii) Pyrite FeS$_2$ ⊥ (100) (iii) Fluorite CaF$_2$ ⊥ (111)

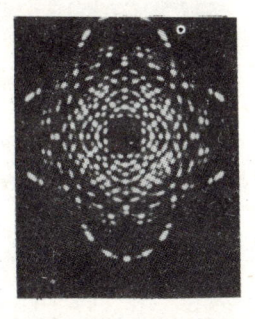

(b) *Tetragonal* mineral: Apophyllite ⊥ (001)

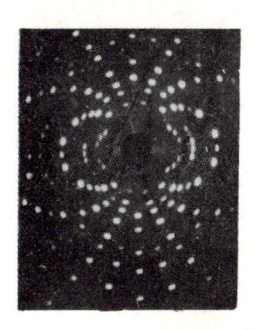

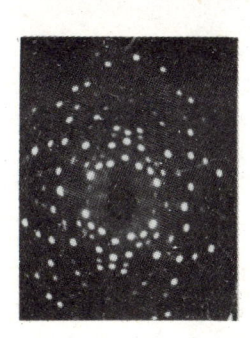

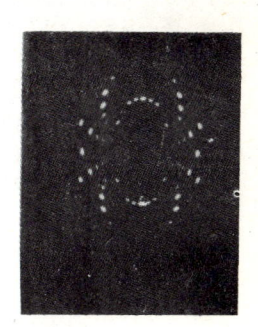

(c) *Orthorhombic* minerals: Anhydrite (i) ⊥ (100) (ii) ⊥ (010) (iii) ⊥ (001)

(d) *Hexagonal* minerals: Carborundum (i) (Moissanite) SiC ⊥ (0001)
(ii) Jeremejevite, Al$_6$ (OH)$_3$ (BO$_3$)$_5$ ⊥ (0001)

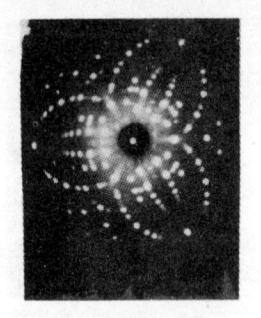

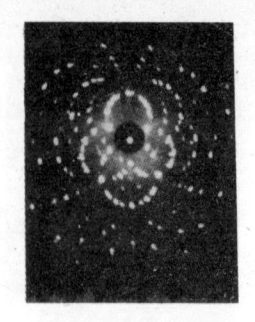

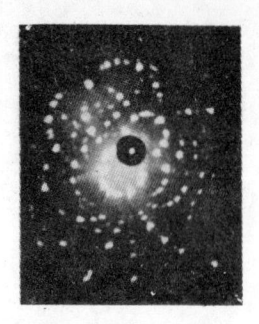

156 *Fundamentals of Optical, Spectroscopic and X-ray Mineralogy*

(e) *Trigonal* minerals: (i) Calcite ⊥ (0001) (ii) Calcite ⊥ (10$\bar{1}$1), (iii) Dolomite ⊥ (10$\bar{1}$1)

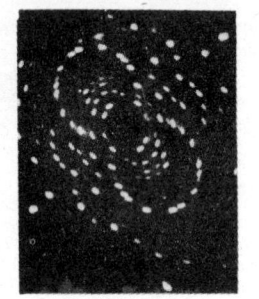

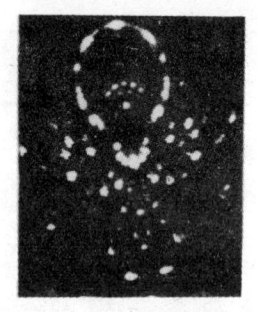

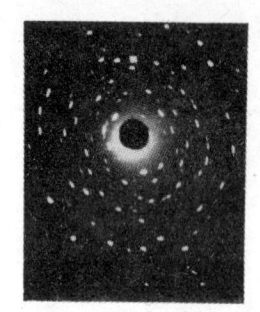

(f) *Monoclinic* minerals: (i) Gypsum: ⊥ (010) (ii) Orthoclase ⊥ (001) (iii) Muscovite ⊥ (001)

The Laue spots on the transmission pattern occur when they lie at the points of intersection of a series of ellipses of tautozonal curves arising from the family of planes which are parallel to a given direction in the crystal and therefore constitute a crystal zone.

In addition, since a lattice plane can be parallel to two zones axes, each reflection spot must be common to two reflection ellipses and lies, therefore, at their points of intersection. In the rear reflection Laue pattern, the spots lie at the intersections of the hyperbolas.

For indexing of the reflections in a Laue pattern, usually the stereographic or gnomonic projections are employed. It is however, very difficult to elucidate the atomic arrangement within a crystal from the intensity measurement on Laue spots. It is generally used for finding the symmetry and the orientation of a crystal.

7.19.2 Scattering and ordered layer lines

When an X-ray falls on a series of regularly spaced atoms, a set of parallel wave fronts are produced about each atom. These make a set of spherical scattered waves of the same frequency and wavelength.

The common tangent to any succession of spherical wave fronts of scattered waves coincides with the reinforcement of the scattered waves; this is commonly marked as diffraction. A series of patterns emerge from such reinforcement or diffraction (Fig. 7.16)

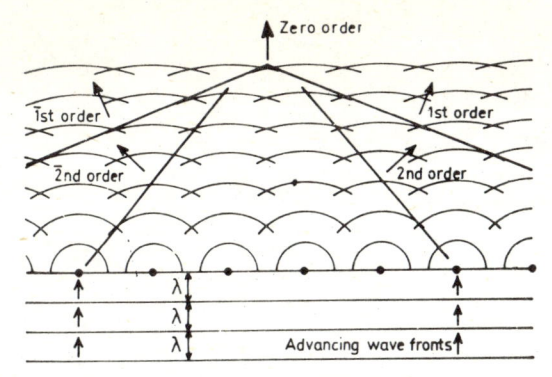

Fig. 7.16 The patterns of diffracted waves in succession to form different order layers (0-order, 1st order, 2nd order etc.).

The reinforcement occurring parallel to the original wavefront gives *zero-order* diffracted beam.

The common tangent with a scattered wave-crest difference of one, between successive atoms, forms the wave front of the *first-order* diffracted beam. Thus, diffracted beams representing 2,3,4 ...n wavelength (wave-crest) differences in phase between the wavelets generated from neighbouring atoms, give the *second-order, third-order* to *nth-order* diffraction.

On the lower side of the zero-order beam a set of corresponding negative orders of diffraction is formed (minus first-order, minus second-order, and so on).

7.20 RECIPROCAL LATTICE

The concept of reciprocal lattice was developed by Ewald. This is a geometrical concept.

In this a plane (*hkl*) is represented by a line drawn normal to the plane, and the length of the line is an inverse of the inter-planar spacing (d_{hkl}) of the plane. The real lattice plane is in a position to satisfy the Bragg diffraction conditions, when the corresponding point in the reciprocal lattice (whose origin is at 0,0,0) lies on a circle (reflecting) of radius $1/\lambda$.

When the normals are drawn to the planes from a common origin, the terminal points of these normals constitute a lattice array (Fig. 7.17a). This is called the reciprocal lattice because the distance of the points from the origin are reciprocal to the *d*-spacings of those planes. The reciprocals are represented by stars, *s. In three dimensions the lattice vectors are:

$$a^* = \sigma_{100} = 1/d_{100}$$
$$b^* = \sigma_{010} = 1/d_{010}$$
$$c^* = \sigma_{001} = 1/d_{001}$$

Similarly, the reciprocal interaxial angles are represented by $\alpha^*, \beta^*, \gamma^*$.

The geometrical concept of reciprocal lattice is presented in Fig. 7.17 (b).

Let S be sample plane diffracting an x-ray beam at an angle θ and the film plate is placed at a distance r from the sample. The normal to the S-plane intersects the film at R, which is the reciprocal lattice-point.

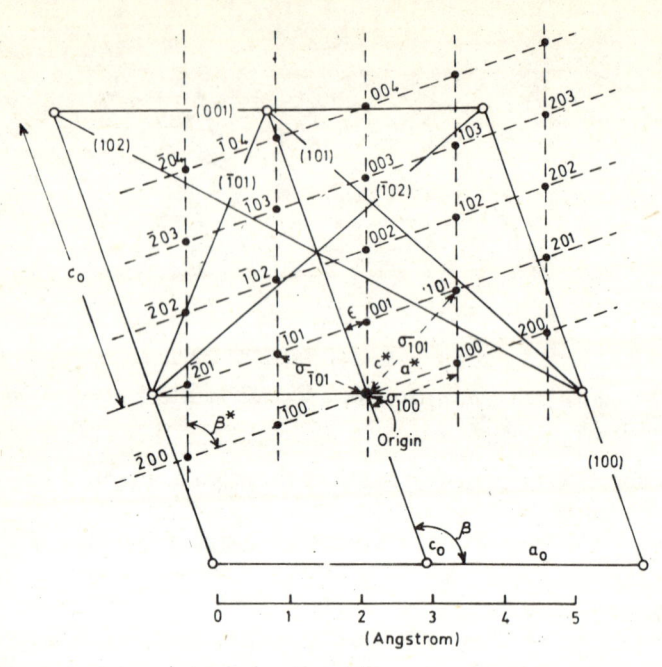

158 *Fundamentals of Optical, Spectroscopic and X-ray Mineralogy*

Fig. 7.17(a) The relation between the cell lattice and the reciprocal lattice. It should be imagined that the reciprocal lattice continues in all directions.

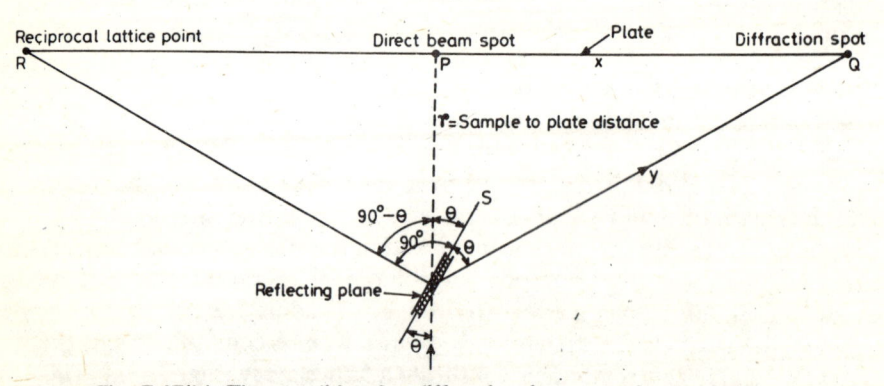

Fig. 7.17(b) The condition for diffraction from a reciprocal lattice.

On the photographic film the distance between the diffraction spot (Q) of a crystallographic plane and the direct beam spot (P) can be determined. Let x be the distance between P and Q.

$$\text{Then, } x = r \tan 2\theta$$
$$y = r \sec 2\theta$$
$$= r \sec [\tan^{-1} \frac{x}{r}]$$

From the above equation, for each reflection spot Q, the corresponding reciprocal lattice point R can be obtained. From the distribution of the points, R's, lattice constants, Miller Indices of the reflecting spots etc. are obtained.

7.21 MOVING CRYSTAL METHOD
Several types of x-ray photography of moving crystals are widely used:

7.21.1 Rotation Photographs
In this method, monochromatic x-rays are made to impinge on a small crystal, rotated around any of its principal crystal axes, which is also a zone axis. Each lattice plane in the zone on rotation reaches the appropriate angle for reflection and produces a single spot on the cylindrical film, whose axis coincides with the rotation axis. The reflection cones intersect the film in a set of circles. When the film is unrolled, the reflection spots appear in straight *layer-lines*.

All reflections from the zonal planes will lie on a layer line, stretching perpendicular to the axis of rotation (Fig. 7.18). If this zone axis is along c-axis (or as in the figure) the reflections on the layer line through the direct beam would have indices as *hko* and this line is called the *equator* or *zero-layer* line. The *hkl* reflections distributed parallel to this equator at some definite distances above and below correspond to 1 = 1; 1 = − 1; 1 = 2; 1 = −2, and so on. Thus, *hk* 2 reflections lie on the second layer line and so on.*

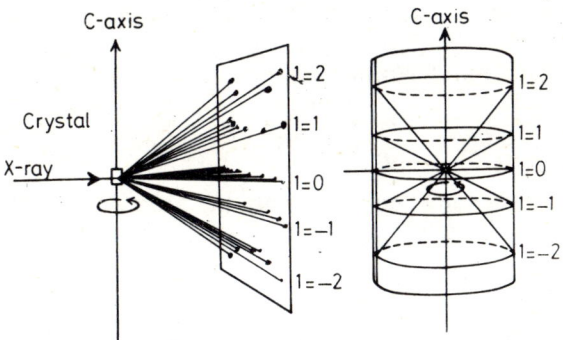

Fig. 7.18 Rotation photograph of single crystal, rotated around its principal symmetry axis. The diffracted beams are recorded on a flat plate (left) or on a cylindrical film (right). The images of the diffracted beams lie on layer lying corresponding to l = 0, 1, −1, 2, −2, etc. and the spots of all values of h and k.

To find out the repeat (or cell-) distance, *d*, in the rotation direction, the following relations are used:

$$a = \frac{h}{\sin \mu} \text{ and } \tan \mu = \frac{P}{R}$$

where R = camera radius
P = distance between a layer line and the equatorial line
h = order (index) number of the layer line
μ = the layer line angle.

*These can be explained easily in terms of a reciprocal lattice rotating or oscillating about an axis and a sphere of reflection.

From measurements on rotation patterns taken around three crystallographic axes the unit cell parameters can be more accurately determined.

A cubic crystal rotation photograph about any one of the axis gives cell size. For tetragonal and hexagonal crystals, two photographs are required, one along the vertical axis and the other along any of the horizontal axes. For orthorhombic, monoclinic and triclinic crystals, three photographs along the three axes are needed along with a knowledge of the symmetry and interaxial angles, determined by optical or other goniometric methods.

The order of a layer line furnishes one of the three Miller indices of each reflection. But this method suffers from the drawback that each layer-line represents a one-dimensional pattern of reflections representing a two-dimensional array of reciprocal points.

In contrast, the oscillation, the Weissenberg and the precision methods present uniquely the diffraction symmetry of the crystal and allow the reflection spots to be properly indexed and help in determining the parameters of the unit cell.

Example: Rotation photograph of quartz (SiO_2) around [0001] in Cuk_α is shown in Fig. 7.19.

We can proceed with the photograph for processing of the equatorial reflections.

The following quadratic form for the hexagonal system is used for indexing

$$\sin^2\theta = \frac{\lambda^2}{4a^2}\left[\frac{4}{3}(h^2 + k^2 + hk) + \left(\frac{a}{c}\right)^2 l^2\right]$$

According to the zonal equation only the reflections $hki0$ occur at the equator. Therefore, the simplified equation is:

$$\sin^2\theta = \frac{\lambda^2}{4a^2}\cdot\frac{4}{3}(h^2 + k^2 + hk).$$

Now, to determine the value of $\sin^2\theta_{(10\bar{1}0)}$, we need the value of a_0 However, from the layer line distances in our photographs we obtain only $c_0 = 5.40$. For the direct calculation of a_0 one ought to take a second photograph about a_0 (as the rotation axis). However, from the axial ratio c/a, obtained from goniometric measurements, a_0 can be obtained without any further photographs. For quartz, $c_0/a_0 = 1.099$. From this we obtain,

$$a_0 = 4.904$$
$$\sin^2\theta_{(10\bar{1}0)} = 0.0328$$

Dividing all values of $\sin^2\theta$ by this quantity, we obtain for every reflection, the corresponding sum of $h^2 + k^2 + hk$.

In the following table 7.2.1 it is denoted by S (i.e. $h^2 + k^2 + hk$.). Now the corresponding indexes of $hki0$, can be read off from the table 7.2.2.

7.21.2 Oscillation Photograph

For removing the ambiguity in indexing rotation photograph spots, the crystal is

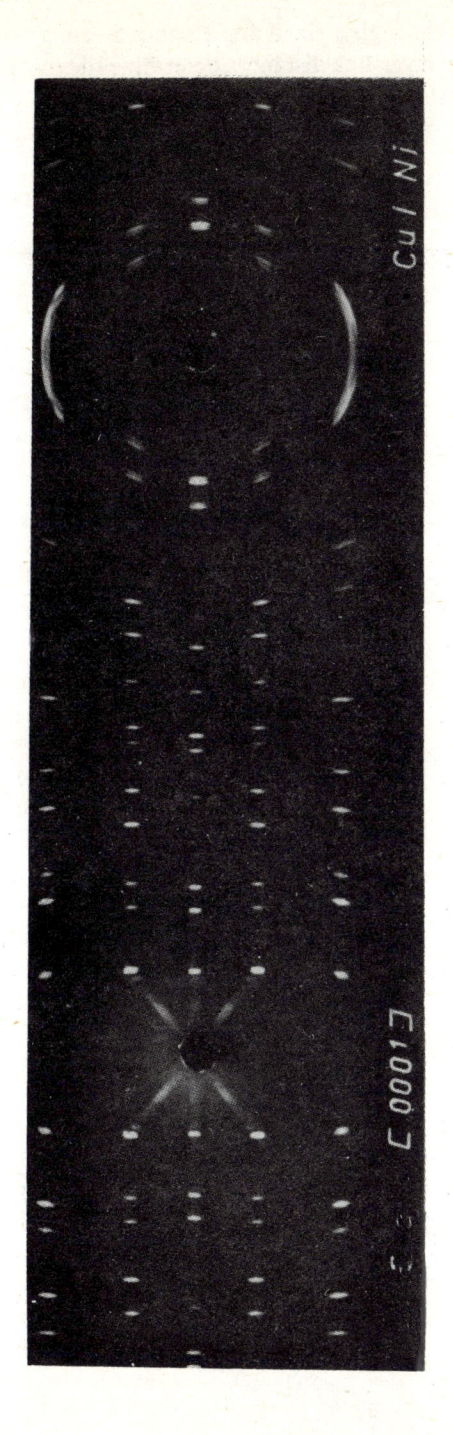

X-ray Optics   161

Fig. 7.19   Rotation photograph of quartz around [0001] in CuK$_\alpha$ (with nickel filter).

made to oscillate through a fixed angle using a camera with an oscillation-rotation cam. The layer-line distances determine the translation period along the lattice-line parallel to the rotation-axis. Translation of identical points not only occurs along the crystal axis but also along other crystallographic directions.

Table 7.2.1. Measurement on a film of rotation photograph of quartz

No.	I	mm	mm/2	θ	$\sin^2\theta$	S	hkiO
1.							
2.							
3.							
4.							
etc.							

Table 7.2.2. Quadratic Form

Cubic System $h^2+k^2+l^2$	h,k,l	Tetragonal h^2+k^2	Prismatic h,k	Hexagonal S	4/3S	Prismatic h,k,i
1	1,0,0	1	1,0	1	1.3	1,0,$\bar{1}$
2	1,1,0	2	1,1	3	4.0	1,1,$\bar{2}$
3	1,1,1	4	2,0	4	5.3	2,0,$\bar{2}$
4	2,0,0	5	2,1	7	9.3	2,1,$\bar{3}$
5	2,1,0	8	2,2	9	12.0	3,0,$\bar{3}$
6	2,1,1	9	3,0	12	16.0	2,2,$\bar{4}$
8	2,2,0	10	3,1	13	17.3	3,1,$\bar{4}$
9	3,0,0	13	3,2	16	21.3	4,0,$\bar{4}$
	2,2,1	16	4,0	19	25.3	3,2,$\bar{5}$
10	3,1,0	17	4,1	21	28.0	4,1,$\bar{5}$
11	3,1,1	18	3,3	25	33.3	5,0,$\bar{5}$
12	2,2,2	20	4,2	27	36.0	3,3,$\bar{6}$
13	3,2,0	25	5,0	28	37.3	4,2,$\bar{6}$
14	3,2,1		4,3	31	41.3	5,1,$\bar{6}$
16	4,0,0	26	5,1	36	48.0	6,0,$\bar{6}$
17	4,1,0	29	5,2	37	49.3	4,3,$\bar{7}$
	3,2,2	32	4,4	39	52.0	5,2,$\bar{7}$
18	4,1,1	34	5,3	43	57.3	6,1,$\bar{7}$
19	3,3,1	36	6,0	48	64.0	4,4,$\bar{8}$
20	4,2,0	37	6,1	49	65.3	7,0,$\bar{7}$
21	4,2,1	40	6,2			5,3,$\bar{8}$
22	3,3,2	41	5,4	52	69.3	6,2,$\bar{8}$
24	4,2,2	45	6,3	57	76.0	7,1,$\bar{8}$
25	5,0,0	49	7,0	61	81.3	5,4,$\bar{9}$
	4,3,0	50	7,1	63	84.0	6,3,$\bar{9}$
26	5,1,0		5,5	64	85.3	8,0,$\bar{8}$
	4,3,1	52	6,4	67	89.3	7,2,$\bar{9}$

1	2	3	4	5	6	7
27	5,1,1	53	7,2	73	97.3	8,1,$\bar{9}$
	3,3,3					
29	5,2,0	58	7,3	75	100.0	5,5,$\bar{10}$
	4,3,2	61	6,5	76	101.3	6,4,$\bar{10}$
30	5,2,1	64	8,0	79	105.3	7,3,$\bar{10}$
32	4,4,0	65	8,1	81	108.0	9,0,$\bar{9}$
			7,4	84	112.0	8,2,$\bar{10}$
33	5,2,2					
	4,4,1	68	8,2	91	121.3	6,5,$\bar{11}$
						9,1,$\bar{10}$
34	5,3,0	72	6,6			7,4,$\bar{11}$
	4,3,3	73	8,3	93	124.0	
35	5,3,1	74	7,5	97	129.3	8,3,$\bar{11}$
etc.						

If an oscillation photograph of a rocksalt cube is taken with the surface diagonal 110 as the rotation axis, one would expect a value for P equal to $a_o\sqrt{2}$. Therefore $P = 5.63\sqrt{2} = 7.96$. Actually, one obtains the value for P as 3.99, which leads us to conclude that there is already an identical point at the centre of the layer. Thus, it must be a face-centred cell. A photograph about a space-diagonal would also give the value $a_o\sqrt{3}$.

On the other hand, for a body-centred cell, the photograph about the surface diagonal would give the full-value and that along the body diagonal would give half-value.

A theoretical attempt to construct a space-lattice from using the translation of identical points gives the fourteen lattice types, known as Bravais lattice.

7.21.3 Weissenberg Camera

In the Weissenberg methods the crystal is rotated through 180° and back again continuously, and at the same time the cylindrical camera containing the film moves at a constant speed backwards and forwards, parallel to the axis of rotation; the movements of camera and film being so synchronised that a given position of the camera corresponds accurately to a definite angular position of the crystal in its rotation (Fig. 7.20a, b). The coordinates of a spot on the film then give both the angle of reflection and the position of the reflecting plane, and this allows the spot to be indexed without ambiguity. A metal cylinder, provided with an equatorial slit a few millimeters wide, is interposed between the crystal and the film in such a position that the spots of only one layer line pass through it. The Weissenberg photograph records all the spots belonging to that layer line, spread out into a characteristic pattern on the single film, and with a little experience they can readily be indexed.

164 *Fundamentals of Optical, Spectroscopic and X-ray Mineralogy*

Fig. 7.20a. Weissenberg photographs of diopside, along equatorial plane (target: copper, filter: nickel) with axis of rotation along 001.

7.21.4 Lattice Line Template for Weissenberg Indexing

Zero-level Weissenberg films often show a set of reciprocal-lattice curves. Photographs of suceeding levels also show a similar set of curves in that region. Commercially available template of lattice-lines is used as a guide for constructing the reciprocal-lattice line curves on a transparent paper placed over the photographs. Because the curves are related from film to film the lattice is resolved into a set of parallel planes.

X-ray Optics 165

Fig. 7.20b. Weissenberg photographs of diopside, along equatorial plane (target: copper, filter: nickel) with axis of rotation along 010.

Starting with one axis of reciprocal lattice the films are shifted in positions over the template and a set of curves are drawn through the reflections of each film. Each set of curves through the reflections resolve the levels into a set of parallel rows. Each reciprocal point can then be referred to three axes, one of which passes through the symmetry-equivalent points in each of the levels and the other two on the zero-level. After appropriately labelling the reflection spots (from the reciprocal points) are indexed.

7.21.5 Precession Camera

In this device the crystal, photographic plate and screen perform a sinuous rhythm in such way that those spots, for instance, with a definite value of l and all values of h and k are recorded as rectilinear net. This method is particularly suitable for crystals with large unit cells and consequently numerous values of the indices.

While the Laue, Rotation-oscillation and Weissenberg patterns are records of a distorted reciprocal lattice, the precision camera gives an undistorted reciprocal lattice.

7.21.6 Buerger Precession Method

For producing undistorted symmetry-true photographs of reciprocal lattice, level by level on a flat film, Buerger (1942, 1964) developed a precession method. In this technique the crystal, placed in the path of X-ray beam, is given an unusual motion, which produces remarkably well-produced reflection spots. (Fig. 7.21).

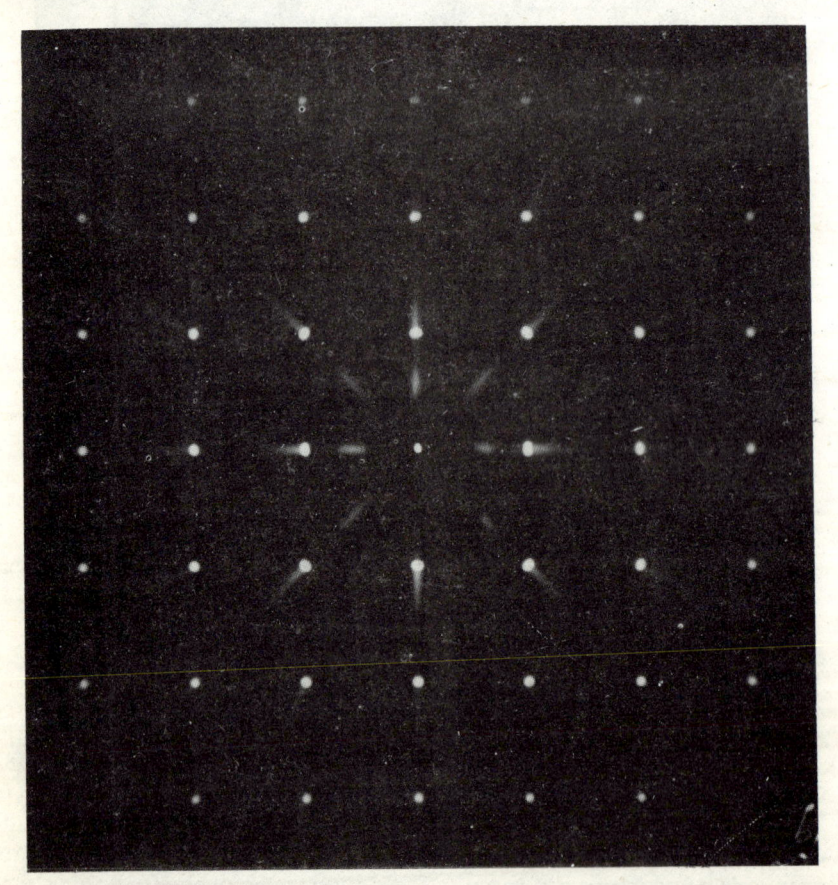

Fig. 7.21 Buerger precession photograph of NaCl; equatorial, with (100) as the precession axis.

Buerger's precession method is superior to Weissenberg's method in indexing a crystal when its reciprocal cell is small. For this reason, it is chosen as the most suitable technique for single-crystal studies.

Symmetry and lattice-type determination

All symmetries are generated by four major operations: (1) rotation, (2) reflection, (3) inversion and (4) translation. A combination of the first three yields 32 distinct point groups. Combining (4) with those 32 we have 230 space groups. These are known as the *Federov space group*. If spin is also brought into consideration we get 1685 space groups called the *Schubnikov space group*.

In X-ray diffraction studies the symmetry axes and planes can be identified by noting regular absence of diffracted wave-orders; the presence or absence of symmetry centres can be determined by statistical survey of intensities.

A number of possible reflection lines are systematically absent in the X-ray diffractogram of some minerals. These are due to the type of the lattice that characterises the mineral.

The *lattice types* as indicated by systematic absences of lines are as follows:

Absent hkl reflection lines	Lattice type
When $(k + l)$ is odd	A — centred
When $(h + l)$ is odd	B — centred
When $(h + k)$ is odd	C — centred
When $(h + k + l)$ is odd	I — centred
When indices are mixed odd and even	F — centred

7.22 STRUCTURE SYMMETRY

7.22.1 Symmetry of Continuum

For determining the structure of the crystals the symmetry of the cell must be known apart from the metric. The symmetry determination is done initially by the exterior morphology (or continuum) and then from the crystal structure. The latter is done by Laue's method. The morphological or continuum studies give the symmetries belonging to 32 classes. For detailed interpretation, the studies on crystal morphology, optics, piezo- and pyro-electricity behaviour etc. are necessary.

As is well-known only 2-, 3-, 4- and 6- fold rotation axes are possible for crystals with translational periodicities. These are represented as ◊, ▲, ■, ● respectively. The four- and six-fold roto-reflection axes, reflection planes and centre of symmetry are denoted by ▰, ◆, *m* and *i* respectively. Observation on a simple cube shows the combination of four 6- fold roto-reflection axes, three 4- fold and six 2- fold rotation axes; 9(3 + 6) mirror planes and the centre of symmetry. This is the highest possble symmetry. All the symmetry elements are shown in seven cube blocks in Fig. 7.22.

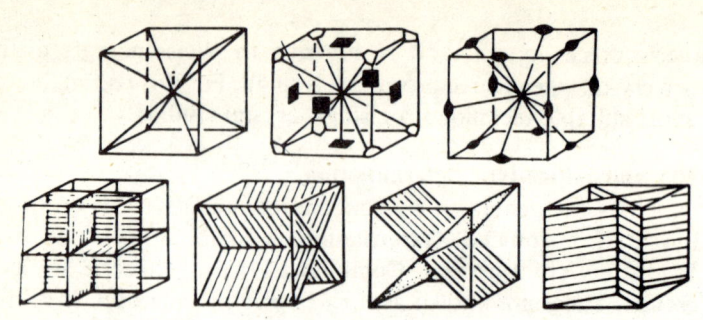

Fig. 7.22 Symmetry elements in a cubic block.

Similarly, all the symmetry elements can be combined with one another which gives rise to 32 possible combinations i.e. 32 classes of crystals which are grouped in seven systems based on axial ratios and angles; viz., cubic, tetragonal, hexagonal (also rhombohedral), orthorhombic, monoclinic and triclinic.

The minimum symmetry elements for each crystal system are as follows:

Cubic	—	Four 3-fold rotation axis
Tetragonal	—	One 4-fold rotation or rotoinversion axis
Orthorhombic	—	Three perpendicular 2-fold rotation axis
Trigonal	—	One 3-fold rotation or rotoinversion axis
Hexagonal	—	One 6-fold rotation or rotoinversion axis
Monoclinic	—	One 2-fold rotation or rotoinversion axis
Triclinic	—	None.

7.22.2 Symmetry of Discontinuum and Omission Laws

As we have a rotation axis in continuum case there is a rotation plus screw axis in a *discontinuum*. In discontinuum, the mirror plane corresponds to mirror plus glide plane.

Screw axis is obtained only when points of some structural value come to an identical position after rotation and gliding. Reflection and gliding make the *gliding mirror plane*.

The order of gliding distances correspond to the interatomic distances of a few angstroms. Hence, they are so small that in a macrocrystal the screw axes and the gliding mirror planes appear as rotation axis and mirror planes. They can only be recognised by X-ray studies (using omission or extinction laws).

Symbols of screw axes are 2_1, 3_1, 3_2, 4_1 etc. (Fig. 7.23) with displacement of $\frac{1}{2}, \frac{1}{3}, \frac{2}{3}, \frac{1}{4}$ etc. along the rotation axes. The gliding mirror planes are: $a, b, c, \ldots n$, with the gliding components: $\frac{1}{2}a_o$, $\frac{1}{2}b_o$, $\frac{1}{2}c_o$, $\frac{1}{2}a_o + \frac{1}{2}b_o$, etc.

Omission Laws

1. Serial Omissions

It is only in reflections at planes normal to screw axis the omissions occur. For example, through a 3-fold screw-axis parallel to the c-axis the lattice distance is split into three equal distances. This tri-section occurs because

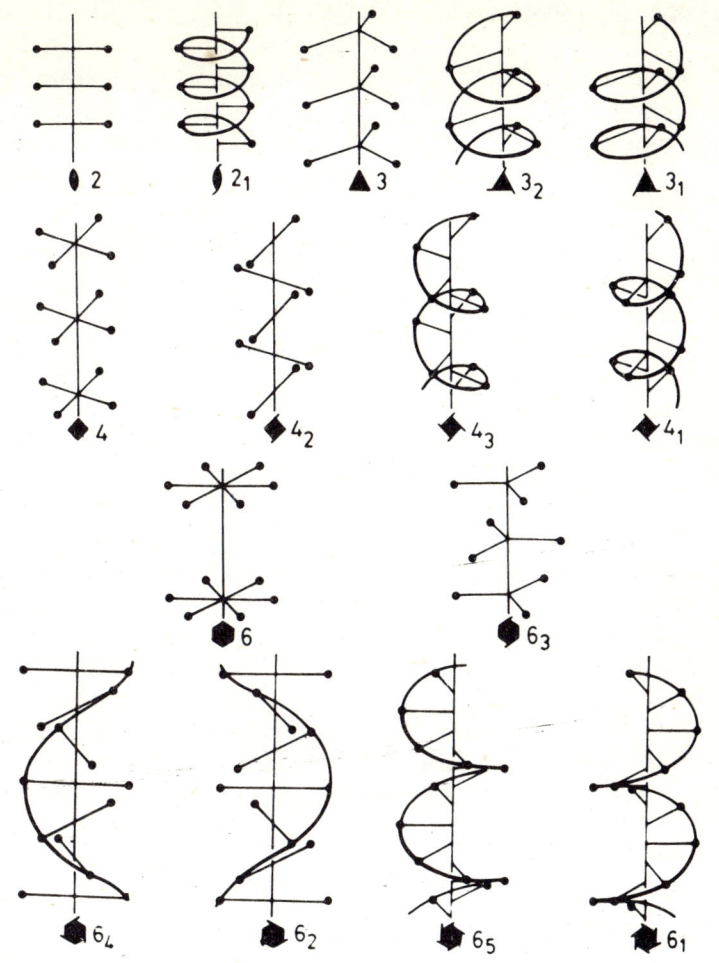

Fig. 7.23 The types of symmetry operations.

only the reflections 003, 006 and 009 are possible for all reflections *00l* (*l* = 3*n*). For a two-fold screw-axis parallel to C, only the reflections 002, 004 and 006 appear (*l* = 2*n*). And finally, for a 4-fold screw-axis (4₁) only reflections 004, 008 etc. (*l* = 4*n*) occur. This, for a screw-axis a definite series of reflections are seen.

2. Zonal omissions

Gliding mirror planes are manifested in the appearance of omissions in zones perpendicular to the mirror planes. Thus, a gliding mirror plane parallel to (010) causes omission only for the reflection of the zone *h0l*. If the gliding component is equal to $\frac{1}{2}a_0$ then the distance between lattice-planes appear no longer as a_0 but $\frac{1}{2}a_0$, which means that all reflections of *h0l* with odd *h* are missing.

Gliding mirror plane 001 : omissions in the zone *hk0*
Gliding mirror plane 010 : omissions in the zone *h0l*
Gliding mirror plane 100 : omissions in the zone *0kl*

The following reflections are omitted

h = odd .... gliding component $a_0/2$
k = odd .... gliding component $b_0/2$
l = odd .... gliding component $c_0/2$
$h + k$ = odd .... gliding component $a_0/2 + b_0/2$
$h + l$ = odd .... glidding component $a_0/2 + c_0/2$
$k + l$ = odd .... gliding component $b_0/2 + c_0/2$

3. Integral Omissions

This corresponds to those omission laws which apply, without restriction, to the entire manifold of all the reflecting lattice planes hkl; they allow one to determine the translation lattice.

hkl with any value present corresponds to ... Primitive cell
hkl with $h+k+l = 2n$ (even) value present corresponds to ... Body-centred cell
hkl with $h+k = 2n$, $h+l = 2n$, ... corresponds to ... Face-centred cell.

Thus, from the omissions it is possible to derive whether the rotation-axis or the screw-axis, mirror-planes or glide-mirror planes are present; whether the cell is a primitive cell or body-centred or face-centred cell. When one considers all these extra-possibilities in combining the symmetry elements, one obtains the discontinuum 230 space groups (in place of 32 crystal classes in continuum cases).

The characteristic omissions in each space group are arranged in International Tables in the form of Determinative Tables. This helps one to infer the space group very easily from the observed reflections.

7.23 SCATTERING POWER OF A SET OF ATOMS

In the case of a series of atoms with specified interatomic distances the X-ray diffraction principle works as in the case of diatomic molecules. However, the atoms in the sequence can collectively contribute to the reflection pattern.

The radiation reflected from each atom gets enhanced when the phase differences are λ, 2λ, 3λ, etc., corresponding to the diffraction cones of 1st, 2nd, 3rd .... order. From a single atomic sequence, however, it is not possible to be seen.

When crystalline substances are exposed to X-rays the diffraction pattern observed on a photographic plate consists of hyperbolic dark lines. When this is taken in a cylindrical film the sectional lines of the diffracted cones parallel to the film form layer lines. The vertical distances from the equator (e) gives the diffraction angle.

For measurement and calculation of interatomic distances we have layer lines with the following relationships:

1st order layer line : $\lambda = P \sin \mu_1$,
2nd order layer line : $2\lambda = P \sin \mu_2$,
nth order layer line : $n\lambda = P \sin \mu_n$

and $\tan \mu_1 = \dfrac{e_1}{R}$

$\cot \mu_2 = \dfrac{e_2}{R}$

$\tan \mu_n = \dfrac{e_n}{R}$

where, P = characteristic periodic distance along the rotation axis in Å
λ = wavelength of X-ray in Å
e_1, e_2, e_n = distances of the layer lines of 1st, 2nd and nth order from the equator.
μ_1, μ_2, μ_n = angle between the equatorial plane and the diffracted rays of 1st, 2nd, nth order.
R = distance of the film (radius of the X-ray camera)

7.24 X-RAY INTENSITIES AND ATOMIC ORDERING

While indexing the lines we have seen that each reflection causes a definite (relative as well as absolute) degree of darkening on the film, which appears with a definite intensity viz., very strong (v.s), strong(s), medium(m), weak(w) etc.

Now, to answer questions such as which factors control the intensities, how they can be measured from the atomic distribution in the cell etc., the following aspects are considered.

7.24.1 Atomic Scattering or Form Factor

Let us consider the scattering power of each atom and its decay with increasing incident angle. The amplitude of wave scattered by each atom is called the atomic scattering or form factor, f. It gives the intensity of scattering as a function of the angle of scattering and is represented by a function, $\sin \theta/\lambda$ (θ = half of the angle of scattering).

Along the direction of the primary ray, the scattering from one single electron is equal to one. From hydrogen atom it is 1, from one helium atom 2, one lithium atom 3, one Li^+-ion 2, one sodium ion 10. Thus heavier atoms scatter more strongly than lighter atoms. (Table 7.3).

The decay of the scattering power with increasing θ (f-curves) are obtained from measurements and exact analyses of the intensities are calculated with atomic models.

The graphical values are given in the International Tables of Crystal Structure Determination and are presented in the following pages for a few atoms. These values are plotted on a mm graph paper to be read off from an arbitrary value of $\sin \theta/\lambda$ (Fig. 7.24).

The atomic form factor is a continuous function (Table 7.3). On the other hand, the surface factor and the structure factor are discontinuous functions.

7.24.2 Surface Factor

When we study a cubic crystal we see there are 6 cubic faces with the same

Table 7.3. Atomic Scattering (form) factor

Sin θ/λ [Å^{-1}]	0.0	0.1	0.2	0.3	0.4	0.5	0.6	0.7	0.8	0.9
H	1.0	0.81	0.48	0.25	0.13	0.07	0.04	0.024	0.015	0.01
Li$^+$	2.0	1.94	1.76	1.52	1.27	1.03	0.82	0.65	0.51	0.40
Li	3.0	2.22	1.74	1.51	1.27	1.03	0.82	0.65	0.51	0.40
B^{+3}	2,0	1.98	1.92	1.82	1.70	1.57	1.42	1.27	1.13	1.0
C	6.0	5.13	3.58	2.50	1.95	1.67	1.54	1.43	1.32	1.22
N^{+3}	4.0	3.77	3.23	2.64	2.17	1.87	1.68	1.56	1.46	1.37
N	7.0	6.20	4.60	3.24	2.40	1.94	1.70	1.55	1.44	1.35
O	8.0	7.25	5.63	4.09	3.01	2.34	1.94	1.71	1.57	1.46
Na$^+$	10.0	9.55	8.39	6.93	5.51	4.33	3.42	2.77	2.31	2.00
Al^{+3}	10.0	9.74	9.01	7.98	6.82	5.69	4.69	3.86	3.20	2.70
Al	13.0	11.23	9.16	7.88	6.77	5.69	4.71	3.88	3.21	2.71
Si^{+4}	10.0	9.79	9.20	8.33	7.31	6.26	5.28	4.42	3.71	3.13
F$^-$	10.0	9.11	7.13	5.19	3.79	2.89	2.32	1.97	1.75	1.60
Ca^{++}	18.0	16.93	14.40	11.70	9.61	8.25	7.38	6.75	6.22	5.70
K$^+$	18.0	16.68	13.76	10.96	9.04	7.86	7.11	6.51	5.94	5.39
Cl$^-$	18.0	16.02	12.20	9.40	8.03	7.28	6.64	5.97	5.27	4.61
K	19.0	16.73	13.73	10.97	9.05	7.87	7.11	6.51	5.95	5.39
Ca	20.0	17.33	14.32	11.71	9.64	8.26	7.38	8.75	6.21	5.70
Ti	22.0	19.41	16.07	13.20	10.83	9.12	7.98	7.22	6.65	6.19
Cr	24.0	21.93	18.37	15.01	12.22	10.14	8.72	7.75	7.09	6.58
Mn	25.0	22.61	19.06	15.84	13.02	10.80	9.20	8.09	7.32	6.77
Fe	26.0	23.68	20.09	16.77	13.84	11.47	9.71	8.47	7.60	6.99
Co	27.0	24.74	21.13	17.74	14.68	12.17	10.26	8.88	7.91	7.22
Ni	28.0	25.80	22.19	18.73	15.56	12.91	10.85	9.33	8.25	7.48
Cu	29.0	27.19	23.63	19.90	16.48	13.65	11,44	9.80	8.61	7.76
Zn	30.0	27.92	24.33	20.77	17.42	14.41	12.16	10.37	9.04	8.08
W	74.0	69.07	61.58	54.59	48.27	42.83	38.12	34.06	30.46	27.36
Cs	55.0	51.27	44.54	38.72	33.80	29.80	26.46	23.69	21.34	19.36
Pb	82.0	77.24	68.45	60.48	53.56	47.77	42.85	38.69	35.13	32.08
O^{-2}	10.0	8.0	5.5	3.8	2.7	2.1	1.8	1.5	1.5	1.4
Ti^{+4}	18.0	17.0	14.4	11.9	9.9	8.5	7.85	7.3	6.7	6.15

crystal parameters; similarly octahedrons have 8 such faces and rhombohedrons have 12 faces.

If the probabilities of the proper reflecting positions are equal (only a chosen orientation may be different), we can see that the reflection probability of all the cubic surfaces to that of octahedral surfaces is in a ratio of 6:8. This implies, in general, that the intensities of reflections are directly proportional to the corresponding number of faces.

This also applies to powder photographs which have no chosen directions. For rotation photographs about [001], for example, one may note that the number of surfaces of the form (hko) lie in the zone [001]. For a cubic crystal (and the reflection hoo) this is 4; for a rhombdodecahedron (and the reflection hk0) this is also 4; for pyramidal cubes (and the reflection hk0) this is 8.

Example: The primitive cell: For a primitive elementary cell, there is no reason for intensity variation due to structure, because only the atoms at the corners

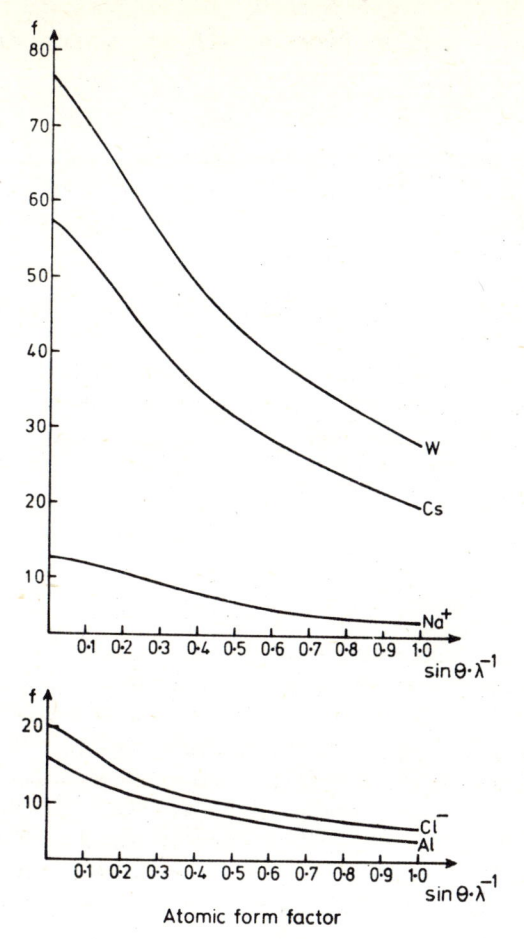

Fig. 7.24 The plot on a mm. graph paper of the value of f against sin θ/λ.

of the cell scatter coherently i.e., with the same phase. Thus, for such a simple lattice only the atomic form factor, *f*, and the surface factor need be considered for calculating the intensities.

As an example, one can consider the photographs of KCl because K^+ and Cl^- respond similarity to X-rays and thus it simulates the primitive cell. This has been discussed in Section 7.24.4.

7.24.3 Structure Factor
Intensities of reflections depend essentially on the distribution of the atoms in the unit cell.

In the contiguous primitive translation lattices there appear phase-differences of scattered waves, which strongly influence the original scattering.

In general, the summation of all the basis points is given by the formula

$$S_K = \exp\{2\pi i . (hx_j + ky_j + lz_j)\},$$

where index, j, is the number of one particular atom in the basis, and similarly for all kinds of atoms, so that one gets,

$$S = \Sigma S_K$$

This quantity is called the *geometrical structure amplitude*. In general it is a complex quantity. Multiplying it by its complex conjugate one obtains a real quantity,

$$S S^* = S^2.$$

The *'structure factor'*, which is directly proportional to the intensity is given by $F = \Sigma f_K S_K$, where f_K = atomic form factor. The geometrical structure factor for the all the space groups are tabulated in the International Tables.

These values are multiplied by the atomic form factor, f, characteristic of each element, and one obtains

$$I \sim F^2 \sim (f_\nu A)^2 + (f_\nu B)^2$$

where, $A = \cos 2\pi (hx_\nu + ky_\nu + lz_\nu)$

and $B = \sin 2\pi (hx_\nu + ky_\nu + lz_\nu)$

7.24.4 Study of Halite Structure (NaCl, KCl and CsCl)

Sodium chloride structure can be described as one in which Na+ ions form cubic face-centred lattice and the Cl⁻ ions are arranged likewise. Figure 7.25 shows that there are two sets of planes containing both types of ions. These are planes which are occupied by sheets of identical atoms, each containing an equal number of sodium and chlorine atoms viz. the set (200), (400), (600) etc. and also (220), (440), (660). Reflections or orders from these would be expected to fall off regularly in intensity. In planes (111) sheets are alternately occupied by sodium or chlorine atoms. Since for (111) there is a path difference of one wavelength for the strongly reflecting chlorine planes, the waves reflected from the weaker sodium planes half-way between them will be opposite in phase. The order of (111) will be weak, since the chlorine contribution would partly affect the sodium contribution. In the case of (222), however, both the contributions would be in phase and the order of reflection would be strong. In such space lattices the orders with even indices ($h+k+l$ = even), such as (200), (220) and

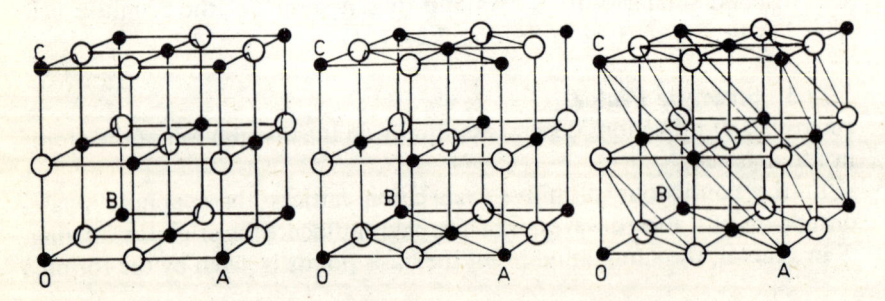

Fig. 7.25 The three major reflection planes (hoo, hko and hkl, where h=k=l) of a sodium chloride crystal.

(222) should form a strong sequence, whereas those with odd indices ($h+k+l$ = odd), such as (111), (113) and (333) would be weak.

For potassium chloride, on the other hand, the scattering powers of the K-atoms and Cl-atoms are so near that the order of (111) is too weak to be observed. It is because the resolving power of atoms and ions depends on the number of electrons. K^+ ion (18 electrons) and Cl-ion (18 electrons) behave similarly with respect to X-rays. So one gets the impression as if KCl contains a cell with a value of $\frac{1}{2} a_o$. For the same reason, the first layer line for NaCl is weak, whereas it is strong for galena (PbS).

Structure Factor
For structure determination the knowledge of the symmetry class of the crystal and the intensities of X-ray reflections are required. The intensities depend on several factors; of these the scattering power factors of the atoms and ions are important. The scattering powers are dependent on the number of electrons in the atom or ion. Previously, we considered the case of NaCl. Let us now consider the structure of CsCl, which has the crystallographic properties:

Crystal class: cubic, hexoctahedral O_h:

Cell dimension, a_o = 4.11 Å and Z = 1.
Omissions, after indexing are found as follows:
(hkl) (okl) and (hhl) are present in all orders.

The powder x-ray diffraction of CsCl is shown in Fig. 7.26.

Fig. 7.26 X-ray diffraction pattern of cesium chloride (in Cu K_α with Ni filter) showing the relative intensities of lines.

The structure may be visualised as composed of two primitive lattices, one consisting only Cs^+, the other Cl^--ions. The two lattices are displaced with respect to each other by one-half the cube diagonal i.e., if a Cs^+ ion is placed at the origin of the unit cell, the coordinates of Cl-ion are $\frac{1}{2}, \frac{1}{2}, \frac{1}{2}$.

Space group: absence of each omission indicates the space group $O_h^1 - Pm\,3m$.

Point positions: since one Cs and one Cl atom are present, only one singly 1-numbered positions can be occupied, which are the points (000) and $\frac{1}{2}, \frac{1}{2}, \frac{1}{2}$.

Thus, CsCl lattice has similar atom coordinates as a body-centred cubic lattice with the only difference that cell vertices and cell centres are occupied by different types of atoms.

The phase difference of the waves, scattered from the Cs-lattice and from the Cl-lattice with indices hkl measured in radian, is $2\pi \cdot \frac{1}{2}(h+k+l)$. When $(h+k+l)$ is an odd number the two waves destroy each other, and when the sum is even they reinforce each other. But still complete omissions are not observed. It is because the scattering power of Cs$^+$ and Cl$^-$ ions are not equal. If we designate the amplitudes of the Cs-scattered waves as f_{cs} and that of the Cl-scattered wave as f_{Cl}, then we can proceed with the following relation of intensity.

$$I = \nu F^2 \text{ where } F = f_{Cl}A_{Cl} + f_{Cs} \cdot A_{Cs}$$

$$A = \Sigma \cos 2\pi (hx + ky + lz)$$

The respective atomic coordinates are : 1 Cl in 000, 1 Cs in $\frac{1}{2}, \frac{1}{2}, \frac{1}{2}$.

$$A_{Cl} = \cos 2\pi (h0 + k0 + l0) = \cos 0 = 1$$
$$A_{Cs} = \cos 2\pi (h\tfrac{1}{2} + k\tfrac{1}{2} + l\tfrac{1}{2}) = \cos \pi(h+k+l)$$
$$= +1 \text{ for } h+k+l \text{ (when even)}$$
$$= -1 \text{ for } h+k+l \text{ (when odd)}.$$

The structure factor, designated as F, for CsCl, thereby becomes

$$F = f_{Cs} + f_{Cl} \text{ for } (h+k+l) \text{ even}$$
and $F = f_{Cs} - f_{Cl}$ for $(h+k+l)$ odd.

7.25 Electron Density Distribution: A Fourier Analysis

In this analysis of the x-ray pattern, the structure is treated not as a cluster of atoms but as a continuous electron distribution capable of scattering x-rays. The positions of atoms are the sites of high electron density. The density map shows the distribution of atoms on a particular plane.

Since the crystal structure is periodic in three dimensions Fourier Series is employed for this analysis because this mathematical expression is employed for analysing variables that change periodically. The density distributions are mapped by adding together the terms of a Fourier Series. Each element of the series is a set of electron sheets that vary periodically in density. When the amplitudes and phases of these sheets in all directions are determined, their sum results in a plot of the density distribution.

Analysis of electron density distribution has enabled locating the major elements of high atomic weight. Thus iron in a silicate mineral can be located in relation to the lighter elements aluminium and magnesium. But often it becomes impossible to distinguish between ions of similar scattering factors or atomic numbers viz., iron and manganese; and also between ions of different valencies viz., Fe^{2+} and Fe^{3+}.

The electron density map (Fourier Series summation) of the mineral diopside, $CaMg[SiO_3]_2$, projected on 010 plane is shown in Fig. 7.27. Contour lines drawn through points of equal density are shown in 7.27 (a). Using projections on other planes the nature and configuration of atoms can be deciphered, knowing that the space group is C_2/c [Fig. 7.27(b)].

7.25.1 Site Occupancy Derivation

Site-occupancy studies are generally carried out using X-rays or thermal neutrons. For X-rays the scattering factor, f, is a function is sin θ/λ.

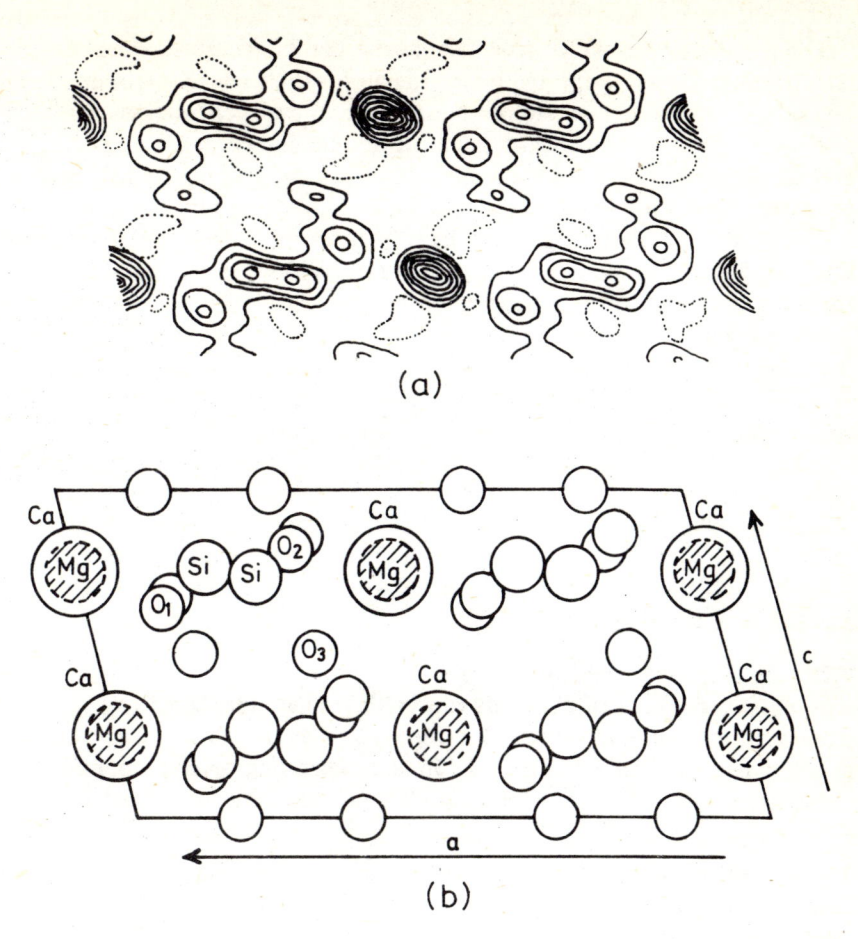

Fig. 7.27 Fourier series summation electron density map of diopside, $CaMg(SiO_3)_2$ projected on (010). Equal densities are contoured (a), and the relative configuration of atoms are derived from the electron density distribution (b) [From W.L. Bragg. *Z. fur Kristall*, 70, 488, 1929.].

Earlier site-occupancies were evaluated on Fourier projections by electron counts or peak hight values at the sites involved in disorder. A satisfactory solution, however, involves determination of site-populations from least-squares refinement of diffraction data.

From a diffraction experiment the total scattering power S_j at each site occupied by an atom is

$$S_j = \sum_i f_i\, a_{ij}$$

Where, S_j is the scattering factor at the jth site in the cell, f_i is the scattering factor of the ith atom, and a_{ij} are the site occupancies of the j-th site by the i-th atom. When the site is completely occupied by the scattering species,

$$\sum_i a_{ij} = 1.$$

The precision with which site-occupancies can be measured depends on the difference in scattering factors of the atoms involved. X-ray scattering factors are a function of atomic number. It, therefore, becomes easy to distinguish between elements of disparate atomic number (viz., Fe and Mg; Fe and Al etc.). Using bulk chemical constraints Mg/Fe site occupanies are determined with standard deviations of ~ 0.005 atoms.

Determination of Al/Si ordering by refinement of high-angle X-ray data is possible. At high values of Sin θ/λ, the relative difference in X-ray scattering factors between Al and Si increases, reaching ~ 15% at Sin θ/λ ~ 0.8 – 0.9. This high-angle x-ray diffraction method is reported to give Al/Si site-occupancy values in some minerals, viz., aluminous orthopyroxene (Kosoi et al 1974). For Al/Si ordering study in felspars, this method has been found to be inadequate (Brown et al., 1974; Phillips and Ribbe, 1973).

Quantitative information on ordering can be obtained from the powder diffraction data when the structure is not too complex. In powder diffraction scattering intensities of individual atoms overlap and hence cannot be measured directly. By using the profile-fitting refinement procedure the structure refinement and site-occupancies are determined (Nord, 1977; Nord and Stefanidis, 1980).

7.26 ELECTRON DIFFRACTION

An impinging stream of electrons can be diffracted like light waves. Electrons with mass m and velocity v behave like waves with wavelength $\lambda = h/mv$, where h is Planck's constant. In an accelerating potential applied to electrons in volts, $\lambda = \sqrt{\frac{150}{v}}$ Å But the electron beams are not only influenced by the electrons in the shells but also by the nucleii of the atoms. The diffracted intensity is ~ 10^8 times higher than that obtained by X-rays. But they are more strongly absorbed and are transmitted only through thin layers.

The electrons are beamed by an accelerating potential in kilo volts. At 63 kilo volts the electrons would have wavelength equal to about .05Å. But often higher potentials are used to produce energetic electrons.

7.27 NEUTRON DIFFRACTION

The position of hydrogen atoms in a mineral structure can be most conveniently determined by neutron diffraction. This is useful in studies of hydrated and OH- bearing minerals. The underlying principle is that excited neutrons are diffracted by crystal lattice like any other elementary particles. Neutrons produced in an atomic pile have a wide range of velocities. From the thermal neutrons, which are in thermal equilibrium with the surroundings, a narrow beam of monochromatic neutrons can be produced by means of reflection from a crystal and appropriate collimation. The monochromatic neutrons can be used for diffraction studies of minerals. Neutrons have spins. Hence spin-spin interactions peculiar to magnetic materials are more conveniently studied by neutron diffraction.

X-Ray Spectroscopy:
This topic has been left out of the scope of this present volume. Enthusiastic readers are advised to look into a review article dealing with its principle and applications by D.S. Urch, X-ray Spectroscopy and Chemical Bonding in Minerals, pp. 31-59 (in Chemical Bonding & Spectroscopy in Mineral Chemistry edited by F.J. Berry & D.J. Vaughan, Chapman & Hall, 1985).

References and Selected Reading (also *vide* Appendix VA)

Advances in X-ray Analysis—Proceedings of the Annual Conferences (vols. 1-29, 1985), Plenum Press, USA.

A.S.T.M. Index to the X-ray powder data file. A.S.T.M. Diffraction Data Sales Dept., 1916 Race Street, Philadelphia, Penn. 19103, USA.

Azaroff, L.V. and Buerger, M.J. (1958) The Powder Method in X-ray Crystallography, McGraw-Hill, New York.

Barinskii, R.L. (1967) Study of Chemical bond by X-ray spectra. *Journal of Structural Chemistry,* **8**, 805-16.

Battey, M.H. (1972) Mineralogy for Students, Oliver & Boyd Edinburgh, 323 p.

Berry, I.G. (1974) ed. Selected Powder Diffraction Data for Mineralogy JCPDS, Swanthmore, P.A. USA.

Berry, L.G. and Thompson, R.M. (1962) X-ray Powder Data for Ore Minerals: The Peacock Atlas, Geological Society of America.

Brown, G.E. Hamilton, W.C. Prewitt, C.T., and Sueno, S. (1974) Neutron diffraction study of Al/Si ordering in sanidine: a comparison with X-ray diffraction data. In W.S. Mackenzie and J. Zussman (eds.) The Felspars: Manchester University Press.

Buerger, M.J. (1942) X-ray Crystallography, John Wiley & Sons, New York.

Buerger, M.J. (1964) The Precession Method in X-ray Crystallography, John Wiley and Sons, New York.

Cullity, B.D. (1956) Elements of X-ray Diffraction, Addison-Wesley, Reading, Mass.

Djurle, S. (1958) An X-ray study of the system Cu-S *Acta chimica Scandinavia,* **12**, 1415-26.

Frueh, A.J. (1988) Some applications of X-ray Crystallography to geological thermometry. *Journal of Geology,* **66**, 218-23.

Graham, A.R. (1969) Quantitative determination of hexagonal and monoclinic pyrrhotites by X-ray diffraction, *Canadian Mineralogist,* **10**, 4-24.

Hall, S.R. and Stewart, J.M. (1973) The crystal structure refinement of chalcopyrite, $CuFeS_2$. *Acta Crystallographica,* **29B**, 579-85.

Henry, N.F.M., Lipson, H. and Wooster, W.A. (1960) The interpretation of X-ray Diffraction Photographs, 2nd ed., MacMillan.

Hutchison, C.S. (1974) Laboratory Handbook of Petrographic Techniques, John Wiley & Sons, New York, 527 p.

Kaelbe, E.F. (1967) (ed.) Handbook of X-rays. McGraw-Hill, New York.

Klug, H.P. and Alexander, L.E. (1954) X-ray Diffraction Procedures, Wiley, New York.

Kostov, I (1968) Mineralogy, Oliver and Boyd, England, p. 587.

Lonsdale, K. (general editor) International Tables for X-ray Crystallography Vols. 1-3, International Union of Crystallography, Kynoch Press, Birmingham, England.

Mirkin, L.I. (1964) Handbook of X-ray Analysis of Polycrystalline Materials (Transl. by J.E.S. Bradley), Consultants Bureau, New York.

Mitra, S. (1973). Orthopyroxenes from Sukinda ultramafites and the nature of the parental magma. *Acta Mineralogica—Petrographica, Szeged,* **21**(1), 87-106.

National Bureau of Standards (1950) Tables for Conversion of X-ray Diffraction Angles to Interplanar Spacings, U.S. Govt. Printing Office, Washington, D.C.

Nord, A.G. (1977) The cation distributions in $Zn_2(PO_4)_2$ determined by X-ray profile-fitting refinements. *Materials Research Bulletin,* **12**, 563-568.

Nord, A.G., and Stefanidis, T., (1980) The cation distribution in two (Co, Mg)$_3$ (PO$_4$)$_2$ solid solutions. *Zeitschrift fur Kristallographie, 153*, 141-149.

Norrish, K. and Taylor, R.M. (1962) Quantitative Analysis by X-ray Diffraction. *Clay Minerals Bulletin, 5*, 98.

Nuffield, E.W. (1966) X-ray Diffraction Methods, John Wiley, N.Y.

Parrish, W. and Mack, M. (1963) Data for X-ray Analysis, vols. 1-3. Charts for the solution of Bragg's Equation. Phillips Technical Library.

Peiser, H.S. Rooksby, H.P. and Wilson, A.J. C. (ed.) (1955) X-ray Diffraction by Polycrystalline Materials, Institute of Physics, London.

Phillips, M.W. and Ribbe, P.H. (1973) The structures of monoclinic potassium-rich felspars. *American Mineralogist, 58* 263-270.

Stout, G.H. and Jensen, L.H. (1968) X-ray Structure Determination, McMillan, London.

Sorem, R.K. (1960) X-ray diffraction technique for small samples. *American Mineralogist, 45*, 1104.

Zussman, J. (1967) X-ray diffraction. In Physical Methods in Determinative Mineralogy (ed. J. Zussman), Academic Press, N.Y. 261-334.

[Additional references cited in Appendix VA]

Allen, W.C. (1966) An x-ray method for defining composition of a magnesium spinel. *American Mineralogist, 51*, 239-243.

Borg, I.Y. and Smith, D.K. (1968) Calculated powder patters. 1 Five plagioclases. *American Mineralogist, 53*, 1709-1723.

Desborough, G.A. and Rose, H.J. (1968) X-ray and chemical analysis of orthopyroxenes from the lower part of the Bushveld Complex, South Africa, *U.S. Geological Survey Professional Paper*, 600-B, B1-6.

Hamilton, D.L. and Edgar, A.D. (1969) The variation of the $\overline{2}01$ reflection in plagioclase. *Mineralogical Magazine, 37*, 16-25.

Himmelberg, G.R. and Jackson, E.D. (1967) X-ray determinative curve for some orthopyroxens of composition Mg$_{48-85}$ from stillwater complex, Montana, *U.S. Geological Professional Paper, 575-B*, B101-B102.

Hormann, P.K. and Morteani, G. (1969) On a systematic error in the x-ray determination of the iron contents of chlorites and biotites: a discussion. *American Mineralogist, 54*, 1491-1494.

Hutchison, C.S. (1972) Alpine type chromite in North Borneo, with special reference to Darvel Bay, *American Mineralogist, 57*, 835-856.

Jones, J.B., Nesbitt, R.W. and Slade, P.G. (1969) The determination of orthoclase content of homogenized alkali felspar using the 201 x-ray method. *Mineralogical Magazine, 37*, 489-496.

Kuellmer, F.J. (1960) X-ray intensity measurement on perthitic materials I: Theoretical considerations. *Journal of Geology, 67*, 648-660.

MacGregor, I.D. and Smith, C.H. (1963) The use of chrome spinels in petrographic studies of ultramafic intrusions, *Canadian Mineralogist, 7*, 403-412.

Orville, P.M. (1967) Unit-cell parameters of the microcline—low albite and the sanidine—high albite solid solution series, *American Mineralogist, 52*, 55-86.

Stevens, R.E. (1944) Compositions of some chromites of the Western Hemisphere. *American Mineralogist, 29*, 1-34.

Winchell, H. and Tilling, R. (1960) Regressions of physical properties on the compositions of clinopyroxenes. *American Journal of Science, 258*, 529.

Yoder, H.S. and Engster, H.P. (1955) Synthetic and natural muscovites. *Geochim et Cosmochim Acta, 8*, 225.

Yoder, H.S. and Sahama, Th.G. (1957) Olivine x-ray determination curve. *American Mineralogist, 42*, 475-491.

Appendices

APPENDIX I (*CHAPTER III*)

The Symbols for Spectroscopic States of Ions

A spectroscopic or Russel-Saunder's state is designated ^rL, where r is the electron spin multiplicity and L is the total atomic orbital angular momentum.

$L = \Sigma l$

when $l = 0$, the spectroscopic state is S
$ = 1$, the spectroscopic state is P
$ = 2$, the spectroscopic state is D
$ = 3$, the sepctroscopic state is F
$ = 4$, the spectroscopic state is G
$ = 5$, the spectroscopic state is H
$ = 6$, the spectroscopic state is I

and $r = (2\Sigma S + 1)$

where ΣS is the number of unpaired electrons, having spin, $S = \frac{1}{2}$ for each electron.

For no unpaired electrons, $\Sigma S = 0$ and $r = 1$: singlet state
for one unpaired electron, $\Sigma S = \frac{1}{2}$ and $r = 2$: doublet state
for two unpaired electrons, $\Sigma S = 1$ and $r = 3$: triplet state
for three unpaired electrons, $\Sigma S = \frac{3}{2}$ and $r = 4$: quartet state
for four unpaired electrons, $\Sigma S = 2$ and $r = 5$: quintet state
for five unpaired electrons, $\Sigma S = \frac{5}{2}$ and $r = 6$: sextet state.

The symbols used to designate the spectroscopic states are made following the group theoretic nomenclature as stated below.

The symbols a(A), b(B), e(E) and t(T) designate the nature of degeneracy. The small and capital letters refer to one-electron and multi-electron systems of symmetry properties:

- a, A = non-degenerate; symmetry with respect to the principal axis of symmetry; no change of sign of the wave function when rotated through $2\pi/n$.
- b, B = non-degenerate; antisymmetric with respect to the principal symmetry axis; change of sign of wave function when rotated through $2\pi/n$.
- e, E = two-fold degenerate.
- t, T = three-fold degenerate.

The superscripts $r = 1, 2, 3$, etc. refer to spin-multiplicities.
The subscripts g, u, $_1$, $_2$ etc. refer to the result of operations involved viz.

g   = (from the German word, *gerade,* meaning straight) no change of sign of wave-function upon inversion i.e., symmetric under the operation of inversion through the centre of symmetry.

u   = (from the German word, *ungerade,* meaning not straight) change in sign of wave-function upon inversion i.e., anti-symmetric under the operation of inversion at the centre of symmetry.

1   = mirror planes parallel to a symmetry axis.

2   = diagonal mirror planes.

Examples:

a_{1g} = s orbitals.

t_{1u} = p orbitals.

t_{2g} = d_{xy}, d_{xz}, d_{yz}.

e_g = $d_{z^2}, d_{x^2-y^2}$

APPENDIX II (CHAPTER IV)
OPTICAL PROPERTIES OF COMMON ORE MINERALS

Name & Composition	Colour & pleochroism	Anisotropism	R%	~VHN	Distinguishing characters & elemental (mineral) associations
(1)	(2)	(3)	(4)	(5)	(6)
Acanthite Ag$_2$S	light greenish grey	distinct	30-31%	20-60	twin lamellae common (Ag-Au-As-Bi-Pb-Sb)
Argentite Ag$_2$S (high temp. form)	light greenish grey	isotropic	30%	20-60	Cu-Pb-S, Au
Arsenopyrite Fe Ag S	white	distinct	52%	1050-1130	rhomb. shape, twinning, zoning, (Au-Bi-Te-Sn, W)
Bismuth Bi	bright white	distinct	60-65%	10-20	tarnishes brown, multiple twinning (Co-Ni-As-S, Au-Bi, Te)
Bismuthinite Bi$_2$S$_3$	bluish white	v. strong	40-50%	70-220	fibrous habit, straight extinction (Co-Ni-As-S, Au-Bi-Te, Mo, Sn, W)
Bornite Cu$_5$FeS$_4$	light pink to brown	v. weak	22%	100	tarnishes to blue and purple, often intergrowths with chalcopyrite (Cu-Fe-S)
Bournonite CuPbSbS$_3$	light bluish grey	weak	35-37%	130-210	twinning common (Pb-Sb-S, Cu-Fe-S)
Boulangerite Pb$_5$Sb$_4$S$_{11}$	light greenish grey	distinct	38-41%	90-180	anisotropy stronger than bournonite (Pb-Sb-S, Cu-Fe-S)
Braunite Mn$_7$SiO$_{12}$	light brownish grey	weak	20-22%	880-1190	often magnetic, mostly in metamorphic rocks (Mn-Fe-Si-O)
Bravoite (Fe, Ni, Co) S$_2$	light grey to brownish white	isotropic	31-54% Ni, Co → Fe	670-1540	colour zoning
Carrollite Co$_2$CuS$_4$	pinkish white	isotropic	43%	350-570	much like linnaeite (Co-Cu-Fe-S)
Chalcocite Cu$_2$S	light bluish grey	multi coloured but weak	32%	70-100	twinning common (Cu-Fe-S)
Chalcopyrite CuFeS$_2$	yellow	weak	42-46%	190-220	greenish yellow and softer with reference to pyrite (Cu-Fe-Ni-Zn-S)
Chromite FeCr$_2$O$_4$	grey	isotropic (weakly anisotropic)	12%	1200-1210	octahedral, non-magnetic with reference to magnetite (Fe-Ti-O)

184 Fundamentals of Optical, Spectroscopic and X-ray Mineralogy

1	2	3	4	5	6
Cinnabar HgS	light bluish grey	moderate	28-29%	50-100	red internal reflections, twinning (Hg-Sb-S, Fe-S)
Cobaltite $CoAsS$	pinkish white	weak	53%	1180-1230	cubic, cleavage distinct, (Cu-Fe-S, Co-Ni-As-S)
Copper Cu	metallic pink with tarnish	isotropic	81%	120-140	low scratching hardness (Cu-O, Cu-Fe-S)
Covellite CuS	light grey to blue, pleochroic	very strong	7-22%	70-80	pleochroic plates (Cu-Fe-S)
Cubanite $CuFe_2S_3$	light grey (pinkish)	isotropic	35		Cu-Fe-S
Cuprite Cu_2O	light grey (bluish)	strong	25-30	180-220	deep internal reflection
Digenite Cu_9S_5	light grey (bluish)	isotropic	22	60-70	Cu-Fe-S
Enargite Cu_3AsS_4	light grey (pinkish)	strong multi-coloured	25-29	130-580	cleavage (110)
Galena PbS	white	isotropic	43	70-80	triangular cleavage pits
Gersdorffite $(Ni, Co, Fe)AsS$	white (pinkish)	isotropic	47-54	520-910	zoning common, triangular pits due to (100)
Goethite $HFeO_2$	grey	distinct	17	770-820	Botroydal, colloform, red to brown internal reflections
Gold Au	bright yellow	isotropic	74	50	very bright
Graphite C	dark grey to grey	strong	6-16	10	bireflectance strong, cleavage flakes often deformed
Hematite Fe_2O_3	light grey	strong	25-30	920-1060	weak bireflectance, red internal reflection
Ilmenite $FeTiO_3$	light pinkish grey	moderate	18-21	520-700	lamellar twinning often present
Iron Fe	bright white	isotropic	65	120-290	commonly isolated grains or laths
Jacobsite $(Mn, Fe, Mg)(Fe, Mn)_2O_4$	grey	isotropic	19	720-750	strongly magnetic
Jamesonite $Pb_4FeSb_2S_8$	light grey	strong	36-41	70-130	twin lamellae common
Lepidocrocite $FeO(OH)$	grey	v. strong	10-19	690-780	internal reflections of red & brown colours
Limonite $FeO(OH)$	grey	v. strong	10-19	690-780	red to brown internal reflections
Limonite (FeO, OH, nH_2O)	Bluish grey	strong	16-19	690-820	strong internal reflection of brown and red colours

Appendices

1	2	3	4	5	6
Linnaeite Co_3S_4	pinkish white	isotropic	45-50	350-570	Cleavage (100) common
Pentlandite $(Fe, Ni)_9S_8$	yellowish white	isotropic	44	200-230	triangular cleavage pits
Proustite Ag_3AsS_3	light bluish grey	strong	25-28	110-140	bireflectance strong, red internal reflection
Psilomelane $Mn_8O_{16}(O,OH)_6$ with Ba, Mn, Fe etc.	light grey to grey	strong	15-30	200-810	fine crystalline aggregates
Pyrargyrite Ag_3SbS_3	light bluish grey	strong	28-31	50-130	bireflectance distinct, red internal reflections
Pyrite FeS_2	yellowish white	isotropic	54	1030-1240	often ideomorphic
Pyrolusite MnO_2	light yellowish grey	v. strong	30-36	80-1500	coarse to cryptocrystalline
Pyrrhotite $Fe_{1-x}S$	brownish white	strong	40	230-320	magnetic
Rutile TiO_2	light bluish grey	strong	20-23	1070-1210	internal reflections common
Scheelite $CaWO_4$	dark grey	distinct	10	290-460	white internal reflections & rhombohedral cleavage
Sphalerite $(Zn, Fe)S$	Grey	isotropic	17	190-210	colourless to red internal reflections
Spinel $Mg Al_2O_4$	dark grey	isotropic	8	860-1650	octahedral
Stannite Cu_2SnFeS_4	light greenish grey to brownish grey	strong	28	140-330	triangular pits common
Stibnite Sb_2S_3	light grey to white	v. strong	30-40	40-110	distinct bireflection
Tetradymite Bi_2Te_2S	yellowish white	distinct	50-52	30-50	hexagonal forms
Tetrahedrite $(Cu, Ag)_{10}(Sn, Fe, Hg)_2(Sb, As)_4S_{13}$	light grey	isotropic	31	320-370	
Titanomagnetite $(Fe, Ti)_3O_4$	brownish grey	isotropic	17	720-730	
Todorokite $(H_2O) Mn (O,OH)_{16}$	light grey	strong	20		column or fibrous habit
Ulvospinel Fe_2TiO_4	brownish grey	isotropic	17		often intergrown in titanomagnetite
Loellingite $Fe As_2$	Yellowish white	v. strong	55	370-1050	twinning common

1	2	3	4	5	6
Mackinawite (Fe, Ni, Co, Cu)S	light pinkish grey	v. strong	22-45	50-60	parallel lamellae common
Maghemite γ-Fe$_2$O$_3$	light bluish grey	isotropic	26	360-990	magnetic
Magnetite Fe$_3$O$_4$	light grey	isotropic	21	530-600	magnetic, octahedra
Manganite MnO·OH	brownish grey	strong	15-21	370-800	red internal reflections common
Marcasite FeS$_2$	yellowish white	strong	49-55	940-1290	radial aggregates of twins, orthorhombic
Millerite NiS	yellow	strong	52-56	190-380	twinning common
Molybdenite MoS$_2$	grey to white	v. strong	15-37	20-30	flaky, often hexagonal
Niccolite NiAs	pinkish white	v. strong	52-58	330-460	twinning common

APPENDIX IV *(CHAPTER VII)*

COMPUTER OUTPUT FOR CELL-SIZE DETERMINATION

PROGRAM COHEN CASE		= 12/196/01
ENSTATITE		
TWO THETA CORRECTION		= 0.000
REJECTED LINES. ERROR		= 0.00400
NONE		
REJECTED LINES. ERROR		= 0.00360
NONE		
REJECTED LINES. ERROR		= 0.00320
NONE		
REJECTED LINES. ERROR		= 0.00280
NONE		
REJECTED LINES. ERROR		= 0.00240
NONE		
REJECTED LINES. ERROR		= 0.00200
NONE		
REJECTED LINES. ERROR		= 0.00160
NONE		
REJECTED LINES. ERROR		= 0.00120
NONE		
REJECTED LINES. ERROR		= 0.00080
NONE		
LINE USED		

H	K	L	D OBS	D CALC	DELATA D	SIN SQ THETA OBS	CALC	DIFF
1	2	0	6.35271	6.34361	0.00910	0.01985	0.01991	—0.00006
2	0	0	4.41401	4.42390	—0.00990	0.04112	0.04094	0.00018
2	1	1	3.30579	3.31240	—0.00661	0.07332	0.07302	0.00029
2	4	0	3.17029	3.17233	—0.00203	0.07972	0.07961	0.00010
2	2	1	3.17029	3.15925	0.01104	0.07972	0.08027	—0.00056
2	3	1	2.94410	2.94531	—0.00121	0.09244	0.09236	0.00008
1	6	0	2.87348	2.87013	0.00335	0.09704	0.09726	—0.00023
1	5	1	2.82834	2.82577	0.00257	0.10016	0.10034	—0.00018
2	4	1	2.70822	2.70772	0.00050	0.10924	0.10928	—0.00004
3	1	1	2.53816	2.54008	—0.00191	0.12437	0.12418	0.00019
0	2	2	2.49773	2.49853	—0.00080	0.12843	0.12835	0.00008
3	4	0	2.47412	2.47524	—0.00112	0.13089	0.13077	0.00012
2	5	1	2.47412	2.47276	0.00136	0.13089	0.13103	—0.00014
3	6	1	1.95823	1.95900	—0.00077	0.20894	0.20878	0.00016
2	9	1	1.73291	1.73428	—0.00138	0.26681	0.26638	0.00043
5	5	1	1.52284	1.52199	0.00085	0.34549	0.34588	—0.00039

PROGRAM COHEN CASE 12196701
NELSON—REILEY EXTRAPOLATION
DIFFRACTOMETER COS THETA X COT THETA EXTRAPOLATION
WEIGHTING FUNCTION COSEC SQUARED TWO THETA
PARAMETER S. DEV
b = 8.84968 ± 0.00511 Å
a = 18.20741 ± 0.01140 Å
c = 5.19721 ± 0.00469 Å

QEW = 9.68E-07
SLOPE = 8.91E-05   3.94E-04

APPENDIX III (CHAPTER VI)
Micron-Wave number Conversion Table

	0	1	2	3	4	5	6	7	8	9	1	2	3	4	5	6	7	8	9
1.0	1.0000	.9901	.9804	.9709	.9615	.9524	.9434	.9346	.9259	.9174	9	18	27	36	45	55	64	73	82
1.1	.9091	.9009	.8929	.8850	.8772	.8696	.8621	.8547	.8475	.8403	8	15	23	30	38	45	53	61	68
1.2	.8333	.8264	.8197	.8130	.8065	.8000	.7937	.7874	.7813	.7752	6	13	19	26	32	38	45	51	58
1.3	.7692	.7634	.7576	.7519	.7463	.7407	.7353	.7299	.7246	.7194	5	11	16	22	27	33	38	44	49
1.4	.7143	.7092	.7042	.6993	.6944	.6897	.6849	.6803	.6757	.6711	5	10	14	19	24	29	33	38	43
1.5	.6667	.6623	.6579	.6536	.6494	.6452	.6410	.6369	.6329	.6289	4	8	13	17	21	25	29	33	38
1.6	.6250	.6211	.6173	.6135	.6098	.6061	.6024	.5988	.5952	.5917	4	7	11	15	18	22	26	29	33
1.7	.5882	.5848	.5814	.5780	.5747	.5714	.5682	.5650	.5618	.5587	3	7	10	13	16	20	23	26	30
1.8	.5556	.5525	.5495	.5464	.5435	.5405	.5376	.5348	.5319	.5291	3	6	9	12	15	18	20	23	26
1.9	.5263	.5236	.5208	.5181	.5155	.5128	.5102	.5076	.5051	.5025	3	5	8	11	13	16	18	21	24
2.0	.5000	.4975	.4950	.4926	.4902	.4878	.4854	.4831	.4808	.4785	2	5	7	10	12	14	17	19	21
2.1	.4762	.4739	.4717	.4695	.4673	.4651	.4630	.4608	.4587	.4566	2	4	7	9	11	13	15	17	20
2.2	.4545	.4525	.4505	.4484	.4464	.4444	.4425	.4405	.4386	.4367	2	4	6	8	10	12	14	16	18
2.3	.4348	.4329	.4310	.4292	.4274	.4255	.4237	.4219	.4202	.4184	2	4	5	7	9	11	13	14	16
2.4	.4167	.4149	.4132	.4115	.4098	.4082	.4065	.4049	.4032	.4016	2	3	5	7	8	10	12	13	15
2.5	.4000	.3984	.3968	.3953	.3937	.3922	.3906	.3891	.3876	.3861	2	3	5	6	8	9	11	12	14
2.6	.3846	.3831	.3817	.3802	.3788	.3774	.3759	.3745	.3731	.3717	1	3	4	6	7	9	10	11	13
2.7	.3704	.3690	.3676	.3663	.3650	.3636	.3623	.3610	.3597	.3584	1	3	4	5	7	8	9	11	12
2.8	.3571	.3559	.3546	.3534	.3521	.3509	.3497	.3484	.3472	.3460	1	2	4	5	6	7	9	10	11
2.9	.3448	.3436	.3425	.3413	.3410	.3390	.3378	.3367	.3356	.3344	1	2	3	5	6	7	8	9	10
3.0	.3333	.3322	.3311	.3300	.3289	.3279	.3268	.3257	.3247	.3236	1	2	3	4	5	6	7	9	10
3.1	.3226	.3215	.3205	.3195	.3185	.3175	.3165	.3155	.3145	.3135	1	2	3	4	5	6	7	8	9
3.2	.3125	.3115	.3100	.3096	.3086	.3077	.3067	.3058	.3049	.3040	1	2	3	4	5	6	7	8	9
3.3	.3030	.3021	.3012	.3003	.2994	.2985	.2976	.2967	.2959	.2950	1	2	3	4	5	5	6	7	8
3.4	.2941	.2933	.2924	.2915	.2907	.2899	.2890	.2882	.2874	.2865	1	2	2	3	4	5	6	7	8
3.5	.2857	.2849	.2841	.2833	.2825	.2817	.2809	.2801	.2793	.2786	1	2	2	3	4	5	6	6	7
3.6	.2778	.2770	.2762	.2755	.2747	.2740	.2732	.2725	.2717	.2710	1	1	2	3	4	5	5	6	7
3.7	.2703	.2695	.2688	.2681	.2674	.2667	.2660	.2653	.2646	.2639	1	1	2	3	4	4	5	6	6
3.8	.2632	.2625	.2618	.2611	.2604	.2597	.2591	.2584	.2577	.2571	1	1	2	3	3	4	5	5	6
3.9	.2564	.2558	.2551	.2545	.2538	.2532	.2525	.2519	.2513	.2506	1	1	2	3	3	4	4	5	6
4.0	.2500	.2494	.2488	.2481	.2475	.2469	.2463	.2457	.2451	.2445	1	1	2	2	3	4	4	5	5
4.1	.2439	.2433	.2427	.2421	.2415	.2410	.2404	.2398	.2392	.2387	1	1	2	2	3	3	4	5	5
4.2	.2381	.2375	.2370	.2364	.2358	.2353	.2347	.2342	.2336	.2331	1	1	2	2	3	3	4	4	5
4.3	.2326	.2320	.2315	.2309	.2304	.2299	.2294	.2288	.2283	.2278	1	1	2	2	3	3	4	4	5
4.4	.2273	.2268	.2262	.2257	.2252	.2247	.2242	.2237	.2232	.2227	1	1	1	2	3	3	3	4	4
4.5	.2222	.2217	.2212	.2208	.2203	.2198	.2193	.2188	.2183	.2179	0	1	1	2	2	3	3	4	4
4.6	.2174	.2169	.2165	.2160	.2155	.2151	.2146	.2141	.2137	.2132	0	1	1	2	2	3	3	4	4
4.7	.2128	.2123	.2119	.2114	.2110	.2105	.2101	.2096	.2092	.2088	0	1	1	2	2	3	3	3	4
4.8	.2083	.2079	.2075	.2070	.2066	.2062	.2058	.2053	.2049	.2045	0	1	1	2	2	2	3	3	4
4.9	.2041	.2037	.2033	.2028	.2024	.2020	.2016	.2012	.2008	.2004	0	1	1	2	2	2	3	3	4
5.0	.2000	.1996	.1992	.1988	.1984	.1980	.1976	.1972	.1969	.1965	0	1	1	2	2	2	3	3	3
5.1	.1961	.1957	.1953	.1949	.1946	.1942	.1938	.1934	.1931	.1927	0	1	1	1	2	2	3	3	3
5.2	.1923	.1919	.1916	.1912	.1908	.1905	.1901	.1898	.1894	.1890	0	1	1	1	2	2	3	3	3
5.3	.1887	.1883	.1880	.1876	.1873	.1869	.1866	.1862	.1859	.1855	0	1	1	1	2	2	2	3	3
5.4	.1852	.1848	.1845	.1842	.1838	.1835	.1832	.1828	.1825	.1821	0	1	1	1	2	2	2	3	3

	0	1	2	3	4	5	6	7	8	9
5.5	.1818	.1815	.1812	.1808	.1805	.1802	.1799	.1795	.1792	.1789
5.6	.1786	.1783	.1779	.1776	.1773	.1770	.1767	.1764	.1761	.1757
5.7	.1754	.1751	.1748	.1745	.1742	.1739	.1736	.1733	.1730	.1727
5.8	.1724	.1721	.1718	.1715	.1712	.1709	.1706	.1704	.1701	.1698
5.9	.1695	.1692	.1689	.1686	.1684	.1681	.1678	.1675	.1672	.1669
6.0	.1667	.1664	.1661	.1658	.1656	.1653	.1650	.1647	.1645	.1642
6.1	.1639	.1637	.1634	.1631	.1629	.1626	.1623	.1621	.1618	.1616
6.2	.1613	.1610	.1608	.1605	.1603	.1600	.1597	.1595	.1592	.1590
6.3	.1587	.1585	.1582	.1580	.1577	.1575	.1572	.1570	.1567	.1565
6.4	.1563	.1560	.1558	.1555	.1553	.1550	.1548	.1546	.1543	.1541
6.5	.1538	.1536	.1534	.1531	.1529	.1527	.1524	.1522	.1520	.1517
6.6	.1515	.1513	.1511	.1508	.1506	.1504	.1502	.1499	.1497	.1495
6.7	.1493	.1490	.1488	.1486	.1484	.1481	.1479	.1477	.1475	.1473
6.8	.1471	.1468	.1466	.1464	.1462	.1460	.1458	.1456	.1453	.1451
6.9	.1449	.1447	.1445	.1443	.1441	.1439	.1437	.1435	.1433	.1431
7.0	.1429	.1427	.1425	.1422	.1420	.1418	.1416	.1414	.1412	.1410
7.1	.1408	.1406	.1404	.1403	.1401	.1399	.1397	.1395	.1393	.1391
7.2	.1389	.1387	.1385	.1383	.1381	.1379	.1377	.1376	.1374	.1372
7.3	.1370	.1368	.1366	.1364	.1362	.1361	.1359	.1357	.1355	.1353
7.4	.1351	.1350	.1348	.1346	.1344	.1342	.1340	.1339	.1337	.1335
7.5	.1333	.1332	.1330	.1328	.1326	.1325	.1323	.1321	.1319	.1318
7.6	.1316	.1314	.1312	.1311	.1309	.1307	.1305	.1304	.1302	.1300
7.7	.1299	.1297	.1295	.1294	.1292	.1290	.1289	.1287	.1285	.1284
7.8	.1282	.1280	.1279	.1277	.1276	.1274	.1272	.1271	.1269	.1267
7.9	.1266	.1264	.1263	.1261	.1259	.1258	.1256	.1255	.1253	.1252
8.0	.1250	.1248	.1247	.1245	.1244	.1242	.1241	.1239	.1238	.1236
8.1	.1235	.1233	.1232	.1230	.1229	.1227	.1225	.1224	.1222	.1221
8.2	.1220	.1218	.1217	.1215	.1214	.1212	.1211	.1209	.1208	.1206
8.3	.1205	.1203	.1202	.1200	.1199	.1198	.1196	.1195	.1193	.1192
8.4	.1190	.1189	.1188	.1186	.1185	.1183	.1182	.1181	.1179	.1178
8.5	.1176	.1175	.1174	.1172	.1171	.1170	.1168	.1167	.1166	.1164
8.6	.1163	.1161	.1160	.1159	.1157	.1156	.1155	.1153	.1152	.1151
8.7	.1149	.1148	.1147	.1145	.1144	.1143	.1142	.1140	.1139	.1138
8.8	.1136	.1135	.1134	.1133	.1131	.1130	.1129	.1127	.1126	.1125
8.9	.1124	.1122	.1121	.1120	.1119	.1117	.1116	.1115	.1114	.1112
9.0	.1111	.1110	.1109	.1107	.1106	.1105	.1104	.1103	.1101	.1100
9.1	.1099	.1098	.1096	.1095	.1094	.1093	.1092	.1091	.1089	.1088
9.2	.1087	.1086	.1085	.1083	.1082	.1081	.1080	.1079	.1078	.1076
9.3	.1075	.1074	.1073	.1072	.1071	.1070	.1068	.1067	.1066	.1065
9.4	.1064	.1063	.1062	.1060	.1059	.1058	.1057	.1056	.1055	.1054
9.5	.1053	.1052	.1050	.1049	.1048	.1047	.1046	.1045	.1044	.1043
9.6	.1042	.1041	.1040	.1038	.1037	.1036	.1035	.1034	.1033	.1032
9.7	.1031	.1030	.1029	.1028	.1027	.1026	.1025	.1024	.1022	.1021
9.8	.1020	.1019	.1018	.1017	.1016	.1015	.1014	.1013	.1012	.1011
9.9	.1010	.1009	.1008	.1007	.1006	.1005	.1004	.1003	.1002	.1001

APPENDIX V A (CHAPTER VII)

CHARACTERISATION OF MINERAL GROUPS BY X-RAY POWDER DIFFRACTION

For X-ray identification and characterisation of mineral powders the diffractogram is run with a slow scan of 1/4° or 1/8° per minute (with time constant 4 seconds, 1° slit with 0.003 in receiving slit) at the specific 2θ ranges where the characteristic peaks occur. The peak positions are measured at 2/3 peak heights.

A guideline of the studies are presented in the following section with examples of some major rock-forming silicate mineral groups.

FELSPARS

Alkali Felspars

In the range of 21-23° 2θ (in CuK_α radiation) $\bar{2}01$ reflections of low temperature alkali felspars occur between 20.95 and 22.4° (Fig. A.1). The potassic members (microcline) occur at lower angle than the sodic members. In an unmixed state the pure albite phase will show peak at about 1.0° (~0.19) Å away from the pure orthoclase (microcline) phase (Kuellmer, 1960). Good perthites show such distinct peaks.

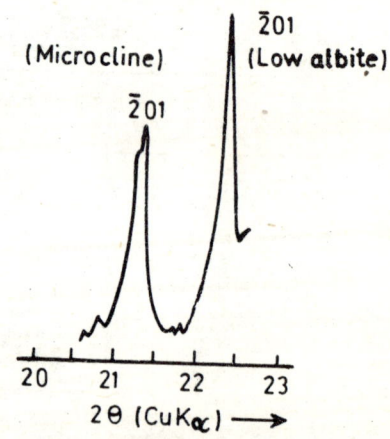

Fig. A-1 The relative positions of $\bar{2}01$ peaks in microcline and low albite in CuK_α

Using the 010 line of the internal standard of potassium bromate (K BrO$_3$) X-ray determinative curves for synthetic (Orville, 1967, p. 75) and natural (Jones et al, 1969) microcline ~ low albite have been prepared.

For high temperature alkali felspars these lines appear between 20-31° $2\theta(CuK_\alpha)$.

With increase in Al/Si disorder in the alkali-felspar the spacings of hkl and $h\bar{k}l$ approach each other and eventually coalesce as a single hkl reflection when the mineral becomes monoclinic. Monoclinic sanidine has a single 131 reflection. In triclinic microcline or orthoclase the 131 peak position is substituted by a doublet of 131 an 1$\bar{3}$1.

The triclinicity (Δ) of felspars can also be determined from the d-spacings of these two lines:

$$\Delta = 12.5\,(d_{131} - d_{1\bar{3}1})$$

When $\Delta = 0$, the alkali-felspar is monoclinic.
When $\Delta = 1$ the alkali-felspar is triclinic.

Plagioclase Felspars

Plagioclase compositional change can be determined by $2\bar{0}1$ reflections (Hamilton and Edgar, 1969). A guide to the indexing of plagioclase X-ray powder patterns which include high and low albite (An_o), oligoclase (An_{29}), bytownite (An_{80}) and anorthite (An_{100}) has been prepared by Borg and Smith (1968).

GARNETS

(JCPDS cards 2–1008, 3–0801, 10-288, 10-367):
The most intense line of 420 occur at 2θ between 38.41 to 40.86 (in CoK_α). The d-value of the line will give a direct measure of the cell-parameter (rather, cell-edge), a, as:

$$a\,(\mathring{A}) = 4.472\,d_{420}$$

At high angle between 104.80 to 114.43 (in CoK_α) 10, 4, 0 occur. The cell-edge, a, in angstrom can be measured as follows:

$$a = 10.77\,d_{10,4,0}$$

CARBONATES

In a scanning in the 2θ range between 26° to 34° (in CuK_α) most dominant peaks for calcite and dolomite occur at 29.43° and 30.98° respectively. The relative heights of these two peaks give a direct measure of the calcite-dolomite ratio.

$Mg^{2+} - Ca^{2+}$ diadochy in dolomitic carbonates can be determined using the calcite 104 peak and using the 111 line of the silicon internal standard.

CLAY GROUP

Clay minerals fall into five groups. These are
1. Kandides (kaolinite, dickite, nacrite, halloysite and metahalloysite)
2. Illites (illite, hydro-micas, phengites, glauconite, celadonite and brammallite)
3. Smectites (montmorillonite, nontronite, hectorite and saponite)
4. Vermiculite
5. Palygorskites (palygorskite, attapulgite and sepiolite).

Kandites

X-ray diffraction patterns are characterised by 7.1 Å (10Å only in halloysite) basal spacings (d_{001}) and glycol treatment has no effect except on halloysite. The varieties of kandites have the following x-ray characteristics.

Kaolinite: (a) ordered species has $d_{001} \sim 7$Å and $d_{002} \sim 3.575$ Å in ordered mounts. d_{060} occurs at ~ 1.50 Å. The structure collapses on heating to 550°C.

(b) disordered species show broadening of the 001 and 002 peaks at $\sim$ 7 Å and 3.755 Å.

Dickite and *nacrite:* peak positions are similar to kaolinite but less intense.

Halloysite: The structure has a single layer of water molecules between its structural sheets; consequently stacking is disordered and d_{001} is increased from ~ 7.2Å to 10Å. Glycol treatment increases this spacing to ~11 Å. No other kandite shows this response.

Illites

They have strong d_{001} reflection at 10Å which does not show any effect by heating to 550°C or on glycol treatment. But when it is interlayered with montmorillonite it may show glycol expansion.

The *d*-spacings of 1M polytype has basal spacings as: d_{060} at ~1.50 Å and d_{020} at ~4.4 Å.

Brammallite is an illite with sodium occuring as interlayer cation.

Glauconite has 1M *d*-reflections as in illites (as in muscovites).

Celadonite gives the same lines as 1M muscovites but chemical tests can prove this species.

Hydro-micas (muscovites, biotites) have similar lines, but spacings differ: d_{060} in muscovite (di-octahedral) occurs at ~1.50Å while in biotite (tri-octahedral) it is at ~ 1.53–1.55Å. In CuK_α the intensities of the former are greater than those of the latter. The basal reflection of d ~5Å is strong for most di-octahedral and weak for most tri-octahedral micas.

Smectites

They have variable d_{001}; mostly ≅ 15Å (for Ca) and take two layers of glycol to swell to 17-18 Å.

In smectites the basal spacings can vary over a wide range. Water is readily adsorbed between the structural layers. The number of layers of adsorbed water molecules is controlled by the interlayer cations. Na cations can generate one ($d_{001} \simeq 12.5$Å), two (15.5 Å), three (19Å) or more layers per cell; but calcium cations (in Ca-montmorillonite) can hold two layers of water per cell ($d_{001} \simeq 15.5$Å).

The dehydrated (at 500°C) collapsed-state smectites may give d_{001} ~9.6Å and become similar to pyrophyllite. Hydration as well as glycol-treatment, however, may lead to increasing the basal spacings to ~21Å.

Vermiculite

The basal spacing d_{002} occurs at ~14.4 Å but on dehydration successive changes occur (accompanied by an endothermic peak on a DTA curve) which result in decreasing the basal spacings to 13.8, 11.6 and 9Å.

Its exfoliation on sudden heating to above 300°C may cause expansion thirty-fold in the $\perp$(001) direction.

Palygorskites

In palygorskite reflections occur at 10.5, 4.5, 3.23 and 2.62Å. Above 400°C the 10.5 Å peak is replaced by a broad band and the structure collapses at ~800°C.

In *sepiolite* characteristic peaks occur at about 12.6, 4.31 and 2.62Å. Dehydration above 300°C makes the 12Å peak diffused out while the intensity of 7.6 Å peak increases. A new peak at 9.8 Å appears.

CHLORITES

The high Mg and low Fe chlorites are characterised by 14Å basal d_{001} spacings. Fe-rich chlorites show weak first and third basal reflections but strong second and fourth reflections.

Dioctahedral chlorite has d_{060} at 1.496 Å while that in normal trioctahedral chlorite occurs at 1.53-1.54 Å.

Some chlorites on glycolation change d_{001} between 28Å to 32Å, and on heating to 450°C collapse to ~13.8Å. These are of intermediate variety between chlorite and montmorillonite or vermiculite.

MIXED LAYER CLAY MINERALS

A series of reflections on the low-angle side of the 10Å spacing and broad humps would indicate mixed layering of mica. Mixed layering of montmorillonite, vermiculite and chlorite is indicatd by reflections on the high-angle side of the 14Å spacing. Heat treatment and glycolation may help to bring out the identifying peaks from the diffused humps and shoulders.

For the identification of clay groups of minerals in a soil sample the following steps may be followed.

1. Treat the sample with 9% hydrogen peroxide to remove all organic compounds.

2. Treat the residue with ammonium oxalate-oxalic acid solution of pH 3.3 (acidity should be specific to avoid acid attack on chlorites). Boil gently for a few minutes with ammonium chloride and wash the residue. Treat the residue with glycerol and take x-ray photograph. Vermiculite, chlorite and montmorillonite present will show lines at 11Å, 14Å and 17.7Å.

Heat treatment of the samples to 500°C followed by XRD may offer clues to the identification as follows:

Name and formula	Layer spacings (Å) approximately	
	Original	After heating to 500°C
Kaolinite $(Al_2(Si_2O_5)(OH)_4$	7.2	Vanishes
Micas $KAl_2(AlSi_3O_{10})(OH)_2$	10	10
Chlorites $Mg_3(AlSi_3O_{10})(OH)_2 \cdot Mg_2 Al(OH)_6$	14	14
Vermiculite $(Fe, Mg)_3 (Al, Si)_4 O_{10} \cdot (OH)_2 \cdot 4H_2O$	14	10
Montmorillonite $(Mg, Ca) Al_2Si (Si_4O_{10})(OH)_8$	9.4–10	9.4–10

At 500°C kaolinite structure is destroyed leaving amorphous residue. The others, however, lose adsorbed H_2O but not much of OH groups present in the structure.

OLIVINES (JCPDS: 7-74, 7-164).

Yoder and Sahama (1957) determined the relationship of the forsterite (Mg_2SiO_4) component in olivines from the *d*-spacing of 130 peak thus:

$$\text{Forsterite (Fo) mol\%} = 4233.91 - 1494.59\ d_{130}.$$

This has an error range of 3-4 mol. per cent. It was later revised by Agterberg (1964) as:

$$\text{Fo mol \%} = 4088.89 - 1442.44\ d_{130}$$

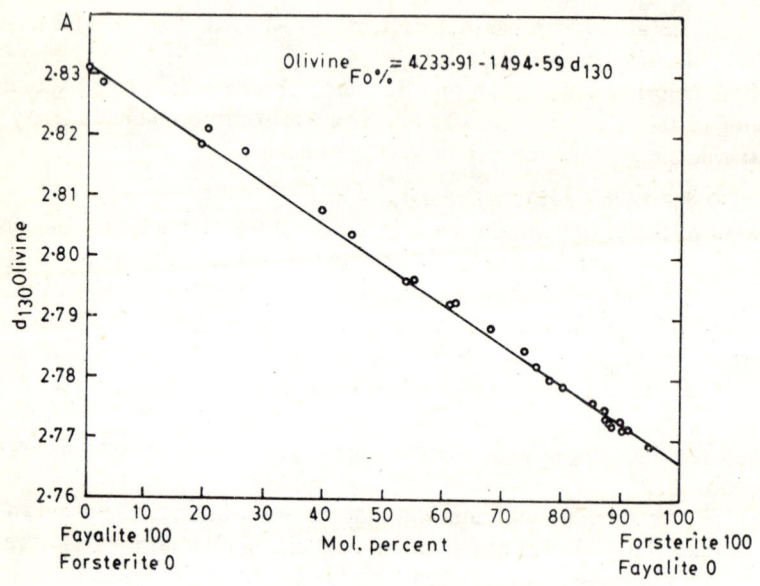

Fig. A-2   The d_{130} olivine determinative curve.

PYROXENES (JCPDS: 7-216; 4-0857)

Orthopyroxene

i) Using 131 and 111 (LiF) peaks.

The scan is made between 34.5 to 40.2° 2θ (in CuK_α) with LiF internal standard (its strong 111 peak appears at 38.75° 2θ).

The difference ($\triangle 2\theta$) in 2θ values of 131 peak of pyroxene and LiF gives a measure of Mg-ionic per cent thus:

$$\triangle 2\theta = 4.112 - 0.009132\ \text{Mg} \quad \text{(Himmelberg \& Jackson, 1967)}$$

or, Mg ionic % = $\dfrac{4.112 - \triangle 2\theta}{0.009132}$

ii) Using 060 peak

The 060 peak of orthopyroxenes occurs at (i) 63.20° 2θ (in CuK_α), *vide* Fig. 7.9. $d = 1.47$Å in enstatite (*En* % about 94), at (ii) 62.36° 2θ (CuK_α, $d = 1.4878$Å) in hypersthene (*En* % about 52) [*vide* US Geol. Soc. Memoir 122, 1969, p. 275]. The position is accurately determined by finding the position of the 220 peak of LiF internal standard which occurs at 65.56° $2\overset{\circ}{A}$ ($d = 1.424$Å).

Desborough and Rose (1968) found the following relationship for orthopyroxenes in the range En_{87} to En_{76} as:

$$\text{Mol. } En\% = \frac{2\theta_{060} - 61.31}{0.02015} \pm 0.3\%$$

CLINOPYROXENES

The composition of clino-pyroxenes of $CaMgSi_2O_6 - CaFeSi_2O_6$ (Di – Hd) series can be determined to an accuracy of ± 5 mol % Hd, using the reflections of 220, 13$\bar{1}$, 150 and 310 as:

$$d_{220} = 3.2329 + 0.000416\ X$$
$$d_{13\bar{1}} = 2.5640 + 0.000206\ X$$
$$d_{150} = 1.7538 + 0.000194\ X$$
$$d_{310} = 2.9505 + 0.000439\ X$$

where X is the mol. % of $CaFeSi_2O_6$ (Hd).

Linear regression graphs of the compositions of clinopyroxenes against cell parameters are available (Winchell and Tilling, 1960).

MICAS

The polymorphs of micas are distinguishable by the spacings between 4.4 and 2.6 Å in powder patterns. Disordered micas of 1M type show absence of *hkl* reflections.

The di- and tri-octahedral micas can be distinguished by using the 060 reflections. In dioctahedral micas the 060 occurs at about 1.50Å and in the trioctahedral micas it lies between 1.53 and 1.55Å. The former shows very strong 5Å basal reflection while the same is weak in most of the trioctahedral micas (lepidolite and glauconite are exceptions).

By powder diffraction study of synthetic and natural muscovites Yoder and Eugster (1955) distinguished the patterns of 1M, 2M and 3T polymorphs of dioctahedral micas. However, the powder patterns of 1M and 3T polymorphs are identical in the case of tri-octahedral micas.

K-micas in general show lesser basal (001) *d*-spacings in comparison to Ca- and Na- micas.

In biotite micas (of phlogopite-annite series) Fe/Mg occupancy in octahedral coordination was studied by Franzini & Schiaffino (1965) using a graph relating the intensity ratio $(I_{004} + I_{006})/I_{005}$. Scope of this method has been discussed later by Hormann and Morteani (1969).

SPINELS

The cell edge, a, in spinels determined by X-ray method may be usefully employed to determine the end members as (MacGregor and Smith, 1963):

$$\% \text{ FeO} = 33.7\ (a - 8.00) + 11.6$$
$$\% \text{ MgO} = 18.2 - 34.3\ (a - 8.00)$$
$$\% \text{ Cr}_2\text{O}_3 = 96.0\ (a - 8.00) + 22.3$$
$$\% \text{ Al}_2\text{O}_3 = 45.4 - 104.8\ (a - 8.00)$$

They also found the relationship of a to the refractive index of chromespinels as $n = 1.7008\ a - 12.0332$.

Stevens (1944) supplemented by Hutchison (1972) calculated the Al and Cr cations in unit cells a (in Å) as:

Cr cations = $58.46 \times a - 473.88$
Al cations = $407.3 - 48.6 \times a$.

The intensity ratios of 220/111 in conjuction with a (Å) have been used for determining the composition of Mg-spinels in the series $MgAl_2O_4$ – $MgCr_2O_4$ – $MgFe_2O_4$ by Allen (1966).

References on X-ray studies of some common minerals[1]
Some of the works related to the X-ray study for characterisation of some common minerals are listed below as a ready reference.

Aluminium silicates
Quantitative X-ray examination of alumino-silicates (W. Johnson and K.W. Andrews) *Trans. Brit. Ceram. Soc. 61* (1952), 724.
[mullite, kyanite, andalusite, sillimanite].

Amphiboles
The crystal chemistry of amphiboles (E.J.W. Whittaker) *Acta. Cryst. 13,* (1960), 291.
[Amphiboles in general]
Some observations on igneous ferrohastingsites (G. Borley and M.T. Frost) *Min. Mag, 33* (1963), 646 [ferrohastingsites].
Amphiboles from younger granites of Nigeria: Part II x-ray data (M.T. Frost) *Min. Mag. 33* (1963), 377.
[arfvedsonites and ferrohastingsites]
The effect of Mg^{2+} – Fe^{2+} substitution on the cell dimensions of cummingtonite (K. Viswanathan and S. Ghose) *Am. Mineralogist, 50* (1965), 1106–1112.
[Cummingtonite]
X-ray crystallographic properties of the cummingtonite-grunerite series (C. Klein and D.R. Waldbaum) *J. Geol., 75* (1967), 379-392.
[Cummingtonite-grunerite]

Analcite
Some natural analcite solid solutions (J.F.G. Wilkinson) *Min. Mag. 33* (1963) 498.

Apatite
X-ray diffraction analysis techniques to monitor composition fluctuations within the mineral group, apatite (H.C.W. Skinner) *Appl. Spectr., 22* (5), 412-414.

Aragonite
The determination of calcite: aragonite ratio in mollusc shells by x-ray diffraction: (T.T. Davies and P.R. Hooper) *Min. Mag, 33* (1963), 608.

Arsenopyrite
Arsenopyrite crystal-chemical relations (N. Morimoto and L.A. Clark) *Am. Mineralogist, 46* (1961), 1448.

---

[1] X-ray diffraction data of all minerals and inorganic compounds are obtained from J.C.P.D.S. Powder Data File, published and regularly revised by the Joint Committee on Powder Diffraction Standards, 1601 Park Lane, Swarthmore, Pennsylvania, 19081, U.S.A.

Carbonates
Determinazione della composizione delle fasi krigonali nel sistema $MgCO_3 - FeCO_3 - CaCO_3$ mediante la diffrazione del raggi (G.L. Morelli *Rend. Soc. Mineral, Haliana, 23* (1967), 315-332.
[X-ray determination of composition in the isomorphous compounds in $MgCO_3 - FeCO_3 - CaCO_3$ system]

X-ray determination of dolomite/calcite ratios of a carbonate rock (C.B. Tennant and R.W. Berger) *Am. Mineralogist, 42,* (1957), 23-29.

Accuracy of Calcite/dolomite ratios by x-ray diffraction and comparison with results from staining techniques (R.P. Gensmer and M.P. Weiss) *Journal of Sedimentary Petrology, 50* (1980), 626-629.

Chalcopyrite
The crystal structure refinement of chalcopyrite, $CuFeS_2$ (S.R. Hall and J.M. Stewart) *Acta Crystallographica, 29B,* (1973), 579-85.

Clay group
Clay minerals: a guide to their X-ray identification (D. Carroll) *Geol. Soc. Am., Spec. Paper, 126* (1970), 80.

Crystal structure of clay minerals and their x-ray identificationm (G.W. Brindley and G. Brown, eds.) Mineralogical Society, London, (1961).

Chlorite
A new review of chlorites (M.H. Hey) *Min. Mag. 30* (1954), 277.

X-ray identification of chlorite species (G.W. Brindley and F.H. Gillery) *Am. Mineralogist 41* (1956), 169.

Semi quantitative analysis of chlorites by X-ray diffraction (R. Schoen) *Am. Mineralogist 47* (1962), 1384.

Determination of the heavy atom content in chlorite by means of the X-ray diffractometer (W. Petruk) *Am. Mineralogist, 49* (1964), 61.

On a systematic error in the X-ray determination of iron-content of chlorites and biotites: a discussion (P.K. Hormann and G. Morteani) *Am. Mineralogist, 54,* 1491-1494.

Felspars
Variations in X-ray powder diffraction patterns of plagioclase felspars (J.R. Smith and H.S. Yoder) *Am. Mineral, 41* (1956). 632-647.

The microcline-sanidine stability relations (J.R. Goldsmith and F. Laves) *Geochim. Cosmochim. Acta, 5* (1954), 1-19.

X-ray intensity measurements on perthitic materials I; Theoretical considerations (F.J. Kuellmer) *J. Geol. 67* (1959), 648-660.

X-ray intensity measurements on perthitic materials II: Data from natural alkali Felspars (F.J. Kuellmer) *J. Geol. 68* (1960), 307-323.

Diagrams for the determination of plagioclases using X-ray powder methods (H.U. Bambauer, M. Corlett, E. Eberhard and K. Viswanathan), *Schweiz. Mineral. Petrog. Mitt., 47* (1967), 333-349.

Unit-cell parameters of the microcline-low albite and the sanidine-high albite solid solution series. (P.M. Orville). *Am. Mineral, 52* (1967), 55-86.

X-ray and optical study of alkali felspars II: an X-ray method of determining the composition and structural state from measurement of 2θ values for three reflections (T.L. Wright) *Am. Mineral, 53* (1968), 88-104.

X-ray and optical study of alkali felspars II: determination of composition and structural state from refined unit-cell parameters and 2V (T.L. Wright and D.B. Stewart) *Am. Mineral, 53* (1968), 38-87.

Calculated X-ray powder patterns for silicate minerals (I.Y. Borg and D.K. Smith) *Geol. Soc. Am. Mem., 122* (1969), 896.

Calculated powder patterns, Part II, Six potassium felspars and barium felspars (I.Y. Borg and D.K. Smith) *Am. Mineral, 54* (1969), 163-181.

Modal analysis of granitoids by quantitative x-ray diffraction. (P.D. Maniar and G.A. Cooke), *Am. Mineral, 72,* (1987), 433-437.

The determination of the orthoclase content of homogenised alkali felspar using the $\bar{2}\ 0\ 1$ X-ray method (J.B. Jones, R.W. Nesbitt and P.G. Slade) *Min. Mag. 37*(1969), 489-496.

A new X-ray method to determine the anorthite content and structural state of plagioclases (K. Viswanathan) *Contrib. Mineral. Petrol. 30* (1971), 332-335.

Garnets

Physical properties of the end members of the garnet group (B.J. Skinner) *Am. Mineral, 41* (1956), 428-436.

The composition and physical properties of garnet (H. Winchell) *Am. Mineral. 43* (1958), 595-600.

Ilmenite

Contribution a l'etude de la series ilmenite-geikielite (B. Cervelle). *Bur. Rech. Geol. Minieres Bull. 6* (1967), 1-26.

Loellingite

X-ray method for rapid determination of sulphur and cobalt in loellingite (L.A. Clark) *Can. Mineralogist, 7* (1962), 306-311.

Nepheline

Determination of the composition of natural nephelines by an X-ray method (J.V. Smith and Th. G. Sahama) *Min. Mag. 30* (1954), 439.

Pyrite

The cobaltiferous pyrite series (J.F. Riley) *Am. Mineral, 53* (1968), 293-295. [Pyrite (FeS_2)-Cattierite (CoS_2) linear relation with d_{511} plots]

Pyroxenes

Regressions of physical properties on the compositions of clino-pyroxenes (H. Winchell and R. Tilling) *Am. J. Sci., 258* (1960), 529.

X-ray determinative curve for some orthopyroxenes of composition, Mg_{48-85} from the Stillwater complex, Montana (G.R. Himmelberg and E.D. Jackson), *U.S. Geol. Surv. Prof. Paper, 575-B* (1967), B 101-102.

X-ray and chemical analysis of orthopyroxenes from the lower part of the Bushveld Complex, S. Africa (G.A. Desborough and H.J. Rose Jr.) *U.S. Geol. Surv. Prof. Paper, 600-B,* (1968), B 1-6.

Calculated X-ray powder patterns for silicate minerals (I.Y. Borg and D.K. Smith) *Geol. Soc. Amer. Memoir 122* (1969), 896.

Pyrrhotite

Measurement of the metal content of naturally occurring, metal-deficient, hexagonal pyrrhotite by an X-ray spacing method (R.G. Arnold and L.E. Reichen) *Am. Mineral, 47* (1962), 105.
[Lattice spacings of d_{102} correlated with x in Fe_{1-x} S of pyrrhotite]

Quantitative determination of hexagonal and monoclinic pyrrhotites by X-ray diffraction (A.R. Graham) *Canad. Mineral, 10* (1969), 4-24.

Sheet silicates

Synthetic and natural muscovites (H.S. Yoder and H.P. Eugster) *Geochim. Cosmochim. Acta, 8* (1955), 225.

Coexisting muscovite and paragonite in pelitic schists (E-AN Zen and A.L. Albee) *Am. Mineralogist, 49* (1964), 904-925.

On X-ray determination of the iron-magnesium ratio of biotites (M. Franzini and L. Schiaffino) *Z. Krist. 122* (1965), 100-107.

On a systematic error in the x-ray determination of the iron-content of chlorites and biotites: a discussion (P.K. Hormann and G. Morteani) *Am. Mineral., 54* (1969), 1491-1494.

X-ray powder diffraction identification of illitic materials (J. Srodon) *Clay and Clay Minerals, 32,* (5), (1984), 337-349.

Scapolite

Studies on scapolite (B.J. Burley, E.B. Freeman, and D.M. Shaw) *Can. Mineral. 6* (1961), 670.
[Composition related to $2\theta_{400} - 2\theta_{112}$ in scapolites]

Serpentine

The characterisation of serpentine minerals by X-ray diffraction (E.J.W. Whittacker and J. Zussman) *Min. Mag. 31* (1956), 107.
[identification of chrysotile, lizardite and antigorite]

Determination of olivine and serpentine in Kimberlite by x-ray diffraction (M. Hodgson and A.W.L. Dudeney) *Analyst, 109,* (1984), 1129-1133.
[Relative proportion of olivine and serpentine determined with relative precision of 6%]

Soil

Quantitative determination of calcite and dolomite in soils by x-ray diffraction (M. Ulas and M. Sayin) *J. of Soil Sci, 35,* (1984), 685-691.

Quantitative determination of goetite and hematite in kaolinitic soils by x-ray diffraction (N. Kampf and U. Schwertmann) *Clay Mineral., 17* (1982), 359-363.

Identification of soil iron oxide minerals by differential x-ray diffraction (D.G. Schulze) *Soil Science, Soc. of Am. J., 45,* 437-440.

Spinels
An X-ray method for determining composition of a magnesium spinel (W.C. Allen) *Am. Mineralogist, 51,* 239-243.

Sulphides
Measurement of disorder in Zinc and Cadmium sulphides (M.A. Short and E.G. Steward) *Am. Mineral, 44* (1959), 189.
[Proportion of cubic and hexagonal (wurtzite) structures]
  Effect of FeS on the unit cell edge of sphalerite: A revision (B.J. Skinner, P.B. Barton and G. Kullerud) *Econ. Geol. 54* (1959), 1040.
[Fe-Zn substitution determination]
  The crystal structure refinement of chalcopyrite, $CuFeS_2$ (S.R. Hall and J.M. Stewart) *Act. Crystall, 29B,* (1973), 579-85.

Topaz
Optical and X-ray determinative methods for fluorine in topaz (P.H. Ribbe and P.E. Rosenberg) *Am. Mineral. 56* (1971) 1812–1821.

Tourmaline
Bestimmung des Tourmalin-chemismus auf rontgenographischem Wege (E.E. Horn and H. Schulz) *Neues Jahrb. Mineral. Abhand., 108* (1968), 20-35.
[Chemical composition determined]

APPENDIX V B (*Chapter VII*)
A GUIDE TO MINERAL IDENTIFICATION

As a preliminary guide to the identification of mineals by the set of five to six strong lines, starting from very very strong or the strongest as (10) followed by successive less strong lines marked as (9), (8). (7), (6), (5) etc., have been presented with the corresponding *d*-spacings following the classification of minerals as outlined in the sections below.

Classification of Minerals as a Guide to Search

The minerals described in the following have been arranged to twelve classes based broadly on the bond-types. The scheme starts with minerals of predominantly metallic or covalent bond type (elements and sulphides) going on to minerals with isodesmic (oxides, halides), mesodesmic (silicates and borates), and finally to anisodesmic (phosphates, sulphates, carbonates etc.) bond types.

Class I *Elements*
  A. Metals
  B. Semi-metals and non-metals

Class II *Sulphides and Sulphosalts*
  A. Intermetallic compounds
  B. Sulphides and related compounds
    1. Metallic
      (a) Pt – Pd – Ru combination
      (b) Ni – Co – Fe combination
      (c) Mo – W – Sn combination
         Zn – Cu – Pb combination
         Hg – Ag – Au combination
    2. Semi-metallic and Oxysulphides
  C. Sulphosalts.

Class III *Halides*
  A. Fluorides
  B. Chlorides, Bromides and Iodides
  Al-Mg-Fe combination
  Na - Ca - K combibnation
  Cu - Pb – Hg combination

Class IV *Oxides and Hydroxides*
  A. Metallic
    Be – Al – Mg combination
    Fe – Mn – V combination
    Ti – Nb – Zr combination
    Zn – Cu – Pb (U) combination
  B. Semi-metallic and Non-metallic

Class V *Silicates*
  A. Nesosilicates
    Be – Al – Mg combination
    Ti – Nb – Zr combination
    Ca – Na – Mn combination
    Zn – Cu – Pb (U) combination
  B. Inosilicates
  C. Phyllosilicates
  D. Tectosilicates

Class VI *Borates*
Class VII *Phosphates, Arsenates and Vanadates*
    Be – Al – Mg combination
    Fe – Mn – Na combination
    Na – Ca – REE combination
    Zn – Cu – Co – Pb (U) combination

Class VIII *Tungstates and Molybdates*

Class IX *Sulphates*

Class X *Chromates*

Class XI *Carbonates*

Class XII *Nitrates*

X-ray line Intensities of common minerals and related compounds

The order of intensities of characteristic lines of the minerals are put in italics in first brackets with the corresponding d-spacings in angstroms as follows:

Elements

Metals
Platinum [Pt]: (*10*) 1.180, (*9*) 2.27, (*8*) 1.956, 1.384, (*7*) 0.898, 0.875
Iron [Fe]: (*10*) 1.168, (*9*) 2.02, (*8*) 1.430, (*7*) 1.012.
Zinc [Zn]: (*10*) 2.092, (*9*) 1.339, 1.173, (*8*) 1.683, (*7*) 1.128, (*6*) 2.311.
Gold [Au]: (*10*) 2.35, (*9*) 2.03, 1.226, (*8*) 1.437, (*7*) 0.933, 0.909.
Silver [Ag]: (*10*) 2.37, (*9*) 1.232, (*8*) 2.05, 1.443, (*7*) 0.936, 0.912.
Copper [Cu]: (*10*) 1.276, (*9*) 2.085, (*8*) 1.806.
Lead [Pb]: (*10*) 1.494, 1.138, 1.108, (*9*) 2.86, 1.753, (*8*) 2.48.

Semi-metals and Non-metals
Bismuth [Bi]: (*10*) 3.21, 1.435, (*9*) 2.45, 1.480, (*8*) 2.34.
Antimony [Sb]: (*10*) 2.249, 2.151, 1.366, (*9*) 1.765, 1.552.
Arsenic [As]: (*10*) 1.53, 1.867, 2.74, (*9*) 2.04, (*8*) 3.14, 1.76.
Tellurium [Te]: (*10*) 3.22, (*9*) 2.33, (*7*) 2.22, (*6*) 1.82, (*5*) 3.85.
Selenium [Se]: (*10*) 2.975, 2.06, 1.755 (*9*) 1.642, 1.634, 1.424.
Sulphur [α–S]: (*10*) 3.85, (*7*) 3.21, (*6*) 3.10, 2.85, 2.12. 1.90.
Diamond [C]: (*10*) 2.05, (*9*) 0.721, 0.358, (*8*) 1.26, (*7*) 1.072, 0.473, (*6*) 0.813.
Graphite [C]: (*10*) 3.35, (*9*) 1.230, 1.154, (*8*) 1.675, (*6*) 1.541, 1.117.

Sulphides and Sulphosalts

Intermetallic compounds
Michenerite [Pd Bi$_2$]: (*10*) 2.99, (*9*) 2.01, (*8*) 2.73, 1.79, (*3*) 1.46.
Froodite [Pd Bi$_2$]: (*10*) 2.77, (*8*) 1.556, (*7*) 2.97, 2.48, 2.21, (*6*) 1.637.
Orcellite [Ni$_2$ As]: (*10*) 1.977, 1.918, (*4*) 1.810
Oregonite [Ni$_2$ Fe As$_2$]: (v.s.) 2.314, 2.119, (*s*) 1.991, 1.788, 1.757.
Algodonite [Cu$_6$ As]: (*10*) 2.00, (*8*) 2.13, (*5*) 1.54, 1.30, 1.20, 1.105.
α-Domeykite [Cu$_3$ As]: (*10*) 2.046, 1.219, (*8*) 3.03, 2.15, 1.959, 1.882.
β-Domeykite [Cu$_3$ As] (*10*) 2.05, 2.00, 1.174, (*8*) 2.35, 2.21, 1.184.
Cattierite [Co S$_2$]: (10) 1.663, 1063, (*8*) 2.75, 2.46, (*7*) 2.249, (*6*) 1.474.
Novakite [(Cu, Ag)$_4$ As]: (*10*) 1.870, 1.182, (*9*) 1.998, 1.957, (*7*) 1.910.
Paxite [Cu$_2$ As$_3$] (*10*) 3.16, (*8*)3.63, (*7*) 2.77, 2.62, 1.202, (*6*) 1.882.
Dyscrasite [Ag$_3$ Sb]: (*10*) 2.28, (*6*) 2.58, 2.40, 1.765, 1.500, 1.364.
Aurostibite [Au Sb$_2$]: (*10*) 2.003, (*5*) 3.33, 1.213, 0.867, 0.785.

Sulphides and Related Compounds

Metallic
(a) Pt - Pd - Ru Compounds
Cooperite [Pt S]: (10) 3.03, (*8*) 1.93, 1.513, (*6*) 1.77, 1.74, 1.16.
Braggite [(Pt, Pd)S]: (10) 2.92, (*8*) 2.63, 1.75, 1.72, 1.43, 1.40.
Vysotskite [(Pd, Ni)S]: (*10*) 2.91, 2.86, (*9*) 1.031, (*8*) 2.61, 1.717, 1.185.
Sperrylite [Pt As$_2$]: (*10*) 1.788, 1.144, (*9*) 1.05, (*8*) 2.94, 2.10.
Niggliite [Pt (Sn, Te)]: (*10*) 2.15, 2.05, 1.486, 1.205, (*8*) 1.268, (*7*) 1.304.
Moncheite [Pt Te$_2$]: (*10*) 2.93, (*8*) 2.11, (*7*) 2.02, 1.462, 1.282.

Ni-Co-Fe Compounds
Maucherite group
Maucherite [Ni$_{11}$ As]: (*10*) 2.70, 2.02, 1.717, (*9*) 1.447, (*6*) 1.211.
Hauchecornite [Ni$_9$ (Bi, Sb)$_2$S$_8$]: (*10*) 2.79, (*7*) 2.39, 2.30, (*6*) 4.34, 1.861.
Parkerite [Ni$_3$(Bi, Pb)$_2$S$_2$]: (*10*) 2.85, (*9*) 2.33, (*7*) 4.01, 1.645.
Shandite [Ni$_3$ Pb$_2$ S$_2$]: (*10*) 2.78, (9) 3.94, (*8*) 1.611, (*6*) 1.760.
Heazlewoodite [Ni$_3$ S$_2$]: (*10*) 1.822, (*9*) 2.88, (*8*) 1.657, (*5*) 4.10, 2.03, (*4*) 2.38.

Pyrrhotine group
Breithauptite [Ni Sb]: (*10*) 2.84, (*9*) 2.052, 1.959, (*4*) 1.616, 1.533.
Troilite [Fe S]: (*10*) 2.085, (*9*) 1.720, (*8*) 1.328, (*7*) 2.644, (*6*) 2.968.
Smythite [Fe$_3$ S$_4$]: (*10*) 1.732, (*8*) 1.897, (*7*) 1.979, (*6*) 2.56, 2.26, 1.427.
Ni-pentlandite [Ni, Fe)$_9$ S$_8$]: (*10*) 1.775, (*8*) 3.03 (*5*) 1.931, (*4*) 2.90,(*3*) 2.30.
Co-pentlandite [(Co, Ni, Fe)$_9$ S$_8$]: (*10*) 3.01, 1.763, (*8*) 1.918, (*5*) 2.878, 1.018.
Niccolite (Ni As]: (10) 2.627, (*9*) 1.937, (*8*) 1.788, (*7*) 1.320, 1.032, (*6*) 1.070.
Pyrrhotine [Fe$_{1-x}$ S]: (*10*) 2.062, (*8*) 1.045, 2.63, 1.045, (*7*) 1.718, (*6*) 2.97.
Millerite [Ni S]: (*10*) 1.876, (*9*) 2.792, (*7*) 1.821, 1.730, 1.609, (*6*) 1.536.
Linnaeite Group
Linnaeite [Co$_3$ S$_4$]: (*10*) 1.68, (*9*) 2.82, (*8*) 1.82, (*7*) 2.38, (*6*) 1.23, (*5*) 1.37.

Polydymite [Ni$_3$ S$_4$]: (*10*) 1.67, (*9*) 2.85, 2.36, (*6*) 1.82, (*4*) 3.35.
Carrollite [Cu Co$_2$ S$_4$]: (*10*) 1.676, (*9*) 2.87, 1.83, (*7*) 2.39, 1.231, (*6*) 1.182.
Tyrrellite [Cu, Co, Ni)$_3$ Se$_4$]: (*10*) 1.769, (*9*) 2.50, (*7*) 2.88, (*6*) 3.02, 1.926, (*4*) 1.51.
Violarite [Fe Ni$_2$ S$_4$]: (*10*) 2.86, 1.678, (*8*) 2.37, (*7*) 1.83, 0.969, (*5*) 3.36.
Daubreelite [Fe Cr$_2$ S$_4$]: (*10*) 3.03, 1.299, 1.763, (*8*) 3.54, 2.49, 1.92.
Indite [Fe In$_2$ S$_4$]: (*10*) 3.20, (*9*) 1.877, (*8*) 1.085, (*7*) 2.05, (*5*) 1.384.

Pyrite-Cobaltite group
Pyrite [Fe S]: (*10*) 1.629, (*9*) 1.040, (*8*) 2.696, 2.417, (*7*) 2.206, (*6*) 1.908.
Vaesite [Ni S$_2$]: (10) 1.702, (*9*) 1.080, (*8*) 2.809, (*7*) 2.515, (*6*) 1.992, (*5*) 3.24.
Penroseite [Ni Se$_2$]: (*10*) 2.67, 2.45, (*9*) 1.802, (*4*) 1.596, (*3*) 1.659, 1.301.
Cobaltite [Co As S]: (*10*) 2.53, (*9*) 2.29, 1.687, (*7*) 2.82, (*5*) 1.493, 1.078.
Gersdorffite [Ni As S] (*10*) 2.56, 1.725, 1.582, (*8*) 2.84, 2.33, 1.10.
Ullmannite [Ni Sb S]: (*10*) 2.64, (*8*) 2.41, 1.78, (*5*) 4.15, 2.94, 1.64.

Marcasite-Arsenopyrite Group
Marcasite [Fe S$_2$]: (*10*) 2.69, 1.754, (*8*) 2.412, 2.314, (*6*) 1.908, (*5*) 1.593.
Ferroselite [Fe Se$_2$]: (*10*) 2.46, 1.876, (*9*) 2.56, (*8*) 1.443, (*7*) 1.781, 1.688.
Lollingite [Fe As$_2$]: (*10*) 2.589, 1.846, 1.628, (*9*) 2.535, 2.332.
Safflorite [(Co, Fe) As$_2$]: (*10*) 2.578, 1.859, 1.639, (*9*) 2.373, 1.291.
Rammelsbergite [Ni As$_2$]: (*10*) 2.55, (*8*) 2.86, (*7*) 2.37, 2.01, 1.82, 1.73.
Pararammelsbergite [Ni As$_2$]: (*10*) 2.55, (*8*) 2.86, (*7*) 2.37, 2.01, 1.82, 1.73.
Arsenopyrite [Fe As S]: (*10*) 2.662, 1.817, (*9*) 2.443, 2.412, (*8*) 1.629

Skutterudite group
Skutterudite [(Co As$_3$]: (*10*) 2.585, 1.607, (*9*) 1.828, 1.668, 1.404, (*8*) 2.184.
Chloanthite [(Ni, Co) As$_{3-x}$]: (*10*) 2.61, 1.845, (*8*) 1.616, (*7*) 1.688, 1.425, (*5*) 2.182.

Mo – W – Sn Combination
Molybdenite [Mo S$_2$]: (*10*) 6.01, (*8*) 2.27, (*5*) 1.82, (*2*) 1.58, (*1*) 2.50.
Tungstenite [WS$_2$]: (*10*) 2.058, 1.547, (*9*) 3.10, 1.581, 1.103.

Zn-Cu-Pb Combination
Sphalerite group
Sphalerite [β-ZnS]: (*10*) 3.116, (*9*) 1.908, (*8*) 1.630, (*5*) 1.104, (*3*) 1.045.
Wurtzite 2H [α-ZnS]: (*10*) 3.107, 1.902, (*9*) 1.625, (*8*) 1.106, 1.044, (*4*) 3.283.
Wurtzite 4H [α—ZnS]: (*10*) 3.29, 2.91, 1.76, (*9*) 1.90, (*7*) 3.11, 2.27.
Hawleyite [β-CdS]: (*10*) 3.36, 2.06, (*9*) 1.756, (*4*) 1.189, 1.121, 0.983.
Greenockite [α-CdS]: (*10*) 3.167, (*8*) 3.59, 2.071, 1.900, 1.764, (*7*) 1.258.
Cadmoselite [CdSe]: (*10*) 2.13, (*8*) 1.816, (*7*) 3.67, 1.96, (*6*) 1.443, (*5*) 3.24.

Hauerite-Alabandine Group
Hauerite [Mn S$_2$]: (*10*) 3.035, (*7*) 2.715, 2.49, 2.15, 1.832, 1.627.
Alabandine [MnS]: (*10*) 2.603, 1.843, (*7*) 1.165, 1.063, (*5*) 1.504, (*4*) 1.302
β-Alabandine [MnS]: (*10*) 1.996, 1.823, 1.346, (*4*) 3.46, 3.051.
Oldhamite [CaS]: (*10*) 2.84, 2.00, (*9*) 1.268, (*8*) 1.63, 1.158, (*5*) 1.419.

Chalcopyrite Group
Chalcopyrite [Cu Fe S$_2$]: (*10*) 3.03, 1.855, 1.586, (*8*) 1.205, 1.074.
Gallite [Cu Ga S$_2$]: (*10*) 3.06, (*7*) 1.876, (*6*) 1.611, 1.088, (*5*) 1.124, (*4*) 1.898.

Stannite [Cu$_2$ Fe Sn S$_4$]: (*10*) 1.888, (*8*) 3.06, 1.103, (*7*) 1.618, (*6*) 1.607, 1.234.
Bornite [Cu$_5$ Fe S$_4$]: (*10*) 1.924, (*8*) 3.30, 3.165, 2.737, 1.369, 1.117.
Valleriite [Cu$_3$ Fe$_4$ S$_7$]: (*10*) 11.056, 5.739, 3.270, 1.900, 1.889, 1.871, (*8*) 3.811.
Betekhtinite [Pb$_2$ (Cu, Fe)$_{21}$S$_{15}$]: (*9*) 1.815, (*8*) 3.047, (*7*) 2.90, 1.743, (*4*) 3.11, 1.309.

Chalcocite—Covellite Group
Chalcocite [Cu$_2$S]: (*10*) 1.963, 1.868, (*8*) 3.14, 2.38, 1.868, 1.643, (*6*) 2.51.
Digenite [Cu$_{2-x}$ S]: (*10*) 1.966, (*4*) 2.78, (*3*) 3.21, (*2*) 1.677, 1.135.
Berzelianite (Cu$_{2-x}$Se): (*10*) 2.02, (*9*) 3.32 (*8*) 1.726, (*5*) 1.169, (*4*) 1.431, (*3*) 1.314.
Crookesite [(Cu, Tl, Ag)$_2$ Se]: (*10*) 3.29, 2.59, (*8*) 3.00, (*6*) 2.11, (*5*) 1.833, (*4*) 2.32.
Weissite [Cu$_2$ Te]: (*10*) 3.61, 1.994, (*9*) 2.084, (*7*) 1.441, (*6*) 1.806, (*5*) 3.24.
Umangite [Cu$_3$ Se$_2$]: (*10*) 3.57, (*9*) 1.819, (*8*) 1.776, (*5*) 3.20, 3.10, 2.26.
Rickardite [Cu$_{4-x}$ Te$_2$]: (*10*) 2.07, (*6*) 3.35, (*4*) 2.54, 1.984, (*3*) 1.156, (*2*) 2.81.
Vulcanite [Cu Te]: (*10*) 2.03, (*7*) 2.86, (*6*) 3.52, (*4*) 6.94, (*3*) 2.65, 2.32.
Covellite [Cu S]: (*10*) 2.81, 1.89, (*8*) 3.04, 2.72, 1.555, (*7*) 1.73.
Klockmannite [Cu Se] (*10*) 2.87, (*9*) 3.17, (*8*) 1.963, (*6*) 3.34, 1.815, (*5*) 1.619.

Galena Group
Galena [Pb S] (*10*) 2.965, 2.093, 1.324, (*9*) 3.44, 1.780, (*8*) 1.707.

Ag – Au – Hg Compounds
Argentite Group
Acanthite [Ag$_2$ S]: (*10*) 2.58, 2.44, (*9*) 2.37, (*8*) 3.07, 2.81, 2.08, (*7*) 3.40.
Aguilarite [Ag$_4$ SeS]: (*10*) 2.82, (*9*) 2.44, (*8*) 1.99, (*6*) 2.09, (*5*) 2.22, (*3*) 3.11.
Naumannite [Ag$_2$ Se]: (*10*) 2.66, 2.56, (*6*) 2.23, (*4*) 2.00, (*2*) 4.14, 2.42.
Hessite [Ag$_2$ Te]: (*10*) 2.31, (*8*) 2.87, (*7*) 2.25, (*6*) 3.01, 2.14, (*2*) 3.19.
Petzite [Ag$_3$ Au Te$_2$]: (*10*) 2.77, (*5*) 2.11, (*4*) 2.02, (*3*) 2.44, 2.31, 1.897.
Empressite [Ag$_5$ Te$_3$] (*10*) 2.16, (*5*) 2.54, (*2*) 3.03, 2.12, 1.347, (*1*) 1.542.
Stromeyerite [Ag Cu S]: (*10*) 2.61, (*8*) 3.33, 2.07, (*7*) 3.46, 1.99, (*6*) 3.07.
Eucairite [Ag Cu Se]: (*10*) 2.12, (*7*) 2.61, (*5*) 2.88, (*4*) 2.48, (*2*) 2.02, (*1*) 1.202.
Sternbergite [Ag Fe$_2$ S$_3$]: (*10*) 4.25, (*8*) 3.25, 2.79, (*7*) 1.79, (*5*) 2.64, 1.945,
Argentopyrite [Ag Fe$_2$ S$_3$]: (*10*) 3.341, 3.318, (*7*) 1.808, (*5*) 3.62, 1.931, (*4*) 3.11.

Cinnabar Group
Cinnabar [HgS]: (*10*) 3.37, 2.869, (*8*) 3.16, 2.074, 1.980, 1.765.
Metacinnabarite [HgS]: (*10*) 3.40, 2.07, 1.765, (*9*) 1.191, 1.122, (*8*) 2.94.

Semi-metallic
1. *Tetradymite Group*
Wehrlite [Bi$_3$ Te$_2$]: (*10*) 3.22, (*7*) 2.36, (*5*) 2.21, (*4*) 1.402, (*3*) 1.99, 1.606.
Tellurbismuth [Bi$_2$ Te$_3$]: (*10*) 3.21, (*8*) 2.37, (*5*) 2.19, 2.03, 1.486.
Tetradymite [Bi$_2$ Te$_2$ S]: (*10*) 3.10, (*5*) 2.28, 1.64, 1.292, 1.207, (*4*) 2.11.

2. Stibnite Group
Stibnite [Sb$_2$ S$_3$] (*10*) 1.933, 1.687, (*9*) 3.57, 3.045, 2.757, 2.511.
Bismuthinite [Bi$_2$ S$_3$]: (*10*) 3.44, (*8*) 1.91, (*7*) 2.48, (*6*) 1.472, (*5*) 3.03.
Orpiment [As$_2$ S$_3$]: (*10*) 4.775, (*8*) 1.743, (*6*) 2.707, 2.446, (*4*) 2.785, 2.085
Realgar [As S]: (*10*) 3.17, (*7*) 2.93, 2.72, 2.122, 1.855, (*4*) 2.478.

a) Copper Sulphosalts
1. *Tetrahedrite Group*
Tetrahedrite [Cu$_{12}$ Sb$_4$ S$_{13}$]: (*10*) 3.00, 1.839, (*8*) 1.568, (*6*) 2.60, (*4*) 2.45, 1.193.
Tenantite [Cu$_{12}$ As$_4$ S$_{13}$]: (*10*) 2.94, 1.803, (*8*) 1.537, (*6*) 2.55, (*4*) 2.40, 1.170.
Luzonite [Cu$_3$ As S$_4$]: (*10*) 3.01, 1.861, (*9*) 1.591, (*8*) 1.849, (*6*) 1.585, 1.074.

2. *Enargite Group*
Enargite (Cu$_3$ As S$_4$): (*10*) 3.16, (*9*) 2.827, 1.842, (*8*) 1.714, 1.581, (*7*) 2.183.
Chalcostibite [Cu Sb S$_2$] (*10*) 3.10, 2.98, 1.751, (*8*) 2.29, 2.11, 1.818.
Emplectite [Cu Bi S$_2$] (*10*) 3.20, 3.02, (*9*) 2.34, 2.16, (*8*) 1.80, 1.655.

3. *Bournonite Group*
Bournonite [Pb Cu Sb S$_3$]: (10) 2.73, (*8*) 3.86, 2.81, 2.58, 1.98, 1.763.
Seligmannite [Pb Cu As S$_3$]: (*10*) 2.72, (*9*) 3.85, (*8*) 2.57, 1.77, (*6*) 1.66.
Aikinite [Pb Cu Bi S$_3$]: (*10*) 3.67, (*9*) 3.18, (*8*) 2.88, (*7*) 3.58, (*6*) 2.58, (*5*) 1.981.

b) Silver Sulphosalts
1. *Pyrargyrite Group*
Proustite [Ag$_3$ As S$_2$]: (*10*) 3.20, (*9*) 2.53, (*7*) 2.75, (*5*) 1.94 (*5*) 2.27, 2.08.
Pyrargyrite [Ag$_3$ Sb S$_3$]: (*10*) 2.79, 2.55, (*9*) 3.20, (*7*) 3.35. (*6*) 1.680, 1.600.
Polybasite [Ag$_{16}$ Sb$_2$ S$_{11}$]: (*10*) 2.99, (*9*) 3.18, (*8*) 2.87, (*6*) 2.52, 1.886, (*5*) 2.69.
Stephanite [Ag$_5$ Sb S$_4$]: (*10*) 3.05, 2.88, 2.56, (*6*) 2.42, 2.12, 1.86.

c) Lead sulphosalts
Sartorite [Pb As$_2$ S$_4$]; (*10*) 3.48, (*9*) 2.95, 2.76, (*7*) 2.62, 2.10, (*4*) 3.23.
Zinckenite [Pb Sb$_2$ S$_4$]: (*10*) 3.44, (*4*) 2.80, (*3*) 1.823, (*2*) 3.01, 2.13.
Jamesonite [Pb$_4$ Sb$_6$ S$_{13}$]: (*6*) 3.615, (*4*) 2.708, 2.014, (*3*) 2.798, 2.278, 2.227.
Boulangerite [Pb$_5$ Sb$_4$ S$_{11}$]: (*10*) 3.71, (*8*) 2.815, (*7*) 1.861, (*6*) 1.757, (*5*) 3.21, 2.337.

Halides
Flourite [Ca F$_2$]: (*10*) 1.928, (*7*) 3.15, 1.644, 1.113, (*5*) 1.251.
Cryolite [Na$_3$ Al, F$_6$]: (*10*) 1.939, (*9*) 2.75, (*8*) 2.33, 1.568, (*7*) 4.47, 3.87.
Halite [Na Cl]: (*10*) 1.990, 1.259, 1.149, 0.938,(*6*) 1.625.
Sylvite (KCl]: (*10*) 1.403, 1.045, (*9*) 2.225, 1.282, (*8*) 3.158, (*7*) 1.816.
Carnallite [KMg Cl$_3$, 6H$_2$O]: (*10*) 2.60. (*9*) 3.09. (*8*) 3.80, (*7*) 2.87, 2.04, (*6*) 1.90.
Chlorargyrite [Ag Cl]: (*10*) 2.80, 1.97, (*8*) 1.245, (*7*) 1.61, (*6*) 1.67.
Bromargyrite [Ag Br]: (*10*) 2.88, (*9*) 2.03, (*8*) 1.289, (*7*) 1.178, (*6*) 1.66.
Iodargyrite [Ag I]: (*6*) 2.32, 1.968, (*3*) 3.75, 1.704, 1.504.

Calomel [HgCl]: *(10)* 4.16, 3.17, 2.06, 1.97, 1.478, *(8)* 1.73.
Terlinguaite [Hg$_2$O Cl]: *(10)* 3.26, 2.506, *(6)* 2.814, 1.768, *(4)* 4.18, 1.966.

Oxides and Hydroxides
A. Metallic oxides and hydroxides
(a) Be – Al – Mg Association
Chrysoberyl [Be Al$_2$ O$_4$]: *(10)* 1.61, 2.08, *(8)* 3.24, 2.57, 2.33, 2.08, 1.362.
Spinel [Mg Al$_2$ O$_4$]: *(10)* 1.427, 1.053, *(9)* 2.44, 2.02, 1.552, *(7)* 1.231.
Gahnite [Zn Al$_2$ O$_4$]: *(10)* 2.44, *(8)* 1.232, *(7)* 2.85, *(6)* 1.48, *(4)* 1.65, 1.55.
Corundum [Al$_2$ O$_3$]: *(10)* 1.599, *(9)* 2.081, *(7)* 1.374, *(6)* 2.543, 1.401, *(5)* 1.738.
Diaspore [HAlO$_2$]: *(10)* 3.98, *(8)* 2.312, 1.629, *(7)* 2.124, *(4)* 2.558.
Boehmite [AlO(OH)]: *(10)* 6.23, 3.16, 2.34, 1.85, *(5)* 1.656, 1.447.
Gibbsite [Al(OH)$_3$]: *(10)* 4.83, *(6)* 4.337, *(5)* 2.451, 2.374, *(4)* 2.039, 1.797.
Periclase [MgO]: *(10)* 2.108, 1.485, *(9)* 2.43, *(8)* 1.213, *(6)* 1.268, 1.051.
Brucite [Mg (OH)$_2$]: *(10)* 2.361, 1.793, 1.189, *(8)* 4.75, *(7)* 1.372, 1.090.

(b) Fe - Mn - V Compounds
(i) Spinel Group
Magnetite [Fe Fe$_2$ O$_4$]: *(10)* 2.541, *(9)* 1.612, 1.479, *(8)* 1.091, *(7)* 2.098, *(6)* 2.99.
Coulsonite [(Fe, V)$_3$ O$_4$]: *(10)* 2.50, *(9)* 1.597, 1.466, *(8)* 2.07, *(6)* 2.93, *(4)* 4.79.
Magnesioferrite [MgFe$_2$O$_4$]: *(10)* 2.52, 1.479, *(8)* 1.609, *(7)* 2.09, *(6)* 2.17.
Franklinite [Zn Fe$_2$ O$_4$]' *(10)* 2.510, 1.610, 1.480, *(6)* 1.278, 1.091.
Jacobsite [Mn Fe$_2$ O$_4$]: *(10)* 1.507, *(8)* 2.57, 2.13, 1.640, *(7)* 3.01, 1.739.
Chromite [Fe Cr$_2$ O$_4$]: *(10)* 1.42, 2.42, *(9)* 2.84, 2.01, 1.64, 1.55.
Magnesiochromite [Mg Cr$_2$ O$_4$]: *(10)* 2.52, 1.479, *(8)* 1.609, *(7)* 2.09, *(6)* 2.17, 1.204.

(ii) Hematite Group
Wustite [FeO]: *(10)* 2.14, *(8)* 1.51, *(7)* 2.47, *(4)* 1.293, *(2)* 1.238.
Bunsenite [NiO]: *(9)* 2.085, 1.476, 1.261. 1.208, *(8)* 2.42, *(7)* 1.045.
Eskolaite [Cr$_2$O$_3$]: *(10)* 2.665, *(9)* 2.479, 1.675, *(7)* 3.63, *(4)* 2.174, 1.433.
Hematite [α-Fe$_2$O$_3$]: *(10)* 2.689, *(8)* 2.508, 1.688, 1.448 *(7)* 3.67, 2.198, 1.833, 1.481.
Maghemite [β-Fe$_2$O$_3$]: *(10)* 2.506, *(7)* 2.939, 1.471, *(6)* 2.079, 1.600, *(4)* 1.082.
Goethite (HFeO$_2$): *(10)* 4.15, *(7)* 2.433, *(6)* 2.674, 1.709, *(4)* 2.237, 2.175.
Lepidocrocite [α-FeO (OH)]: *(10)* 6.25, *(9)* 3.284, *(8)* 2.467, *(7)* 1.933, *(4)* 1.729, 1.521.

(iii) Bixbyite-Manganite Group
Manganosite [MnO]: *(10)* 2.218, 1.568, *(8)* 2.561, 1.337, 1.280.
Bixbyite [(Mn, Fe)$_2$O$_3$]: *(10)* 2.705, 1.655, *(7)* 1.409, *(5)* 1.074, *(2)* 1.836, 1.517.
Groutite [HMnO$_2$] *(10)* 4.17, *(7)* 2.798, 2.675, 2.369, *(6)* 2.303, 1.692.
Manganite [Mn OOH]: *(10)* 3.38, *(9)* 2.62, 1.661, *(8)* 1.130, *(7)* 2.263, *(6)* 2.406.

Pyrochroite [Mn (OH)$_2$]: (10) 2.44, 1.814, 1.229, (8) 4.62, 1.653, 1.560.

(iv) Braunite-Hausmannite Group
Braunite [Mn^{2+}(Mn^{4+}, Si)$_2$ O$_4$]: (10) 2.72, 1.66, (8) 2.15, 1.42, 1.08, (5) 2.36.
Hausmannite [Mn^{4+} Mn^{2+}O$_4$]: (10) 1.54, (8) 3.08, 2.75, 1.57, 1.44.
Hydrohausmannite [Mn^{4+} Mn^{3+} Mn^{2+} O$_3$ (OH)$_3$]: (10) 4.61, (8) 2.47, 1.92, (5) 3.14, 2.74, 1.53.
Hetaerolite [ZnMn$_2$O$_4$]: (10) 2.44, (8) 2.68, 1.51, 1.11, 1.07, 1.014.
Marokite (CaMn$_2$O$_4$): (10) 2.71, 2.22, (8) 2.29, 2.07.
Chalcophanite [ZnMn$_3$O$_7$. 3H$_2$O]: (10) 6.63, (9) 3.42, (8) 7.28, (7) 4.02, (6) 2.55, 2.22.
Crednerite [Cu$_2$Mn$_2$O$_5$]: (10) 2.49, (9) 2.84, 2.40, (6) 4.88, 3.66, 1.56.

(v) Pyrolusite-Psilomelane Group
Pyrolusite (β-MnO$_2$): (10) 3.096, (8) 1.618, (7) 1.550, (6) 2.396, (5) 1.356, (4) 2.108, 1.964.
Ramsdellite (γ-MnO$_2$): (10) 4.08, (9) 3.10, (8) 2.53, (7) 1.60, (6) 1.64, (5) 2.13, 1.88.
Nsutite (Mn^{4+} Mn^{2+} O$_{2-2x}$ (OH)$_{2x}$]: (10) 3.96, (8) 2.43, 2.13, 1.64.
Cryptomelane [K Mn$_8$ O$_{16}$]: (10) 3.105, (9) 6.86, (8) 4.89, (7) 2.387, (6) 1.528, (5) 2.145.
Hollandite [Ba Mn$_8$ O$_{16}$]: (10) 3.113, (6) 3.46, (5) 1.544, (3) 2.409, 2.198, 1.657.
Coronadite [Pb Mn$_8$ O$_{16}$]: (10) 3.104, (6) 3.47, (5) 1.542, (4) 2.400, 2.205, (2) 1.836.
Quenselite [Pb Mn O$_2$ (OH)]: (8) 3.04, 2.281, 1.518, (7) 2.711, 1.681, (6) 1.364.
Birnessite [(Na, Ca) Mn$_7$ O$_{14}$. 3H$_2$O]: (10) 7.27, (5) 2.44, 1.412, (3) 3.60.
Psilomelane [(Ba, H$_2$O)$_2$ Mn$_5$ O$_{10}$]: (10) 2.191, (8) 3.49, 2.402, (6) 1.403, (5) 1.816, (4) 2.842.
Todorokite [(Ba, Ca) Mn$_3$ O$_7$, H$_2$O]: (10) 9.65, (8) 4.81, (5) 1.331, (4) 3.20, 2.40, 2.216.

(c) Ti-Nb-Zr Compounds
1. Rutile Group
Rutile (TiO$_2$): (10) 1.689, (9) 3.24, (8) 2.488, 1.624, (7) 2.189, (6) 1.362.
Anatase (TiO$_2$): (10) 3.508, (9) 1.887, (7) 1.696, 1.662, 1.447, 1.261.
Brookite (TiO$_2$): (10) 3.22, 1.681, (8) 2.45, 1.356, (6) 3.46, 2.87, 1.619, (4) 2.17, 1.881.
Cassiterite (Sn O$_2$): (8) 1.758, (7) 1.079, (6) 2.631, 1.213, 1.059, (5) 3.333.

2. Columbite—Tantalite (Fe, Mn) (Nb, Ta)$_2$ O$_6$ Group
Columbite: (10) 2.968, (9) 1.457, (7) 3.66, 1.735, (6) 1.767
Tantalite: (10) 2.97, 1.72, 1.458, (7) 3.64, 1.77, 1.74, (6) 1.90

3. Ilmenite Group
Ilmenite FeTiO$_3$: (10) 2.74, (9) 2.53, (8) 1.72, (7) 1.504, 1.465, (6) 1.865.
Geikielite: (10) 2.74, (6) 1.865, (9) 1.715, (8) 2.23, (7) 2.55, (6) 1.86.

4. Perovskite Group
Perovskite, CaTiO$_3$: (10) 2.69, (8) 1.903, 1.552, (7) 1.017, (6) 1.345.

5. *Pyrochlore Group* (Ca, Na)$_2$ (Nb, Ta)$_2$O$_6$, (O, OH, F)
Pyrochlore: (*10*) 2.998, 1.835, (*9*) 1.565, (*8*) 1.192, 1.162, (*7*) 1.059
Betafite: (*10*) 1.802, 1.535, (*8*) 2.91, 1.17; 1.142, 1.043.
Microlite: (*10*) 2.98, (*9*) 1.836, 1.563, (*4*) 2.58, 1.194, 1.165
Samarskite: (*10*) 3.07, 2.92, (*4*) 2.59, 1.84, 1.502, (*3*) 3.68

(d) **Zn-Cu-Pb(U) Association**

1. Zincite-Tenorite Group
Zincite ZnO: (*10*) 2.459, 1.623, 1.491, 1.373, (*9*) 2.79, (*8*) 1.90
Montroydite, CdO: (*10*) 3.848, (*5*) 2.864 (*3*) 5.26, 2.402, 1.762, 1.497.
Tenorite CuO: (*10*) 2.51, (*9*) 2.307, (*7*) 1.852, 1.5, 1.37, (*6*) 1.258
Paramelaconite, CuO: (*10*) 2.49, 1.575, 1.251, (*6*) 2.05, 1.449, 1.403.
Cuprite, Cu$_2$O: (*10*) 2.456, 1.280, (*9*) 1.505, (*8*) 2.130, (*4*) 1.226
Delafossite, CuFeO$_2$: (*10*) 2.51, (*8*) 2.85, 2.23, 1.666, 1.511, 1.436

2. Plattnerite Group
Litharge PbO: (*10*) 3.11. (*8*) 2.80, 1.68, (*7*) 1.87 (*6*) 1.99, 1.55
Plumboferrite, PbFe$_4$O$_7$: (*10*) 2.947, 2.816, 2.648, (*5*) 3.93, 1.684
Magnetoplumbite, PbFe$_{12}$O$_{19}$: (8) 2.76, 2.62, 2.23, 1.66, 1.47, (*6*) 1.38

3. Uraninite—Becquerlite Group
Uraninite, UO$_2$: (*10*) 3.163, 1.934, 1.654, (*8*) 1.255, (*7*) 1.224, (*6*) 1.117
Pitchblende, UO$_2$ with REE: (*10*) 3.09, 1.634, (*9*) 1.9, 1.041, (*8*) 1.245, 1.210, 1.108
Thorianite, ThO$_2$: (*10*) 3.216, 1.964, 1.675, (*8*) 2.776, 2.77, 1.074

4. Curite Group
Curite, 3PbO. 8UO$_3$. 4H$_2$O: (*10*) 6.28, (*9*) 3.97, (*8*), 3.14, (*6*) 2.55, (*5*) 2.10, 1.74

B. Semi-metallic and non-metallic oxides
Senarmontite Group
Arsenolite, As$_2$O$_3$: (*10*) 3.10, 1.068, (*9*) 2.534, 1.951, (*8*) 1.665, 1.547
Claudetite, As$_2$O$_3$: (*10*) 3.19, (*9*) 1.955, 1.550 (*8*) 2.766, 2.545, 1.070
Senarmontite, Sb$_2$O$_3$: (*10*) 1.673, (*9*) 3.21, 1.96, 1.071, (*8*) 1.274, 1.243
Valentinite, Sb$_2$O$_3$: (*10*) 3.08, (*8*) 10.85, 1.792, (*7*) 1.91, 1.509, (*6*) 1.174

Class V Silicates
A. Nesosilicates
(a) Be-Al-Mg combination
1. Beryl Group
Beryl, Be$_3$Al$_2$Si$_6$O$_{18}$: (*10*) 3.238, 2.874, (*8*) 2.146, 1.989, 1.737, 1.515, 1.430, 1.276
Cordierite, Mg$_2$Al$_3$[Si$_5$AlO$_{18}$[: (*10*) 3.00, (*9*) 3.34, (*8*) 8.29, (*7*) 1.685 (*6*) 4.03

2. Andalusite-Topaz Group
Mullite, Al$_6$Si$_2$O$_{13}$: (*10*) 3.42, 3.38, 2.21, (*9*) 2.69, 2.55, (*8*) 2.12
Yoderite, Al$_3$MgSi$_2$O$_8$ (OH): (*10*) 3.50, (*8*) 3.03, (*6*) 2.61, 2.00, 1.82, (*4*) 2.46

Dumortierite, [Al Fe)$_7$BSi$_3$O$_{18}$: (*10*) 5.85, (*8*) 2.09 (*6*) 5.06 3.43, 3.22, 2.91
Andalusite, Al$_2$SiO$_5$: (*10*) 4.53, 2.17, 1.46, (*9*) 2.76, 2.26, (*8*) 3.96
Sillimanite, Al$_2$SiO$_5$: (*10*) 3.32, (*8*) 2.16, 1.498, (*7*) 2.49, 1.677, 1.579, 1.429, 1.323
Kyanite, Al$_2$SiO$_5$: (*10*) 1.95, 1.38, (*8*) 3.33, 3.14, 2.37, (*6*) 1.76
Topaz, Al$_2$SiO$_4$ (F, OH)$_2$): (*10*) 2.96, 1.403, 1.384, (*9*) 3.20, 2.07, 1.65, 1.343
Staurolite, Al$_4$FeSi$_2$O$_{10}$ (OH)$_2$: (*10*) 2.38, 1.396, (*9*) 1.974, (*8*) 3.01, (*5*) 1.516, (*3*) 1.539

3. Olivine Group

Forsterite, Mg$_2$SiO$_4$: (*10*) 2.441, 1.741, (*9*) 2.25, 1.475, 1.347, (*7*) 3.875
Olivine, (Mg, Fe)$_2$SiO$_4$: (*10*) 2.49, (*8*) 2.41, 1.734, (*7*) 2.24, 1.498, 1.468
Fayalite, Fe$_2$SiO$_4$: (*10*) 2.85, (*7*) 1.755, (*3*) 3.71, (*2*) 2.03, 1.508, 1.383
Tephroite, Mn$_2$SiO$_4$: (*10*) 1.81, (*8*) 2.86, 2.55, 1.077, 1.063, (*6*) 1.54
Monticellite: (*10*) 2.65, (*9*) 1.811, 1.59, (*7*) 2.86, (*6*) 1.386, (*5*) 2.39
Kirschsteinite: (*10*) 2.95, (*9*) 2.68, (*8*) 2.60, (*7*) 3.66, (*6*) 1.83, (*4*) 2.41

4. Humite Group

Norbergite, Mg$_2$SiO$_4$ · Mg(F, OH)$_2$: (10) 1.744, (8) 3.08, 2.26, 1.492, (7) 2.66, (*6*) 1.346

(b) Ti-Nb-Zr Combination

1. Zircon Group

Zircon, ZrSiO$_4$: (*10*) 3.29, (*9*) 1.710, (*7*) 4.41 (*6*) 1.644
Thorite, ThSiO$_4$: (*4*) 2.583, (*3*) 2.855, 1.665, (*2*) 1.882, 1.579
Uranothorite, (Jh, U) SiO$_4$: (*5*) 3.23, (*4*) 1.686, (*3*) 1.986, 1.487, 1.285, 1.128
Huttonite, ThSiO$_4$: (*8*) 3.09, (*7*) 2.89, (*6*) 4.23, 3.29, (*5*) 4.71, (*4*) 3.53.

2. Sphene Group

Sphene, CaTiSiO$_5$: (*10*) 2.59, (*9*) 3.21, (*8*) 1.647, 1.498, (*7*) 2.97, 1.135, 1.110

(c) Ca-Mn-Na Combination

3. Melilite Group

Melilite, Ca$_2$ (Al, Mg) (Si, Al)$_2$O$_7$: (*10*) 2.858, (*8*) 1.762, 1.758, (*7*) 2.452, 1.513, 1.434
Gehlenite, Ca$_2$Mg (Si$_2$O$_7$)]: (*10*) 2.847, (*7*) 2.432, 2.407, 2.396, 2.297, 2.287
Akermanite, Ca$_2$Fe (Si$_2$O$_7$)]: (*10*) 2.874, (*8*) 1.763, (*7*) 3.088, 2.480, 2.039, 1.385

4. Epidote Group

Epidote, Ca$_2$Al$_2$FeSi$_3$O$_{12}$(OH): (*10*) 2.90, (*8*) 2.40, 1.64 (*7*) 1.88, (*6*) 1.409
Tuhualite, Na$_4$Fe$_2$ Fe$_3$ Si$_{15}$ O$_{37}$ (OH)$_3$: (*10*) 7.16, (*8*) 2.76
Cherkinite, Ce$_2$FeTi$_2$Si$_2$O$_{12}$: (*10*) 2.702, (*8*) 3.156, 2.163, 1.960, (*7*) 2.850, (*5*) 2.600
Fenaksite, (K, Na)$_4$ Fe$_2$Si$_8$O$_{20}$ (OH, F): (*10*) 3.03, (*7*) 3.55, 3.44, 2.46, (*6*) 2.88, 1.752
Pumpellyite, Ca$_4$Al$_6$Si$_6$O$_{23}$(OH)$_3$ · 2H$_2$O: (*10*) 2.90, (*8*) 3.79, (*6*) 2.74, (*5*) 2.64, 2.45, 2.21

Howieite, Na(Fe, Mn)$_{10}$ (Fe, Al)$_2$ Si$_{12}$ (O, OH)$_{44}$: (s) 9.13, 3.25, 7.90 2.205, (ms) 2.775
Zussmanite, K(Fe, Mg, Mn)$_{13}$ Al$_2$ Si$_{17}$ (O, OH)$_{56}$: (s) 9.56, 4.77, 3.19, 2.516
Deerite, Fe$_{13}$^{II} Fe$_7$^{III} Si$_{13}$ (O, OH)$_{54}$: 9.00-2.99, 2.5-2.63
Julgoldite, Ca$_2$Fe^{II}Fe^{III}$_2$ (SiO$_4$) (Si$_2$O$_7$) (OH)$_2$ H$_2$O: (*10*) 2.95, (*8*) 3.84, (*7*) 4.80, (*6*) 2.568, 2.776
Allanite (Orthite), (Ca, Ce)$_2$ (Al, Fe)$_3$ Si$_3$O$_{12}$ (O, OH): (*10*) 2.94, (*8*) 2.74, (*6*) 3.57, 2.65, 1.65, (*4*) 2.14
Prehnite, Ca$_2$Al$_2$Si$_3$O$_{10}$ (OH)$_2$: (*10*) 3.49, 3.28, 3.05, 2.54, (*7*) 2.81, 1.764

5. *Garnet Group*
Pyrope, Mg$_3$Al$_2$Si$_3$O$_{12}$: (*10*) 1.542, (*9*) 2.583, 1.598, (*8*) 2.886, 1.070, 1.052
Almandine, Fe$_3$Al$_2$Si$_3$O$_{12}$: (*10*) 2. 589, 2.539, 1.071, (*9*) 1.595, 1.259, 1.054
Spessartine, Mn$_3$Al$_2$Si$_3$O$_{12}$: (*10*) 2.60, 1.553, (*9*) 1.61, 1.079, (*8*) 1.262, 1.060
Grossular, Ca$_3$Al$_2$Si$_3$O$_{12}$: (*10*) 2.66, 1.581, 1.101, (*9*) 1.639, 1.291, 1.082
Andradite, Ca$_3$Fe$_2$Si$_3$O$_{12}$: (*10*) 2.707, 1.611, (*8*) 3.026, (*7*) 1.659, 1.099, 1.065
Uvarovite, Ca$_3$ Cr$_2$ Si$_3$ O$_{12}$: (*10*) 1.604, (*9*) 3.020, 2.691, (*8*) 1.665, 1.094, 1.059
Idocrase (Vesuvianite), Ca$_{10}$ (Mg, Fe)$_2$ Al$_4$ [(OH)$_4$/(SiO$_4$)$_5$/(Si$_2$O$_7$)$_2$]: (*10*) 2.75, (*9*) 2.60, (*8*) 2.46, 1.625, (*4*) 1.108, (*3*) 1.296

6. *Tourmaline Group*
Tourmaline, (Na,Ca) (Mg,Fe,Li)$_3$ Al$_6$B$_3$Si$_6$O$_{27}$(OH)$_4$
Elbaite, Na (Li, Al)$_3$ Al$_6$ B$_3$ Si$_6$ O$_{27}$ (OH)$_4$: (*10*) 2.552, (*9*) 2.022, 1.439, 1.393, 1.019, (*8*) 2.923, 1.900, 1.639, 1.344, 1.298, 1.018
Schorl, NaFe$_3$Al$_6$B$_3$Si$_6$O$_7$ (OH)$_4$: (*10*) 2.59, (*8*) 6.5, 3.48, 2.98, 1.66, 1.033, 1.022

7. *Axinite-Danburite Group*
Axinite, Ca$_2$(Fe, Mn) Al$_2$ BSi$_4$O$_{15}$ (OH): (*10*) 2.79, (*8*) 3.45, 3.13, (*6*) 2.99, 2.41, 1.99
Harkerite, Ca (Mg, Al) (Si, B) (O, OH)$_4$. CaCO$_3$: (*10*) 2.61, (*8*) 1.84, (*7*) 2.13, 1.51, (*5*) 5.22, (*4*) 2.84

(d) Zn-Cu-Pb (U) Association
Hemimorphite, Zn$_4$Si$_2$O$_7$ (OH)$_2$.H$_2$O: (*10*) 3.276, 3.080, 1.442, 1.384, (*9*) 2.395, 1.517, (*8*) 1.302
Dioptase, Cu$_6$Si$_6$O$_{18}$, 6H$_2$O: (*10*) 7.5, (*8*) 2.62, (*6*) 4.16, 2.14, 2.05, 1.72
Papagoite, (Ca, Cu, Al) Si$_2$O$_6$ (OH)$_3$: (*10*) 2.874, (*9*) 4.29, 2.204, (*8*) 3.44, 2.795, (*7*) 1.667
Shattuckite, 2CuSiO$_3$, H$_2$O: (*10*) 4.43, (*8*) 3.31, (*7*) 4.95, 3.50, (*6*) 2.36
Coffinite, USiO$_4$(OH)$_4$: (*10*) 3.51, (*9*) 4.67, 2.665, 1.183, (*8*) 2.189, (*7*) 2.002
Uranophane, Ca(H$_3$O)$_2$U$_2$O$_4$(SiO$_4$)$_2$.3H$_2$O: (*10*) 7.89, 3.948, (*9*) 2.981, 2.898, (*8*) 4.78, 3.20.
Ca-Ursilite, Ca$_2$U$_2$O$_4$Si$_5$O$_{14}$. 9H$_2$O: (*10*) 3.37, 3.02, (*9*) 4.56, 1.828, 1.120, (*8*) 1.562
Mg-Ursilite, Mg$_2$U$_2$O$_4$Si$_5$O$_{14}$. 9H$_2$O: (*10*) 4.98, 3.06, (*9*) 2.30, 2.07 (*8*) 4.58, 3.87

Sklodowskite, MgU$_2$O$_4$Si$_2$O$_7$. 6H$_2$O: (*10*) 8.42, (*8*) 4.19, (*7*) 3.27, (*6*) 3.52, 3.00

Cuprosklodowskite, CuU$_2$O$_4$Si$_2$O$_7$, 6H$_2$O: (*10*) 8.18, (*9*) 4.09, (*8*) 2.97, (*7*) 4.82, (*6*) 3.52, 2.21

Orlite, Pb$_3$U$_3$O$_6$Si$_4$O$_{14}$. 6H$_2$O: (*10*)3.23, (*7*) 1.678, (*5*) 6.36, 1.967, 1.849, (*4*) 2.897

Bismutoferrite Group

Bismutoferrite, BiFe$_2$ (SiO$_4$)$_2$ (OH): (*10*) 7.63, 3.87, (*7*) 2.90, (*5*) 3.18 (*4*) 3.58, 2.59

Chapmanite, Sb Fe$_2$ (SiO$_4$)$_2$(OH): (*10*) 7.63, 3.58, (*9*) 3.88, 3.19, (*7*) 2.90, 2.59

B. Inosilicates

1. Pyroxene Group

Enstatite, Mg$_2$Si$_2$O$_6$: (*10*) 3.158, (*7*) 2.864, 2.526, 2.472, 2.105, 1.483

Clinoenstatite, Mg$_2$Si$_2$O$_6$: (*10*) 3.149, 2.859, 1.519, 1.482, 1.468, 1.374

Hypersthene, (Mg, Fe)$_2$Si$_2$O$_6$: (*10*) 3.20, (*8*) 2.890, 1.486, (*6*) 1.599, 1.389, (*5*) 1.304

Diopside, CaMgSi$_2$O$_6$: (*10*) 3.00, 2.523, 1.616, 1.071, (*9*) 1.418, (*8*) 1.322, (*7*) 1.280

Augite, Ca(Mg, Fe, Al) (Si, Al)$_2$) O$_6$: (*10*) 2.98, 2.522, 1.619, 1.412, 1.071, (*8*) 1.324, (*7*) 1.277

Aegirine, NaFeSi$_2$O$_6$: (*10*) 3.012, 2.545, (*7*) 2.916, (*6*) 2.483, (*4*) 2.033, 1.562, 1.385

Jadeite, NaAlSi$_2$O$_6$: (*10*) 2.835, (*7*) 2.919, (*4*) 2.533 (*3*) 2.495, 2.416, 1.966

Spodumene, LiAlSi$_2$O$_6$: (*10*) 2.921, 2.790, (*9*) 1.604, 1.565, 1.461, (*8*) 2.445, 1.212.

2. Amphibole Group

Anthophyllite, (Mg, Fe)$_7$ Si$_8$ O$_{22}$ (OH)$_2$: (*10*) 8.25, 3.23, 2.84, (*9*) 9.1, (*8*) 2.75, 1.610, (*7*) 1.542

Gedrite, (Mg, Fe)$_6$ (Al, Fe) (Si, Al)$_8$ O$_{22}$ (OH)$_2$: (*10*) 3.06, (*8*) 8.27, (*7*) 3.23, (*5*) 8.97, (*4*) 4.48, 2.50

Tremolite, Ca$_2$ Mg$_5$ Si$_8$ O$_{22}$ (OH)$_2$: (*10*) 1.438, (*9*) 1.047 (*8*) 2.71, 1.503, (*6*) 3.12, (*5*) 1.582, 1.293.

Actinolite, Ca$_2$ (Mg, Fe)$_5$ Si$_8$ O$_{22}$ (OH)$_2$: (*10*) 2.705, 1.432, 1.046, (*9*) 3.314, 1.507, (*8*) 2.541, 2.155, 1.642, 1.576, 1.359, 1.292.

Hornblende, NaCa$_2$ (Mg, Fe Al)$_3$ (Si, Al)$_8$ O$_{22}$ (OH)$_2$: (*10*) 2.711, 1.436, 1.049, (*9*) 3.15, 1.504, (*8*) 2.155, 1.645, 1.290, 1.079

Riebeckite, Na$_2$Fe$_3$ Fe$_2$ Si$_8$ O$_{22}$ (OH)$_2$: (*10*) 3.13, (*9*) 2.72, (*8*) 8.42, 4.51 (*6*) 1.661, 1.619, 1.504

Arfvedsonite, Na$_3$ Fe$_4$ Fe Si$_8$ O$_{22}$ (OH)$_2$: (*10*) 3.18, 2.75, 1.450, (*9*) 1.055, (*8*) 1.302, (*7*) 1.632, 1.204

Holmquistite, Li$_2$ (Mg, Fe)$_3$ Al$_2$ Si$_8$ O$_{22}$ (OH)$_2$: (*10*) 8.16, 2.99, (*9*) 4.40, (*8*) 3.61 (*6*) 3.83, 2.53, 2.27

3. Wollastonite-Rhodonite Group
Wollastonite, $Ca_3Si_3O_9$: (*10*) 2.963, (*8*) 3.30, (*7*) 1.705, (*6*) 2.165, 1.594, 1.471, 1.355
Xonotlite, $Ca_6 [Si_6 O_{17}]$ O: (*10*) 3.06, (*9*) 4.24, 3.89, 3.234, 2.029, 1.938
Rhodonite, $MnSiO_3$: (*10*) 2.938, (*9*) 2.968. (*8*) 2.755, 2.595, (*6*) 3.08, 1.554, 1.427.

4. Sepiolite-Palygorskite Group
Sepiolite, $Mg_8 H_6 Si_2 O_{30} (OH)_{10} \cdot 6H_2O$: (*10*) 12.3, 4.29, 2.55, (*7*) 3.75, (*6*) 3.97, 3.35, 3.18, 2.61
Palygorskite, $Mg_3H_2 Si_8 O_{22} (H_2O) \cdot 2H_2O$: (*10*) 10.2, 4.3, 3.25, 2.55, (*4*) 1.25, 1.49

C. Phyllosilicates
1. Talc-Pyrophyllite Group
Talc, $Mg_3Si_4O_{10} (OH)_2$: (*10*) 9.25, 3.104, 1.525, (*6*) 4.64, 2.471, 1.383
Minnesotaite, $Fe_3Si_4O_{10} (OH)_2$: (*10*) 9.53, (*5*) 3.177, (*3*) 2.524, (*2*) 2.005, 1.567, 1.329
Calciotale, $CaMg_2Si_4O_{10} (OH)_2$: (*9*) 9.25, 3.079, (*6*) 1.518, (*5*) 2.469, (*4*) 4.503, (*3*) 1.677
Pyrophyllite, $Al_2Si_4O_{10} (OH)_2$: (*10*) 3.045, (*9*) 1.489, 1.381, 1.365, (*8*) 2.403, (*7*) 1.239

2. Biotite Group
Biotite, $K (Mg, Fe)_3 [AlSi_3O_{10}] (OH, F)_2$: (*10*) 10.00, 3.34, (*8*) 2.63, 1.541, (*6*) 2.44, 1.672, (*4*) 1.363.
Phlogopite: (*10*) 10.00, 3.35, (*9*) 1.533, (*8*) 2.62, 2.435, 2.175, 1.998, 1.669
Lepidomelane: (*10*) 10.1, 3.36, (*8*) 2.65, 2.452, 2.186, 2.006, 1.676, 1.548
Muscovite, $KAl_2[AlSi_3O_{10}] (OH)_2$
$2M_1$ Muscovite: (*10*) 10.04, (*8*) 2.56, (*5*) 5.02, 4.48, 4.46
1 M Muscovite: (*10*) 10.08, 3.36, (*8*) 4.49, 2.56, (*5*) 3.07, 2.58
3 T Muscovite: (*10*) 9.99, 3.33, (*8*) 4.99, (*5*) 2.56
Lepidolite, $KLi_2 Al [Si_4O_{10}] (F, OH)_2$: (*10*) 3.36, 2.58, 2.012, 1.492, (*8*) 4.47, 2.84, 1.695, 1.634
Zinnwaldite, $KFe_2Al [Si_4O_{10}] (F, O, H)_2$L (*10*) 10.00, 3.34, (*8*) 2.62, 2,432, 2.184, 1.995, 1.672
Tainiolite, $KLiMg_2 [Si_4O_{10}]F_2$: (*10*) 1.99, (*8*) 3.33, 1.501, (*7*) 2.595, (*5*) 3.65, 1.65, 1.303
Spodiophyllite, $K (Mg, Fe)_3 [Si_4O_{10}] (F, OH)_2$: (*10*) 3.32, (*8*) 10.09, 2.60, 1.966, 1.520, (*7*) 1.659

3. Illite (hydromica) Group
Illite, $(K, H_2O)Al_2 [(Al, Si)Si_3O_{10}](OH)_2$: (*10*) 9.98, (*8*) 4.47, 2.56, 1.50 (*6*) 3.31, 1.98, 1.29
Glauconite, $(K, H_2O) (Fe, Mg, Al)_2 [Al, Si) Si_3O_{10}] (OH)_2$: (*10*) 2.58, 1.516, (*7*) 3.31, (*5*) 3.67, 2.40, 1.656.
Stilpnomelane, $(K, H_2O) (Fe, Fe, Mg, Al)_3 (Si_4O_{10}) (OH)_2 \cdot 2H_2O$: (*10*) 11.9, (*6*) 4.045, (*5*) 3.036, 2.549, (*4*) 1.576, 1.561

Brammelite: (*10*) 10.2, 4.4, 3.2, 1.49, (*9*) 2.54, (*7*) 1.64.
Hydrobiotite: (*10*) 2.596, 1.533, (*8*) 11.4, (*6*) 1.668, 1.322, (*4*) 2.016

4. Brittle Mica
Margarite, CaAl$_2$ [Al$_2$Si$_2$O$_{10}$](OH)$_2$): (*10*) 3.168, (*5*) 4.902, (*3*) 3.51, (*2*) 2.769, 2.402, 1.608
Chloritoid, (Mg, Fe)$_2$Al$_4$Si$_2$O$_{10}$(OH)$_4$: (*10*) 4.46, (*5*) 2.96, 2.36, 2.30, 1.781, (*4*) 2.62.

5. Chlorite Group (Mg, Fe, Al, Cr. Ni, Mn)$_3$ (Si, Al)$_4$O$_{10}$ (OH)$_2$. (Mg, Fe, Mn)$_3$ (OH)$_6$.
Chamosite: (*10*) 7.04, 3.513, (*9*) 2.796, 2.514, (*7*) 1.551, (*6*) 2.137
Corundophilite: (*10*) 7.03, 3.51, (*9*) 4.68, (*5*) 2.59, 2.54, 2.00
Clinochlore: (*10*) 3.53, 1.535, 1.393, (*9*) 1.998, 1.564, (*8*) 1.220
Daphnite: (*9*) 6.76, (*7*) 1.55 (*5*) 3.47, 2.00, (*4*) 2.55, 2.38
Diabantite: (*10*) 15.0, (*9*) 7.15, (*7*) 3.58, 1.545, (*6*) 4.62, 2.47
Kammererite: (*10*) 15.0, 7.1, 4.7, 3.59, (*9*) 1.546, (*7*) 2.51
Pennine: (*10*) 7.17, 4.78, 3.585, (*6*) 14.3, 2.867, (*4*) 1.579
Preudothuringite: (*10*) 3.505, (*8*) 6.7, (*6*) 1.553, (*5*) 4.69, (*4*) 1.539, 1.518
Ripidolite: (*10*) 2.589, 1.556, (*8*) 7.05, (*7*) 3.536, 2.013, (*6*) 4.68
Thuringite: (*10*) 6.8, 3.48, 1.552 (*9*) 2.59, 2.00 (*7*) 2.26

6. Montmorillonite—Vermiculite Group
Montmorillonite, Al$_2$ [Si$_4$O$_{10}$] (OH)$_2$. nH$_2$O: (*10*) 15.3, 4.50, 3.07, 2.55, 1.497
Beidellite, Al$_2$ [AlSi$_3$O$_{10}$] (OH)$_2$. nH$_2$O: (*10*) 15.1, 4.45, 3.02, 2.60, 2.49, 1.488
Nontronite, Fe$_2^{3+}$ [Si$_4$O$_{10}$(OH)$_2$. n H$_2$O: (*10*) 16.6, 1.519, (*7*) 4.52, 1.310, (*6*) 3.066, 1.258
Saponite, Mg$_3$ (Si$_4$O$_{10}$) (OH)$_2$ n H$_2$O: (*10*) 4.57, 1.527, (*8*) 15.8, 2.65, 1.32, 1.301
Vermiculite, Mg$_3$ [AlSi$_3$O$_{10}$] (OH)$_2$. n H$_2$O: (*10*) 13.7, (*9*) 1.533, (*8*) 2.39, (*6*) 2.55, (*4*) 2.65, 1.321

7. Antigorite-Kaolinite Group
Antigorite-Chrysotile, Mg$_6$[Si$_4$O$_{10}$] (OH)$_8$
Antigorite: (*10*) 2.558, (*9*) 7.36, 3.641, (*8*) 1.583, 1.553, (*7*) 2.186
Chrysotile: (*10*) 3.66, 1.522, (*9*) 7.36, (*8*) 2.571, 2.424, (*7*) 1.300
Garnierite: (*10*) 9.8, (*8*) 1.552, (*6*) 2.65, 2.40, (*4*) 4.5, 2.86
Greenalite: (*10*) 2.57, (*8*) 7.12, 3.56, (*6*) 1.593, (*4*) 2.184, 1.553
Amosite: (*10*) 3.47, (*8*) 6.93, (*7*) 1.920, (*6*) 2.467, (*5*) 1.529, (*4*) 1.339.
Kaolinite: Al$_4$[Si$_4$O$_{10}$] (OH)$_3$
Kaolinite: (*10*) 7.14, 3.57, 1.487, (*8*) 2.338, 1.126, (*7*) 1.283
Dickite: (*10*) 3.592, 2.345, 1.666, 1.322, (*8*) 1.988, (*6*) 1.192, (*1*) 7.17
Nacrite: (*10*) 7.15, 3.59, 2.416, (*8*) 4.42, 1.489, (*6*) 1.372
Anauxite: (*10*) 4.6, 4.08, (*8*) 7.4, 3.59, 1.502, (*6*) 1.67
Donbassite: (*10*) 4.80, 3.536, (*7*) 2.334, (*6*) 2.834, 1.662, 1.496
Halloysite: (*10*) 10.4, 4.41, 1.483, (*7*) 9.7, 2.57, 1.236.
Ferrihalloysite: (*10*) 4.45, 1.488, (*8*) 1.695, (*7*) 3.68, 2.565. (*5*) 7.17

8. Okenite—Tobermorite Group
Okenite, $Ca_3Si_6O_{15} \cdot 6H_2O$: (*10*) 8.87, (*9*) 4.57, 3.56, 3.09, 2.92, (*8*) 1.80
Tobermorite, $Ca_{10}Si_{12}O_{31}(OH)_6 \cdot 8H_2O$: (*10*) 11.3, 2.83, (*8*) 5.55, 1.85

9. Apophyllite—Krauskopfite Group
Apophyllite, $KCa_4[Si_4O_{10}]_2F \cdot 8H_2O$: (*10*) 2.495, (*9*) 7.77, 2.97 (*8*) 3.915, 3.596, (*7*) 4.52
Krauskopfite, $BaSi_2O_5 \cdot 3H_2O$: (*10*) 3.84, (*5*) 6.36, 5.34, (*4*) 3.01

D. Tectosilicates
(a) Without additional anions

1. Quartz Group
Quartz (low), SiO_2: (*10*) 3.343, (*8*) 4.25, 1.818, (*7*) 1.541, (*6*) 2.456, 2.281, 2.127, (*5*) 2.236, 1.979.
Tridymite, SiO_2: (*10*) 4.39, 4.12, (*9*) 3.73, (*7*) 2.49, 1.69, 1.528
Cristobalite (low) SiO_2: (*10*) 4.03, (*8*) 2.481, (*7*) 2.834, 1.876, (*6*) 1.924, 1.687, 1.608, 1.530
Cristobalite (high) SiO_2: (*10*) 4.15, (*9*) 2.53, (*7*) 1.641, (*6*) 1.460 (*5*) 2.07, 1.21
Coesite, SiO_2: (*9*) 3.13, 3.08, 3.47, 3.42, 1.709, 1.692, 1.648, 1.579, 1.279, 1.181
Stishovite, SiO_2: (*10*) 2.959, (*5*) 1.53, (*4*) 1.981, (*3*) 1.235, (*2*) 2.246, 1.478

2. Felspar Group
Plagioclase, $(Ca, Na)(Al, Si)(Al, Si)_2O_8$
Albite (0.5% An): (*10*) 3.179, (*7*) 4.016, 3.660, 3.206, 2.952, (*6*) 3.767, 2.952, (*5*) 2.561
Oligoclase (16.5% An): (*10*) 3.176, (*7*) 4.018, 3.198, (*6*) 3.754, (*5*) 3.651, 2.975, 2.545
Andesine (38% An): (*10*) 3.199, (*8*) 3.169, (*7*) 4.028, (*6*) 3.741, 3.640, 2.528, (*5*) 2.999
Labradorite (64.5% An): (*10*) 3.198, (*8*) 3.170, (*7*) 4.036, 3.747, 3.222, (*6*) 2.522, (*4*) 3.013
Bytownite (81.5% An): (*10*) 3.191, (*7*) 4.021, 3.743, 3.160, (*6*) 3.609, 3.231, 2.513, (*5*) 3.016
Anorthite (93% An): (*10*) 3.197, (*7*) 3.249, 3.164, (*6*) 3.033, 3.611, 2.519, (*5*) 3.030 (*4*) 3.740
Orthoclase, $KAlSi_3O_8$
Orthoclase: (*10*) 3.29, (*9*) 1.81, (*7*) 4.25, 2.98, 2.90, 2.16
Adularia: (*10*) 3.313 (*8*) 3.227, 1.792, (*6*) 4.21, 2.560, 1.490
Microcline: (*10*) 3.22, (*8*) 1.80, (*7*) 2.16, (*6*) 4.18, 1.99, (*5*) 1.459

3. Nepheline Group: $(Na, K)AlSiO_4$
Nepheline, $NaAlSiO_4$: (*10*) 3.02, (*9*) 3.86, (*8*) 4.20, 3.28, (*7*) 2.35, (*6*) 1.563
Kaliophilite, $KAlSiO_4$: (*10*) 3.09, (*8*) 2.59, (*6*) 1.65, (*5*) 2.13, (*4*) 1.93, 1.76
Kalsilite, $KAlSiO_4$: (*10*) 3.97, 3.11, 2.59, (*5*) 1.66, 1.23, 1.19

4. Leucite Group
Leucite, $KAlSi_2O_6$: (*10*) 3.252, (*9*) 3.432, 1.659, (*8*) 1.245, (*7*) 5.3, 2.809.
Pollucite, $(Cs, Na)AlSi_2O_6(H_2O)_{<1}$ (*10*) 3.43, (*8*) 2.925, (*7*) 1.740 (*8*) 2.925, (*7*) 1.740, (*6*) 2.424, (*5*) 1.863 (*4*) 1.356.

(b) With additional anions
1. Scapolite Group
Scapolite, (Na, Ca)$_4$ [Al(Al, Si)Si$_2$O$_8$] (Cl, CO$_3$): (*10*) 3.215, 1.797, (*9*) 3.304, 2.565, (*7*) 2.988, (*5*) 1.277
Cancrinite, (Na, K, Ca)$_{3-4}$ (Al, Si)$_6$O$_{12}$ (SO$_4$, CO$_3$, C$_5$), n H$_2$O: (*10*) 3.16, (*7*) 3.39, 2.46, 1.741 (*6*) 2.394, 2.218

2. Sodalite-Helvine Group
Sodalite, Na$_8$[AlSiO$_4$]$_6$Cl$_2$: (*10*) 3.68, (*9*) 2.60, (*8*) 6.38, 2.13, (*7*) 1.594, 1.466
Nosean, Na$_8$(AlSiO$_4$)$_6$SO$_4$: (*10*) 3.69, (*6*) 2.61, 2.13, (*5*) 2.86, 1.60
Hauyne, Na$_6$Ca$_2$[AlSiO$_4$]$_6$SO$_4$: (*10*) 3.72, (*5*) 2.63, (*3*) 2.88, 2.149, 1.788, 1.612
Lazurite, (Na, Ca)$_8$ [AlSiO$_4$]$_6$(SO$_4$, Cl, S)$_2$: (*10*) 3.74, 2.99, (*9*) 2.53, 1.545, (*7*) 1.422, 1.371
Helvine, (Mn, Fe, Zn)$_8$ [BeSiO$_4$]$_6$S$_2$: (*10*) 3.37, 1.958, (*9*) 1.467, 1.382, 1.279, 1.127

(c) Zeolites
1. Stilbite Group
Stilbite, (Ca, Na$_2$) Al$_2$Si$_7$O$_{18}$. 7H$_2$O: (*10*) 8.81, 4.025, (*8*) 2.993, (*6*) 4.36, (*4*) 3.405, (*3*) 2.761
Heulandite, CaAl$_2$Si$_7$O$_{18}$. 6H$_2$O: (*10*) 3.969, (*9*) 2.963, (*8*) 8.81, (*7*) 5.114 (*6*) 4.670, (*5*) 3.425, (*4*) 1.966

2. Chabazite Group
Chabazite (Ca, Na$_2$) Al$_2$Si$_4$O$_{12}$. 6H$_2$O: (*10*) 9.3, 2.93, (*8*) 4.35, 1.81, (*6*) 3.62, 3.24.
Levyne, CaAl$_2$Si$_4$O$_{12}$. 6H$_2$O: (*8*) 4.04, 2.78, (*7*) 8.12, (*6*) 5.13, 4.26.

3. Analcime Group
Analcime, Na$_2$Al$_2$Si$_4$O$_{12}$. 2H$_2$O: (*10*) 3.43, (*8*) 5.61, 2.925, (*6*) 1.743, (*5*) 2.505, 1.903.
Wairakite, CaAl$_2$Si$_4$O$_{12}$. 2H$_2$O: (*10*) 3.39, (*8*) 5.57, (*6*) 3.42, (*5*) 2.909, (*4*) 2.489, 2.215.
Faujasite, Na$_2$Ca [Al$_2$Si$_4$O$_{12}$]$_2$. 16H$_2$O: (*10*) 15.02, (*9*) 3.75, (*8*) 5.68, 4.35, 3.278, 2.860

4. Natrolite Group
Natrolite, Na$_2$Al$_2$Si$_3$O$_{10}$. 2H$_2$O: (*10*) 2.86, (*9*) 6.58, 3.19, (*8*) 5.89, (*6*) 4.42, (*5*) 4.14
Scolecite, CaAl$_2$Si$_3$O$_{10}$. 3H$_2$O: (*10*) 2.90, (*8*) 5.94, 4.42, (*7*) 6.68, 3.22, (*5*) 4.72
Thomsonite, NaCa$_2$Al$_5$Si$_5$O$_{20}$. 7H$_2$O: (*10*) 4.67, 2.69, (*9*) 6.54, 3.53, 2.85, (*8*) 3.22
Laumontite, CaAl$_2$Si$_4$O$_{12}$, 4H$_2$O: (*10*) 9.52, 4.17, (*8*) 6.82, 3.57
Gismondite, CaAl$_2$Si$_2$O$_8$, 4H$_2$O: (*10*) 7.3, 3.24, 2.73, (*8*) 4.19, (*6*) 4.9, 1.78

5. Phillipsite-Harmotome Group
Phillipsite, (K$_2$ Ca) Al$_2$Si$_4$O$_{12}$. 4.5 H$_2$O: (*10*) 7.64, 6.91, 3.18, (*6*) 4.25,, 4.07, 2.67

Harmotome, $BaAl_2Si_6O_{16} \cdot 6H_2O$: *(10)* 8.11, 7.16, 6.25, 4.07, 3.18, 2.69
Edingtonite, $BaAl_2Si_3O_{10} \cdot 4H_2O$: *(10)* 3.46, 2.69, *(8)* 4.54, 2.96, 2.23, 1.81

Class VI
Borates
Kotoite, $Mg_3B_2O_6$: *(10)* 2.82, *(9)* 3.07, 1.99, *(8)* 2.07, *(7)* 2.18, *(6)* 1.262
Sussexite, $MnBO_2(OH)$: *(10)* 6.31, *(7)* 2.718, *(6)* 2.471, *(5)* 3.313, *(4)* 2.250, 2.017
Pinakiolite, $(Mg_3Mn_3)[O_4/B_2O_6]$: *(10)* 2.70, *(8)* 5.39, 2.50, *(7)* 2.16, 1.99 *(6)* 1.631
Boracite (low), $Mg_3B_7O_{13}Cl$: *(10)* 3.041, 2.071, *(8)* 2.727, 1.767, 1.247, *(7)* 1.583, 1.364.
Hydroboracite, $CaMgB_6O_{11} \cdot 6H_2O$: *(10)* 2.438, *(8)* 1.908, *(6)* 2.210, *(5)* 4.54, *(4)* 1.839
Calciborite, $Ca_5B_8O_{17}$: 3.774, 3.419, 2.629, 2.300, 1.787
Colemanite, $Ca_2B_6O_{11} \cdot 5H_2O$: *(10)* 3.13, *(5)* 5.64, 3.85, 2.55 *(4)* 4.00, 3.29, 2.89
Borax, $Na_2B_4O_7 \cdot 10H_2O$: *(10)* 2.574, *(9)* 4.84, 2.818, *(7)* 6.90, 5.74, 3.91
Hilgardite, $Ca_8(B_6O_{11})Cl_4 \cdot 4H_2O$: *(10)* 2.86, 2.84, *(8)* 2.113, 2.109, 1.985, *(6)* 2.755

Class VII
Phosphates, Arsenates and Vanadates
(a) Be-Al-Mg combination
Wavellite, $Al_3(PO_4)_2(OH)_3 \cdot 5H_2O$: *(10)* 3.437, 3.263, *(8)* 8.58, 2.575, 2.103, 1.987
Strengite, $FePO_4 \cdot 2H_2O$: *(10)* 4.35, *(8)* 5.5, 3.12, 2.54, *(7)* 1.61, 1.479
Scorodite, $FeAsO_4 \cdot 2H_2O$: *(10)* 4.46, *(9)* 5.59, 3.17, *(5)* 3.05, 2.57, *(4)* 3.33
Wagnerite, Mg_2PO_4F: *(10)* 2.99, 2.84, *(9)* 3.15, *(8)* 3.32, *(7)* 2.762, *(6)* 2.713
Newberyite, $MgHPO_4 \cdot 3H_2O$: *(10)* 3.45, *(9)* 3.05, *(6)* 2.57 *(5)* 4.70, 2.80, *(4)* 1.92
Amblygonite, $LiAlPO_4(F, OH)$: *(10)* 3.193, *(9)* 4.688, *(8)* 2.974, *(7)* 3.297, *(5)* 2.388, 1.958
Lazulite, $MgAl_2(PO_4)_2(OH)_2$: *(10)* 3.20, *(9)* 3.24, *(8)* 3.14, 3.08, *(7)* 2.55 *(5)* 1.963
Scorzalite, $FeAl_2(PO_4)_2(OH)_2$: *(10)* 2.54, *(8)* 2.27, *(7)* 2.23, 1.627, *(6)* 2.12, 1.897
Metavauxite, $FeAl_2(PO_4)_2(OH)_2 \cdot 8H_2O$: *(10)* 2.75, *(9)* 4.67, 4.32, *(7)* 5.1, *(4)* 2.65

(b) Fe-Mn-Na Combination
Vivianite, $Fe_3(PO_4)_2 \cdot 8H_2O$: 6.79, 3.25, 3.01, 2.70, 2.53, 2.34
β-Kerchenite, $Fe_4Fe_4(OH)_4(PO_4)_6 \cdot 21H_2O$: 2.91, 2.70, 3.16, 2.42, 2.31, 1.68
Cacoxenite, $Fe_4(PO_4)_3(OH)_3 \cdot 12H_2O$: *(10)* 11.81, *(9)* 3.146, *(6)* 4.892, 2.787, *(5)* 9.73, *(4)* 6.94.
Strunzite, $MnFe_2(PO_4)_2(OH)_2 \cdot 6H_2O$: *(10)* 9.25, *(9)* 5.30, *(5)* 3.30, *(3)* 2.90, 2.77, 2.60

Frondelite, MnFe$_4$ (PO$_4$)$_3$ (OH)$_5$: (10) 3.195, (5) 3.38, 1.598, (4) 3.61, (3) 1.979, 1.849, 1.598

(c) Na-Ca-REE

Apatite, Ca$_5$ (PO$_4$)$_3$, (F, Cl, OH):
Fluorapatite: (10) 2.798, (6) 2.072, 1.838, (4) 2.769, (3) 1.745, 1.720
Chlorapatite: (10) 2.764 , (6) 1.954, 1.840, (4) 2.308, (3) 1.809
Fermorite, (Ca, Sr)$_{10}$ [PO$_4$) AsO$_4$)]$_6$ (F, OH, O)$_2$: (10) 2.866, (9) 2.769, 1.873, 1.749, (6) 1.280, (4) 3.494
Xenotime: PO$_4$: (10) 1.749, (9) 1.422, (8) 3.343, 1.703, 1.233, (7) 1.286
Monazite: (Ce, La, Y, Th) PO$_4$: (10) 3.09, (7) 3.31, 2.88, (6) 4.17, 2.139, 1.746
Englishite, K$_2$Ca$_4$Al$_8$(PO$_4$)$_8$ (OH)$_{10}$. 9H$_2$O: (10) 9.3, (7) 2.86, (6) 1.72, (5) 5.8, 3.03, 2.73
Millisite, NaCaAl$_6$ (PO$_4$)$_4$ (OH)$_9$. 3H$_2$O: (10) 4.84, 4.796, 4.732, 2.979, (8) 2.813, (5) 2.840
Viseite, Na$_2$Ca$_{10}$ (Al$_{20}$Si$_6$ P$_{10}$ H$_{36}$) O$_{96}$. 16H$_2$O: (10) 2.92, (6) 1.74, (5) 3.46, (4) 5.68, (2) 1.555
Crandallite, CaAl$_3$ (PO$_4$)$_2$(OH)$_5$. H$_2$O: (10) 2.95, (5) 2.98, (3) 1.895, (2) 4.86, 3.49, 1.755.
Roselite, Ca$_2$ (Co, Mg) (AsO$_4$)$_2$. 2H$_2$O: 3.11-2.73-3.66-2.07, 1.79-6.32
Arsenosiderite, Ca$_3$Fe$_2$ (AsO$_4$)$_4$ (OH)$_4$. 4H$_2$O: (10) 8.95, 1.486, 1.108, (9) 2.502, (8) 2.953, 1.693

(d) Zn-Cu-Co-Pb (U) Association

Hopeite, Zn$_3$ (PO$_4$)$_2$. 4H$_2$O: (10) 2.85, (9) 1.94, (8) 3.47, 2.63, 2.52, 1.82
Adamite, Zn$_2$AsO$_4$ (OH): (10) 2.451, 1.603, (9) 2.965, 2.686, 1.581, (8) 4.90
Reinerite, Zn$_3$ (AsO$_3$)$_2$: (10) 3.995, (8) 3.203, 2.644, 1.440, 1.436, (6) 1.38
Kottigite, Zn$_3$(AsO$_4$)$_2$8H$_2$O: (10) 3.25, (9) 2.79, (8) 4.60, 2.16, 1.83, 1.62
Erythrite, Co$_3$ (AsO$_4$)$_2$. 8H$_2$O: (10) 3.01, (9) 3.23, (7) 6.85, 2.319, 1.485
Annabergite, Ni$_3$ (AsO$_4$)$_2$. 8H$_2$O: (10) 3.19, (9) 2.998, 1.557, (8) 1.680, 1.649, (7) 1.336
Libethenite, Cu$_2$PO$_4$ (OH): (10) 4.81, 2.71, (9) 5.85, 2.917, (7) 1.451 (5) 1.579
Cornetite, Cu$_3$PO$_4$(OH)$_3$: (10) 3.04, (9) 4.29, (8) 3.17, (7) 2.06, (6) 5.07
Pseudomalachite, Cu$_5$ (PO$_4$)$_2$ (OH)$_4$. H$_2$O: (10) 4.48, (8) 2.39, (6) 2.42, (5) 3.46, 2.23, 1.728
Olivenite, Cu$_2$AsO$_4$ (OH): (10) 2.98, (9) 4.82, (7) 2.47, 2.39, (6) 2.65, 1.575
Euchroite, Cu$_2$AsO$_4$ (OH). 3H$_2$O: (10) 2.79, (8) 7.17, 5.17, (6)4.95, 3.65, 2.89.
Clinoclase, Cu$_3$ AsO$_4$ (OH)$_3$: (10) 3.55, (8) 3.13, (6) 4.3, (5) 2.30, 2.05, (4) 2.53
Pyromorphite, Pb$_5$ (PO$_4$)$_3$Cl: (10) 2.92, (9) 1.52, 1.486, 1.292, 1.268, (7) 1.318
Mimetite, Pb$_5$ (AsO$_4$)$_3$Cl: (10) 3.05, 3.001, (9) 1.563, 1.544, 1.069, (8) 1.983
Vanadinite, Pb$_5$(VO$_4$)$_3$ Cl: (10) 2.958, (8) 2.091, 1.888, 1.990, 1.043, (7) 1.976

Descloizite, PbZn VO$_4$(OH): (*10*) 3.04, (*9*) 2.283, 2.083, (*8*) 3.23, 1.640 (*7*) 1.398
Chalcophyllite, Cu$_{18}$Al$_2$ (AsO$_4$)$_3$ (SO$_4$)$_3$ (OH)$_{27}$. 33H$_2$O: (*10*) 3.76, (*8*) 5.14 (*7*) 2.892, 2.537, (*6*) 6.21, (*5*) 1.489
Turquoise, CuAl$_6$ (PO$_4$)$_4$ (OH)$_8$. 4H$_2$O: (*10*) 3.67, (*9*) 2.90, (*5*) 2.01, (*4*) 4.78, 1.84, 1.81.
Chalcosiderite, CuFe$_6$ (PO$_4$)$_4$ (OH)$_8$. 4H$_2$O: (*10*) 3.77, (*7*) 3.01, (*5*) 6.33, 3.39 (*4*) 3.56, (*3*) 1.55
Plumbogummite, PbAl$_3$ (PO$_4$)$_2$ (OH)$_5$.H$_2$O: (*10*) 2.97, (*8*) 5.70, (*6*) 3.45, 2.20, (*5*) 1.888, 1.448
Dumontite, Pb$_2$ (UO$_2$)$_3$ (PO$_4$)$_2$ (OH)$_4$. 3H$_2$O: (*10*), 3.00, 2.95, (*8*) 4.27, (*7*) 3.48, (*6*) 3.74, 3.31
Parsonite, Pb$_2$ (UO$_2$) (PO$_4$)$_2$. 2H$_2$O: (*10*) 4.19, (*8*) 3.24, (*6*) 3.39, (*4*) 3.92, 1.873, 1.850.
Torbernite, Cu (UO$_2$)$_2$ (PO$_4$)$_2$ 12H$_2$O: (*10*) 3.69, (*8*) 1.556, (*7*) 1.642, (*6*) 1.989, (*5*) 1.419, 1.366
Meta-torbernite, Cu (UO$_2$)$_2$ (PO$_4$)$_2$. 8H$_2$O: (*10*) 8.74, (*8*) 3.66, (*7*) 2.16, (*5*) 2.66, (*4*) 4.33, (*3*) 1.62
Autonite, Ca (UO$_2$/PO$_4$)$_2$. 10-12 H$_2$O: (*8*) 3.59, 1.600, (*6*) 2.60, 2.106, (*5*) 8.10, 1.940, 1.527
Carnotite, K$_2$ (UO$_2$)$_2$ V$_2$O$_8$ 2H$_2$O: (*10*) 6.56, (*7*) 3.12, (*5*) 3.53, (*3*) 4.25, 3.25, 2.156.

Class VIII
Tungstates and Molybdates
Wolframite, (Fe, Mn) WO$_4$: (*10*) 2.917, (*8*) 2.46, 2.18, 1.702, (*6*) 1.758, 1.503
Hubnerite, Mn WO$_4$: (*10*) 2.989, (*8*) 1.783, (*7*) 2.497, 1.721, 1.441, 1.378
Ferberite, FeWO$_4$: (*10*) 2.933, 1.711, (*8*) 2.188, 1.765, 1.507, (*7*) 1.371
Scheelite, CaWO$_4$: (*10*) 3.08, (*9*) 1.589, 1.247, (*8*) 1.923, 1.549, 1.082
Wulfenite, PbMoO$_4$: (*10*) 3.25, (*9*) 1.659, (*8*) 2.030, 1.792, 1.314, 1.053
Lindgrenite, Cu$_3$ (M$_0$O$_4$)$_2$ (OH)$_2$: (*10*) 3.50, (*6*) 4.15, (*4*) 2.67, (*3*) 4.34, 3.58, 3.47.

Class IX
Sulphates
(a) Al-Mg-Na(K) Association
Alunite, KAl$_3$(SO$_4$)$_2$ (OH)$_6$: (*10*) 3.10, 3.03, (*9*) 5.02, (*8*) 2.21, 1.967, 1.822
Potash alum, KAl (SO$_4$)$_2$. 12H$_2$O: (*10*) 4.29, (*9*) 4.03, 3.24, (*6*) 2.78, (*5*) 3.03, 1.91.
Soda Alum, NaAl(SO$_4$)$_2$. 12H$_2$O: (*10*) 4.23, (*7*) 3.65, (*5*) 3.98, (*4*) 1.62, (*3*) 1.90, (*2*) 1.50
Kieserite, MgSO$_4$. H$_2$O: (*10*) 3.38, (*7*) 2.55, (*4*) 4.82, 2.05, 1.67, 1.279
Epsomite, MgSO$_4$. 7H$_2$O: (*10*) 4.22, (*7*) 2.66, (*3*) 5.9, 5.3, 2.96, 2.87
Melanterite, FeSO$_4$. 7H$_2$O: (*10*) 4.86, (*5*) 3.74, (*3*) 5.42, 3.22, 2.73, 2.63
Jarosite, KFe$_3$ (SO$_4$)$_2$ (OH)$_6$: (*10*) 3.06, (*9*) 2.28, 1.976, 1.819, (*8*) 1.538 (*7*) 1.510

Anhydrite, CaSO$_4$: (*10*) 3.49, (*8*) 2.85, 1.64, (*6*) 2.32, 2.20.
Gypsum, CaSO$_4$. 2H$_2$O: (*10*) 4.29, (*7*) 2.87, (*6*) 3.06, 2.68, 2.07, (*5*) 1.79
Celestine, SrSO$_4$: (*10*) 2.042, 1.999, (*8*) 1.595, 1.472, 1.202, 1.145
Baryte, BaSO$_4$: (*10*) 2.106, (*7*) 3.058 (*6*) 3.456, 1.526, 1.259, 1.093
Zinkosite, ZnSO$_4$: (*10*) 4.168, 3.544, 2.650, 2.450, (*8*) 2.615, (*6*) 1.457
Chalcanthite, CuSO$_4$. 5H$_2$O: (*10*) 5.46, (*7*) 3.98, (*6*) 4.71, 4.66, (*5*) 1.884, (*3*) 1.827
Brochantite, Cu$_4$SO$_4$ (OH)$_6$: (*10*) 2.49, (*9*) 3.83, (*7*) 6.20, 5.25, 2.65, (*5*) 1.73
Antlerite, Cu$_3$SO$_4$ (OH)$_4$: (*10*) 2.12, (*9*) 2.55. (*7*) 4.80, 3.58, 1.50, (*5*) 1.63
Linarite, PbCu (SO$_4$) (OH)$_2$: (*10*) 3.11, (*8*) 3.52, 2.30, 2.10, 1.797, 1.571
Plumbojarosite, PbFe$_6$ (SO$_4$)$_4$ (OH)$_{12}$: (*10*) 3.13, 3.01, (*8*) 2.19, 1.955, 1.820, (*7*) 2.49
Anglesite, PbSO$_4$: (*10*) 3.06, 2.06, (*8*) 3.21, 2.02, (*7*) 2.68, 1.61

Class X
Chromates
Crocoite, PbCrO$_4$: (*10*) 3.258, (*9*) 1.965, 1.846, (*8*) 3.008, 2.242, 1.689

Class XI
Carbonates
Magnesite, MgCO$_3$: (*10*) 2.737, 1.697, (*9*) 2.101, (*7*) 1.336, (*6*) 1.935, 1.252
Siderite, FeCO$_3$: (*10*) 2.781, (*9*) 1.733, (*7*) 2.135, (*6*) 3.592, 2.348, 1.966
Rhodochrosite, MnCO$_3$: (*10*) 2.850, (*8*) 1.762, (*7*) 3.65, (*5*) 1.99, (*4*) 1.54,

Calcite, CaCO$_3$: (*10*) 3.03, (*9*) 1.910, (*8*) 1.873, (*7*) 2.28, 2.09, (*6*) 1.600
Vaterite, CaCO$_3$: (*10*) 3.57, 3.28, 2.728, (*8*) 2.062, (*6*) 1.82, (*2*) 1.647
Aragonite, CaCO$_3$: (*10*) 1.975, (*9*) 3.40, (*8*) 1.880, (*7*) 2.364, (*6*) 2.71, 1.806
Strontianite, SrCO$_3$: (*10*) 3.47, (*7*) 2.022, 1.794, (*5*) 2.416, 1.876, 1.267
Witherite, BaCO$_3$: (*10*) 3.72, (*6*) 2.62, (*5*) 2.14, 2.03, 1.94, 1.239
Dolomite, CaMg (CO$_3$)$_2$: (*10*) 2.883, (*6*) 1.785, (*5*) 2.191, (*4*) 2.015, 1.167, 1.110
Smithsonite, ZnCO$_3$: (*10*) 2.748, 1.707, (*8*) 1.413, 1.186, (*7*) 1.495, 1.345
Malachite, Cu$_2$CO$_3$ (OH)$_2$: (*10*) 2.82, (*9*) 1.509, (*8*) 3.63, 2.49, (*6*) 4.87, 1.664
Azurite, Cu$_3$ (CO$_3$)$_2$(OH)$_2$: (*10*) 5.20, 3.67, 3.53, 2.54, (*8*) 2.28, 2.24, 1.825
Cerussite, PbCO$_3$: (*10*) 3.574, (*9*) 3.480 2.487, 2.087, 1.941, 1.865
Bismuthite, (BiO)$_2$CO$_3$: (*10*) 2.943, (*9*) 1.745, 1.616, (*8*) 2.724, 2.134, (*7*) 1.936

Class XII
Nitrates
Soda-nitre, NaNO$_3$: (*10*) 3.03, (*6*) 2.31, 1.89, (*3*) 1.65, 1.461, 1.170
Nitre, KNO$_3$: (*10*) 3.77, (*6*) 3.03, (*5*) 2.66, 2.19, (*3*) 1.96, (*2*) 1.54

Author Index

Adams, J.B. 110, 111, 112
Albee, A.L. 199
Allen, W.C. 196, 200
Andrews, K.W. 196
Arnold, R.G. 199

Bambauer, H.U. 197
Barton, P.B. 200
Berger, R.W. 197
Berry, F.J. 179
Bloss, F.D. 9
Borg, I.Y. 191, 198, 199
Borley, G. 196
Bowie, S.H.U. 95, 98, 99
Bragg, W.L. 177
Brindley, G.W. 197
Brown, G. 197
Brown, G.E. 178
Burley, B.J. 199
Burns, R.G. 61, 67, 68, 98, 109

Cervelle, B. 198
Clark, L.A. 196, 198
Cooke, G.A. 198
Corlett, M. 197
Corroll, D. 197
Cotton, J.A. 59

Datta, S.K. 119
Davies, T.T. 196
De, D. 119
Desborough, G.A. 194, 199
Dudeney, A.W.L. 199

Eberhard, E. 197
Edgar, A.D. 191
Eugster, H.P. 195, 199

Fairbairn, H.W. 47
Franzini, M 195, 199
Freeman, E.B. 199

Gaffey, M. 112
Galopin, R. 98, 99
Gensmer, R.P. 197
Ghose, S. 196
Gillery, F.H. 197
Goldman, D.S. 119, 122

Goldsmith, J.R. 197
Goodenough, J.B. 59
Graham, A.R. 199
Gray, I.M. 95

Hall, S.R. 197, 200
Hallimond, A.F. 98
Hamilton, 191
Henry, N.F.M. 98, 99
Hey, M.H. 197
Himmeberg, G.R. 194, 198
Hodgson, M. 199
Hooper, P.R. 196
Hormann, P.K. 195, 197, 199
Horn, E.E. 200
Huguenin, R.L. 111
Hutchison, C.S. 195

Jackson, E.D. 194, 198
Johnson, K.H. 59
Johnson, W. 196
Jones, J.B. 190, 198

Kampf, N. 199
Klein, C. 196
Kuellmer, F.J. 197
Kullerud, G. 200

Launer, P.J. 115
Laues, F. 197

Mach, M. 136
Maniar, P.D. 198
Millman, A.P. 95, 96
Mitra, S. 82, 98, 139
Morelli, G.L. 197
Morimoto, N. 196
Morteani, G. 197, 199
Murchison, D.G. 99

Nassau, K. 16, 119
Nesbitt, R.W. 198
Nord, A.G. 178

Orcel, J. 98
Orville, P.M. 190, 197

Parrish, W. 136
Petruk, W. 197

Phillips, F.C. 47
Phillips, M.W. 178

Reichen, L.E. 199
Ribbe, P.H. 178, 200
Riley, D.P. 149
Riley, J.F. 198
Rose, H.J. 194, 199
Rosenberg, P.E. 200

Sahama, Th. G. 194, 198
Sayin, M. 199
Schiaffino, L. 199
Schoen, R. 197
Schultz, H. 200
Schulze, D.G. 200
Schwertmann, U. 199
Shaw, D.M. 199
Short, M.N. 200
Singh, D.S. 98
Skinner, B.J. 200
Skinner, H.C.W. 196, 198
Slade, P.G. 198
Smith, D.K. 198, 199
Smith, J.R. 191, 197
Smith, J.V. 193
Srodon, J. 199
Stanton, R.I. 97
Stefanidis, T. 178
Stewart, D.B. 198

Talmage, S.B. 94
Taylor, K. 95
Tennant, C.B. 197
Tilling, R. 198
Troger, W.E. 44

Ulas, M. 199
Urch, D.S. 179

Vaughan, D.J. 98, 179
Viswanathan, K. 196, 197

Waldbaum, D.R. 196
Weiss, M.P. 197
Whittaker, E.J.W. 196, 199
Wilkinson, J.F.G. 196
Willey, H.G. 97
Willey, R.R. 117
Winchell, H. 198
Wood, D.L. 119
Wright, T.L. 198

Yoder, H.S. 194, 197, 199
Young, B.B. 96

Zen, E.A.N. 199
Zussman, J. 199

Mineral Index

Acanthite, 183, 203
Actinolite, 61, 210
Adamite, 216
Adularia, 213
Aegirine, 210
Aquilarite, 203
Aikinite, 204
Akermanite, 208
Alabandine, 202
Albite, low, 188, 215
Alexandrite, 17, 70
Algodonite, 201
Alkali amphiboles, pleochroic colour, 72
Alkali Feldspar, 188
Allanite, 211
Almandine, 211
Alunite, 217
Amazonite, colour, 74
Amblygonite, 217
Amethyst, colour, 74
Amosite, 212
Amphiboles, 194
Analcine, 214
Analcite, 194
Anatase, 93, 113, 206
    — absorption bands, 111
    — internal reflection colour, 93
Anauxite, 212
Andalusite, 67, 194, 208
    — absorption line, 80
    — uv fluorescence, 83
Andesine, 213
Andradite, 20
Anglesite, 218
Anhydrite, 218
    — Laue photograph, 155
Ankerite absorption bands, 129
Annabergite, 216
Anorthite, 213
Anthophyllite, 210
Antigorite, 212
Antimony, 200
Antlerite, 218
Apatite, 194, 216
    — absorption line, 80
    — uv, fluorescence, 83
Apophyllite, 17, 119, 121
    — laue photograph, 155
    — nir bands, 120

Aquamarine, 17, 76, 77
    — charge transfer, 76
Aragonite, 194, 218
    — absorption bands, 124, 125
Arfvedsonite, 194, 210
Argentite, 183, 203
Argentopyrite, 203
Armalcolite, lunar, 51
Arsenates, 200, 215
Arsenic, 200
Arsenolite, 207
Arsenopyrite, 93, 94, 183, 194, 202
    — crystal form, 97
    — polishing hardness, 94
Arsenosiderite, 216
Attapulgite, 189
Andradite, 17
Augite, 27, 111, 112, 210
    — bands subcalcic, 110
    — subcalcite, 51
Aurostibite, 201
Autonite, 217
Axinite, 209
Azurite, 93, 218
    — internal reflection colours, 93

Baryte, 220
Beidellite, 214
Benitoite, pleochroic colour, 72
    — uv fluorescence, 83
Beryl, 17, 209
    — charge transfer, 80
    — uv fluorescence, 83
Berzelianite, 205
Betafite, 209
Betekhtinite, 205
Biotite, 17, 27, 195, 199, 213
    — pleochroic colour, 72
Birnessite, 208
Bismuth, 183, 202
Bismuthinite, 94, 183, 206
    — polishing hardness, 94
    — reflectance colour, 101
Bismuthite, 220
Bismutoferrite, 212
Bixbyite, 207
Boehmite, 207
Boracite, 217
Borates, 202, 217

224 Index

Borax, 217
Bornite, 183, 205
Boulangerite, 183, 206
Bournonite, 206
Braggite, 203
Brammellite, 191, 192, 214
Brass, 16
Braunite, 183, 208
Bravoite, 183
Breithauptite, 203
Brochantite, 220
Bromargyrite, 206
Bromides, 201
Bronzite, 80
Brookite, 44, 206
Brucite, 28, 207
Bunsenite, 17, 207
Bytownite, 215

Cacoxenite, 217
Cadmoselite, 204
Calciborite, 217
Calciotale, 213
Calcite, 220
    — absorption bands, 123, 129
    — colour, 72
    — i.r. bands, 117
    — Laue photograph, 115
    — negative, 25
    — peaks, 191, 197
    — structure, 6, 17
    — uv fluorescence, 83
Calomel, 28, 207
Cancrinite, 216
Carbonates, 202
Carnallite, 206
Carnotite, 219
Carrollite, 183, 204
Cassiterite, 93, 208
    — internal reflection colours, 93
Cattierite, 203
Celadonite, 191, 192
Celestine, 220
Cerussite, 220
    — absorption bands, 124
Chabazite, 216
Chalcanthite, 220
Chalcocite, 183, 205
Chalcophanite, 208
Chalcophyllite, 219
Chalcopyrite, 94, 183, 197, 200
Chalcosiderite, 219
Chalcostibite, 206
Chamosite, 214

Chapmanite, 212
Cherkinite, 210
Chloanthite, 204
Chlorapatite, 218
Chlorides, 201
Chlorites, 191, 197, 214
Chlorargyrite, 206
Chloritoid, 214
Chromates, 202, 220
Chromite,
    — crystal form, 97
    — internal reflection colours, 93
    — JCPDS Card, 154
    — placer, 103
    — powder data, 138
Chrysoberyl, 207
    — absorption line, 80
    — uv fluorescence, 83
Chrysotile, 214
Cinnabar, 16, 28, 93, 184, 205
    — band gap energy, 77, 78
    — colour, 77
    — proustite, internal reflection colour, 93
Claudetite, 209
Clay group, 191, 197
Clinoclore, 214
Clinoclase, 218
Clinopyroxene, 195
    — Apollo, 17, 50
    — spectra, 108
Cobaltite, 184, 202
Coesite, 215
Coffinite, 211
Colemanite, 217
Columbite, 208
Cooperite, 203
Copper, 16, 184, 202
    — colour, 77
Cordierite, 209
    — absorption line, 80
    — charge transfer, 76
    — colour, 75
    — pleochroic colour, 72
Cornetite, 218
Coronadite, 208
Corundophyllite, 214
Corundum, 17, 70, 207
    — absorption line, 80
    — colour, 75, 76
    — uv fluorescence, 83
Coulsonite, 207
Covellite, 93, 184, 205
    — crystal form, 97
    — polishing hardness, 94

— reflectance colour, 101
— spectral reflectance curve, 102
Crandallite, 218
Crednerite, 208
Cristobalite, 215
Crocoite, 220
— charge transfer, 75
Crookesite, 205
Cryolite, 206
Cryptomelane, 208
Cubanite, 93, 184
— reflectance colour, 101
Cummingtonite, 61, 196
Cuprite, 93, 184, 209
Cuprosklodowskite, 212

Daphnite, 214
Daubreelite, 204
Deerite, 211
Delafossile, 209
Descloizite, 219
Diabantite, 214
Diamond, 16, 202
— absorption line, 80
— band gap, 78
— luminescence, 81
— uv fluorescence, 83
— yellow colour, 78
Diaspore, 207
Dickite, 191, 214
Digenite, 184, 204
Diopside, 17, 61, 212
— absorption bands, 110
— absorption line, 80
— electron density map, 176, 177
— fourier series map, 176
— spectra, 108
— weissenberg photograph, 164, 165
Dioptase, 211
Dolomite, 191, 197, 220
— absorption bands, 123, 124
— laue photograph, 156
Domeykite, 203
Donbassite, 214
Dumontite, 219
Dumortierite, 210
Dyscrasite, 203

Edingtonite, 217
Egyptian blue, 17
Elbaite, 211
Emerald, 16, 17, 71
— uv fluorescence, 83
Emplectite, 206

Empressite, 205
Enargite, 93, 184, 206
Englishite, 218
Enstatite, 194, 212
Epidote, 17, 210
— absorption line, 80
Epsomite, 219
Erythrite, 17, 218
Eskolaite, 207
Eucairite, 205
Euchroite, 218

Faujacite, 216
Fayalite, 65, 210
Feldspars, 197, 215
— uv fluorescence, 83
Fenaksite, 210
Ferberite, 219
Fermorite, 218
Ferrihalloysite, 214
Ferroselite, 204
Fitzroyite, 17
Fluorapatite, 218
Fluorides, 201
Fluorite, 16, 206
— absorption line, 80
— colour, 72
— colour centres, 73
— electronic spectra, 82
— luminescence, 81
— laue photograph, 155
— refractive index, 86
— uv fluorescence, 83
Fluosilicate NIR bands, 120
Forsterite, 194, 210
Franklinite, 207
Frondelite, 218
Froodite, 203

Gahnite, 207
Galena, 16, 94, 184, 205
— band gap, 78
— crystal form, 97
— polishing hardness, 94
— refractive index, 86
Gallite, 204
Garnets, 61, 198, 211
Garnet cell parameter, 191
Garnierite, 17, 214
Gedrite, 212
Gehlenite, 210
Geikielite, 208
Gersdorffite, 184, 204
Gibbsite, 28, 207

226 Index

Gismondite, 216
Glass, agglutinitic, 108
Glaucophane colour, 75
    pleochroic colour, 72
Glauconite, 191, 192, 213
Goethite, 184, 207
Gold, 16, 94, 184, 210
    — crystal form, 97
    — polishing hardness, 94
Graphite, 93, 94, 184, 202
    crystal form, 97
Greenalite, 214
Greenockite, 88, 204
    — band gap, 78
Grossular, 211
Groutite, 207
Grunerite, 196
Gypsum, 220
    — bands, 119
    — laue photographs, 156
    — NIR bands, 120
    — plate, 33, 35, 39
    — structure, 119

Halides, 201
Halite, 206
    — colour, 72
    — laue photograph, 155
Halloysite, 191, 192, 214
Harkerite, 211
Harmotome, 217
Hauchecornite, 203
Hauerite, 204
Hausmannite, 208
Hauyne, 216
Hawleyite, 204
Heazlewoodite, 203
Hectorite, 191
Hedenbergites, 108
Helidor, charge transfer, 75
Helvine, 216
Hematite, 93, 184
    — absorption bands, 111
    — absorption colour, 88
    — crystal form, 97
    — internal reflection colour, 89, 93
    — refractive index, 86
Hemimorphite, 211
Hessite, 205
Hetaerolite, 208
Heulandite, 216
Hilgardite, 217
Hollandite, 208

Holmquistite, 212
Hopeite, 218
Hornblende, 212
    — colour, 72
    — vibration direction, 27
Howieite, 211
Hubnerite, 219
Humite group, 210
Huttonite, 210
Hydrobiotite, 214
Hydroboracite, 217
Hydrohausmannite, 208
Hydro-micas, 191, 192
Hydroxides, 201
Hypersthene, 27, 194, 212
Hypersthene, absorption band, 110
    — colour, 72
    — polished spectra, 66, 68

Idocrase, 211
Illite, 191, 192, 213
Ilmenite, 93, 94, 184, 198, 208
    — absorption bands, 111
    — crystal form, 97
    — polishing hardness, 94
Indite, 204
Iodargyrite, 206
Iodides, 201
Iolite charge transfer, 76
Iron, 16, 184, 202
    — reflection light, 51

Jacobsite, 184, 207
Jadeite, 212
Jamesonite, 97, 184, 206
    — crystal form, 97
Jarosite, 219
Jeremejevite, Laue photograph, 155
Julgoldite, 211

Kallophilite, 215
Kalsilite, 215
Kammererite, 17, 214
Kandides, 191
Kaolinite, 191, 193, 194
Kerchenite, 217
Kieserite, 219
Klockmanite, 205
Kotoite, 217
Kottingite, 218
Krauskopfite, 215
Krischsteinite, 210

Kyanite, 17, 196, 210
— colour, 75
— charge transfer, 76
— uv-fluorescence, 83

Labradorite, 215
Lapislazuli, 16, 76
Laumontite, 216
Lazulite, 217
Lazurite, 216
Lead, 202
Lepidocrocite, 184, 207
Lepidolite, 213
Lepidomelane, 213
Leucite, 215
Levyne, 216
Libethenite, 218
Limonite, 184
Linarite, 220
Lindgrenite, 219
Linnacite, 185, 203
Litharge, 209
Loellingite, 185, 198, 204
Luzonite, 206

Mackinawite, 186
Maghemite, 186, 207
Magnesiochromite, 207
Magnesioferrite, 207
Magnesite, 220
— absorption, 124
Magnetite, 16, 186, 207
— absorption bands, 111
— crystal form, 97
— placer, 103
— polishing hardness, 94
Magnetoplumbite, 209
Malachite, 17, 93, 220
Manganophyllite, 67
Manganite, 186, 207
Manganosite, 207
Marcasite, 93, 186, 204
— crystal form, 97
Margarite, 214
Marokite, 208
Maucherite, 203
Melanterite, 219
Melilite, 210
Mesolite, NIR bands, 120, 121
Metacinnabarite, 205
Metahalloysite, 191
Metavauxite, 216
Micas, 193, 195, 119

Michenerite, 203
Microcline, 17, 190, 215
Microlite, 209
Millerite, 186, 203
— reflectance colour, 101
Millisite, 218
Mimetite, 218
Minnesotaite, 213
Moissanite, Laue photograph, 155
Molybdates, 202, 219
Molybdenite, 93, 186, 204
— band gap, 78
— polishing hardness, 94
— reflectance colour, 101
Monazite, 218
Moncheite, 203
Moonstone, 16
Monticellite, 210
Montmorillonite, 191, 193, 214
Montroydite, 209
Mullite, 196, 209
Muscovite, 28, 195, 199, 213
— Laue photograph, 156

Nacrite, 191, 214
Natrolite, 216
Naumannite, 205
Nepheline, 198, 215
Newberyite, 217
Niccolite, 93, 186, 203
— polishing hardness, 94
Nigglite, 203
Ni-pentlandite, 203
Nitrates, 202, 220
Nitre, 220
Nontronite, 191, 214
Norbergite, 210
Nosean, 216
Novakite, 203
Nsutite, 208

Okenite, 215
Oldhamite, 204
Oligoclase, 215
Olivenite, 218
Olivines, 17, 61, 64, 65, 194, 210
— polarised spectra, 66
Opal, 83
Orcellite, 203
Oregonite, 203
Orlite, 212
Orpiment, 88, 206
— band gap, 78

Orthite, 211
Orthoclase, 215
   — Laue photograph, 156
Orthoferrosilite, 110
Orthopyroxenes, 61, 194, 199
   — absorption bands, 210
   — atomic structure, 67
   — diffractometer record, 140
   — line positions, 139
   — myrmekitic, 104
   — polarised spectra, 66
Oxides, 201

Palygorskites, 191, 192, 213
Papagoite, 211
Paramelaconite, 209
Pararammelsbergite, 204
Parkerite, 203
Parsonite, 219
Paxite, 203
Pennine, 214
Penroseite, 204
Pentlandite, 185
Periclase, 207
Peridot, 17
   — absorption line, 80
Perovskite, 208
Petzite, 205
Phengites, 191
Phillipsite, 216
Phlogopite, 213
Phosphates, 202, 217
Piemontite, 17, 67
Pigeonite, 61, 110
   — sector zoning, 51
   — spectra, 108
Pinakiolite, 217
Pitchblende, 209
Plagioclase, 190, 215
   — Apollo 17, 50, 51
Platinum, 202
Plumboferrite, 209
Plumbogummite, 219
Plumbojarosite, 220
Pollucite, 215
Polybasite, 206
Polydymite, 204
Potash-alum, 219
Proustite, 185, 206
   — band gap, 78
Prehnite, 211
Pseudomalachite, 218
Pseudothuringite, 214
Psilomelane, 185, 206

Pumpellyite, 210
Pyrargyrite, 93, 185, 206
   — internal reflection colours, 93
Pyrite, 94, 185, 198, 204
   — crystal form, 97
   — diamagnetic, 60
Pyrochlore, 209
Pyrochroite, 208
Pyrolucite, 185, 208
Pyromorphite, 218
Pyrope, 211
Pyrophyllite, 213
Pyrrhotine, 203
Pyrrhotite, 94, 185, 199
   — paramagnetic, 60
   — reflectance, colour, 101
   — reflectance spectra, 109
Pyroxenes, 117, 110, 198, 212
   — iron bearing, octahedral sites, 111
   — lunar, 68, 109
   — structural types, 111

Quartz, 215
   — blue, 17
   — positive, 26
   — smoky, 16, 74
Quenselite, 208

Rammelsbergite, 204
Ramsdellite, 208
Realgar, 206
Reinerite, 218
Rhodochrocite, 17, 220
   — absorption bands, 124
Rhodonite, 17, 213
   — absorption line, 80
Rickardite, 205
Riebeckite, 212
Ripidolite, 214
Roselite, 218
Rubellite, 17
Ruby, 16, 17, 70, 71
   — uv fluorescence, 83
Rutile, 93, 185, 208
   — crystal form, 97
   — internal reflection colours, 93

Safflorite, 204
Samarskite, 209
Saponite, 191, 214
Sapphire, blue, 16, 17, 75, 76
Sartorite, 27, 206
Scapolite, 199, 216
   — uv fluorescence, 83

Index 229

Scheelite, 185, 219
— luminiscence, 81
— uv fluorescence, 83
Schorl, 211
Scolecite, 216
Scorodite, 217
Scorzalite, 217
Selenium, 202
Seligmannite, 206
Senarmontite, 209
Sepiolite, 191, 213
Serpentine, 199
Shandite, 203
Shattuckite, 211
Siderite, 220
Silicates, 202
Silicon, 16
Sillimanite, 196, 210
Silver, 16, 104, 202
— colour, 77
— refractive index, 86
Sklodowskite, 212
Skutterudite, 204
Smectites, 191, 192
Smithsonite, 220
— uv fluorescence, 83
Smythite, 203
Soda-alum, 219
Sodalite, 216
Sodanitre, 220
Sperrylite, 203
Spessartine, 211
Sphalerite, 93, 185, 200, 204
— band gap, 78
— internal reflection colour, 93
— refractive index, 86
— with activators, 81
Sphene, 210
Spinel, 17, 185, 195, 200, 207
— absorption line, 80
— uv fluorescence, 83
Spodiophyllite, 213
Spodumene, 212
— uv fluorescence, 83
Stannite, 185, 205
Staurolite, 210
— colour, 72
Stephanite, 206
Sternbergite, 205
Stibnite, 93, 185, 206
— crystal form, 97
Stilbite, 216
Stilpnomelane, 213
Stishovite, 215

Strengite, 217
Stromeyerite, 205
Strontianite, 220
— absorption bands, 125
Strunzite, 217
Sulphates, 202, 219
Sulphides, 200, 201, 203
Sulphosalts, 200, 202
Sulphur, 202
Sussexite, 217
Sylvite, 206

Tainiolite, 213
Talc, 213
— IR bands, 119
Tantalite, 208
Tellurbismuth, 205
Tellurium, 202
Tenantite, 206
Tenorite, 209
Tephroite, 210
Terlinguaite, 207
Tetradymite, 185, 205
Tetrahedrite, 185, 206
Thomsonite, 216
Thorianite, 209
Thorite, 210
Thuringite, 214
Titanaugite, 17
Titanomagnetite, 185
Ti-phlogopite, 17
Tobermorite, 215
Todorokite, 185, 208
Topaz, 200, 210
— absorption line, 80
— uv fluorescence, 83
Torbernile, 219
Tourmaline, 200, 211
— dichroism, 21
— NIR bands, 121, 122
— pleochroic colour, 72
Tremolite, 212
Tridymite, 215
Troilite, 203
— lunar, 51
Tuhualite, 210
Tungstates, 202, 219
Tungstenite, 204
Turquoise, 17, 219
— uv fluorescence, 83

Ullmannite, 204
Ulvospinel, 185
Umangite, 205

Uraninite, 209
U-minerals, luminescence, 81
Uranophane, 210
Uranothorite, 210
Ursilite, 210
Uvarovite, 17, 210

Vaesite, 204
Valentinite, 209
Valleriite, 93, 205
Vanadates, 202, 217
Vanadinite, 218
— charge transfer, 75
Vaterite, 220
Vermiculite, 192, 193, 199, 214
Vesuvianite, 211
Violarite, 204
Viridine, 17
Viseite, 218
Vivianite, 217
— charge transfer, 76
Vulcanite, 205
Vyrotskite, 203

Wagnerite, 217
Wavellite, 217
Wehrlite, 205
Weissite, 205

Willemite, luminescence, 81
Witherite, 220
— absorption bands, 124
Wolframite, 93, 94, 219
— internal reflection colours, 93
— polishing hardness, 94
Wallastonite, 213
Wulfenite, 219
Wurtzite, 204
Wustite, 207

Xenotime, 218
Xonotlite, 213

Yoderite, 209

Zeolites, 216
Zinc, 202
Zincite, 209
Zinckenite, 206
Zinkosite, 220
Zinnwaldite, 213
Zircon, 210
— absorption line, 80
— uv fluorescence, 83
Zoisite, 17
— absorption line, 80
Zussmanite, 211

Subject Index

Absorption
    bands, 112, 118, 119, 123
    coefficient, 35, 86, 120
    edge effect, 88
    electronic, 82
    error, 149
    frequency, 124
    linear, 132
    lines, 80
    mass, 131, 132
    optical, 131, 132
    spectrum, 4, 72
    sun, 4
    zirconium, 4
Acceptor level, 81
Accessory plates, 23, 30, 33, 35, 41
Activation energy, 81
Activators, 81
Agglutinitic glass, 110
Aluminian chromite, 153
Anisotropic, -tropism, 5, 25, 27, 28, 94, 183
Asteroids, 110
Atmospheric absorption, 109
Atomic form factor, 173
Atomic ordering, 171
Atomic scattering, 171, 172
Aufbau principle, 52
Automated powder diffractometry, 143, 144

Band gap transitions, 72, 79
    — in arsenides, 72
    — in diamond, 78
    — in semiconductors, 77
    — in sulphides, 72
    — in cinnabar, 78
    — in galena, 78
    — in greenockite, 78
    — in molybdenite, 78
    — in orpiment, 78
    — in proustite, 78
Band theory, 12
Band shift, 109
Band structure, 87
    — of metals, 87
    — of semiconductors, 88
Beam trap, 135

Beer-Lambert relationship, 63
Bertrand lens, 23
Bireflectance, 93
Birefringence, 28, 29, 30, 35
Bisectrix, 32
    — acute, 32, 41
    — obtuse, 32, 41
Blue-shift, 61
Bohr equation, 8
Bonding, covalent, 8
    — ionic, 8
    — metallic, 8
Bragg angle, 132, 146
Bragg's diffraction law, 132, 134, 153, 157
Brillouin zone, 88
Buerger precession method, 166, 167

Carbonate bands, 123
Cary spectrometer, 117
Cell lattice, 158
Cell parameter ratio determination, 150
Cell size determination, 189
    — computer output, 189
Characteristic spectrum, 127, 129
Charge transfer, 27, 71, 72, 110, 113
    — oxygen-metal, 74
    — metal-metal, 74
    — transition, 109
Chemical bonding, 179
Chromacity diagram, 89
Chromophores in minerals, 16, 17
    — ion Cr^{3+} in ruby, 70
    — ion Cr^{3+} in emerald, 70
    — ion Cr^{3+} in uvarovite, 70
    — ion Cr^{3+} in kammererite, 70
Cleavage, 98, 102
Clinopyroxenes, 110, 112
    — bands, 112
    — spectra of, 110
Collimators, 135
Colour, 14, 15, 16, 17, 80
    — allochromatic, 17
    — caused by charge transfer, 71
    — caused by electronic processes, 69
    — caused by molecular transition, 76
    — generation, 15
    — idiochromatic, 17
    — of minerals, 15, 16, 68

— of opaques, 86
— of strongly reflecting surface, 89
— spectral, 69
— sensation, 14
Colour centres, 73, 74
— bleaching, 73
— F-band, 73
— F-centres, 73
— F'-centres, 73
— in fluorites, 73, 74
— in halite, 74
— in microcline, 74
— in quartz, 73, 74
— k-band, 73
— v-centres, 73
Colour perception, 89
Common lines of light, 2
Compensator, 35
— berek, 35
Condenser, 23
Conduction band, 79
Conoscopic set up, 23
Continuous spectrum, 127
Coulomb interaction, 53
Crystal field splitting, 56
— parameter, 63
— spectra, 60
Crystal field stabilisation energy, 54 55, 56, 57
— energy difference, 56, 57
— in Cr^{3+}, 55, 56
— strongfield, 56, 57
— weak field, 56, 57
Crystal field theory, 55, 58
— in mineralogy, 59
Crystal form, 97
Crystal habit, 97
Crystal structure determination, 171

Debye-Scherrer camera, 135, 136
Debye-Scherrer powder photographs, 145
Detectors, 141
— Geiger-muller, 141
— scintillation counter, 141
Dichroism, 21, 27
Dielectrics, 85, 92
Dielectrical properties, 85
Diffraction order, 157
Diffraction spot, 158
Diffractometer record, 143
Diffuse component, 109
Diffuse reflection spectra,
mineral mixture, 114
opaques, 113

pyroxenes, 112
Diopside, 110, 112
Dispersion,
biaxial minerals, 44
of reflectance, 102
optic axis, 44
rhombic, 44
uniaxial minerals, 44
Dubious ions in minerals, 67

Eccentricity error, 149
Electromagnetic spectrum,
solar radiation, and black body, 3
visible range, 1
wave length, 2
Electromagnetic theory, 126
Electron density distribution, 176
Electron diffraction, 178
Electron polarization, 27
Electron spin multiplicity, 181
Electronic Structure, in metal, 87
Electronic transition, 69, 110
Electrostatic repulsion, 55
Elongation, negative, 36
— positive, 36
Emission spectrum of iron, 4
Energy level diagram,
Cr^{3+} in emerald, 71
Cr^{3+} in ruby, 71
in a crystal structure, 80
Estimation of 2V, 41, 43, 47
Exchange interaction, 53
Excitation potentials, 129
Extinction angle, 33
— inclined, 33
— parallel, 33
— symmetric, 33
Extraordinary ray, 25, 28
Extrinsic semiconductors, 81

Federov space group, 167
Filters, 99, 130, 137
Fink method, 152
Fink numerical section, 151
Forbidden, 114
Form factor, 171
Fourier analysis, 176
Fraunhofer lines, 1, 4
Fresnel's laws, 109
Fundamental absorption, 81

Galilean satellites, 110
Gamm ray spectroscopy, 109
Gaussian shaped functions, 110

Geiger-Muller counter, 141
Geometrical structure amplitude, 174
Gerade, 182
Gliding mirror plane, 168
Goniometer, 142,
Goniometric method, 150

Hanawalt numerical section, 151
Hanawalt search procedures, 151
Hardness, 94, 95
    — micro indentation, 94, 95
    — polishing, 94
    — scratching, 94
Hedenbergites, 110
Hematite, 113
High spin state, 56, 60
    — pyrrhotite, 60
Hund's rule, 52, 53, 153

Illumination, 10, 12, 46, 92
    — central, 10
    — oblique, 10, 12
Immersion media, 11
Indicatrix, 26, 30
    — of biaxial, 32
    — of uniaxial, 26, 30
    — optical, 30
Indices, 133
    — Miller, 133
Insulator, 77
Inter electron transition, 109
Interference colour, 29, 30, 34, 48, 92
Interference figures, 36, 37, 38, 39, 40
    — acute bisectrix figure, 37, 40
    — biaxial minerals, 37, 40, 41, 42
    — flash figure, 37, 41
    — obtuse bisectrix figure, 37, 41
    — off-centred figure, 37, 42
    — optic axis figure, 37, 41, 42
Internal reflection, 88, 93
Intervalence transition, 111
Integral omission, 170
Isochromes, 35
Isogyres, 36, 37, 41
Isotropic,-tropism, 5, 25, 27, 30, 45

Jahn-Teller effect in minerals, 53

Kalb light line, 94

Lambert's law, 85
Laporte selection rule, 62
Lattice parameter determination, 149

Lattice type determination, 167
Lattice vibration, 117
Laue method, 153, 167
Laue pattern, 134, 153, 156, 166
Laue spot, 153, 156
Layer lines, 159
Ligand field theory, 109
Lorentzian curves, 110
Lorentzian shaped function, 110
Low spin state, 56, 60
Luminescence colours in minerals, 81
Lunar pyroxenes, 110
Lunar surface, Ti-content variations, 112
Lustre, adamantine, 8
    — earthy or dull, 9
    — metallic, 7, 9, 10
    — nonmetallic, 7
    — pearly, 9
    — resinous, 9
    — semivitreous, 8
    — subadamantine, 8
    — submetallic, 8, 9
    — superadamantine, 8
    — supermetallic, 8
    — vitreous, 9

Mallard approximation, 43
Mars, 110
Mare basalt types, 110
Martian soils, 110
Mass absorption, 131
Melatope, 37
Mercury, 109, 110
Metal, 77
Michel-Levy color chart, 31
Microfiche, 151
Microhardness, 95, 96, 97, 102
    — determination, 96
    — types, 96
    — anisotropy, 97
    — variation, 97
Microscopy:
    — lunar samples, 105
    — opaque, 90
    — ore, 91
Microtextures, 49, 50, 51
    — in lunar samples, 50, 51
Mineral identification, guide, 201
Mixed liquid, 11
Molar extinction co-efficient, 63
Molecular centre, 82
Molecular orbital theory, 58, 59
Molecular transition, 76
Monochromatic light, 34

234 Index

Monochromatic neutrons, 178
Monochromatic X-ray, 159
Monochromators, 99, 130
Moon, 109, 110
Multiplex devices, 116
Myrmekitic intergrowth, 104

Napierian base, 132
Nelson-Riley extrapolation
    function, 148, 150
Neutron-diffraction, 178

Objectives, 91, 92
Omission Law, 168
Opaques, 85, 90
    — diffuse reflection spectra, 113
    — microscopy, 90
Optic axial angle, 28, 30, 32, 41
Optic axis, 25, 28, 32, 41
Optic normal, 32
Optic plane, 32
Optic sign, 49
Optic absorption
    — in plane polarised spectra, 64
    — spectra of aquamarine, 76, 77
    — spectra of blue sapphire, 76, 77
Optical absorption spectroscopy, 61
    — electronic transition, 61
    — instrument used, 61
    — rotational transition, 61
    — vibrational transition, 61
Optical depth to grain size ratios, 111
Orbital angular momentum, 181
Ordinary ray, 25
Orthopyroxene, absorption band centres, 112
Orthoscopic set up, 23
Oscillation photographs, 160, 162
Oscillation rotation camera, 162
Oxide minerals, reflection spectra, 113

Parafocussing arrangement, 140, 141
Parafocussing geometry, 140
Parallel slit, 141
Parting, 98
Pauli exclusion principle, 52
Petrofabric analysis, 48
Petrological mapping, 114
Phase difference, 35
Phosphorescence, 81
Photo conduction, 109
Photoconductivity, 81, 109
Photoelectronic emission, 109

Photoemission, 109
Piezoelectricity, 167
Pleochroism, 26, 27, 59, 102, 183
    — for mineral identification, 26
    — in biaxial minerals, 27
Point position, 175
Polarised spectra, 64
    — of olivines, 64, 66
    — of orthopyroxenes, 66, 68
Polaroids, 21
Polarizability of ions, 5
    — atomic, 5
    — electronic, 5
Polarisation colour, 26
Polarization of light, 19, 20, 21, 22
    — plane, 19
    — circular, 19
    — elliptical, 19
    — by absorption, 20, 22
    — uses, 21, 22
Polarizer, 20, 21
    — nicol, 20, 21
Polarizing microscope, 23, 26, 48
Polychromatic x-ray, 153
Polytype characterisation, 152
Powder diffraction data for minerals, 152
Powder diffraction file, 150, 151, 152
Powder diffractometry, 141
Powder photographs, 142
Precession camera, 166
Precession method, 160, 166
Primary X-ray, 126
Primitive cell, 172
Properties of light, (1)
Pyroelectricity, 167
Pyroxenes, 110
    — structural types, 111
    — iron-bearing octahedral sites, 111

Quantum number, 52
Quartz wedge, 29, 33, 34, 35

Racah parameter, 63
Receiving slit, 141
    — tube, 135
Reciprocal angle, 157
Reciprocal lattice, 157, 158
Reflectance, 85, 117
    — spectra, 117
    — oxide minerals, 113
Reflecting colour, 92
Reflection pleochroism, 93, 102
    — spectroscopy, 109
    — specular component, 109

Reflectivity, 98, 101, 102
— measurement, 98, 101
Refractive index, 4, 5, 8, 10, 14, 43
— determination, 110
Relief, 14, 94
— polishing, 94
Remote sensing mineralogy, 109
Retardation, 29
Rotation photographs, 159
Rotation-oscillation pattern, 166
Roundness, 49

Satellite reflection spectra, 109
Saturn rings, 110
Schubnikov space group, 167
Search manuals, 151, 152
— Hanawalt method, 151
Secondary X-ray, 126
Semiconductors, 77, 87
Sequential device, 116
Sign determination, 37, 39, 40, 48
— of uniaxial minerals, 37, 39
— of biaxial minerals, 40
Site occupancy derivation, 176
Site symmetry, 11
Skarn deposits, 110
Soil maturity, 110
Space group, 175
Spatial device, 116
Spectral dispersion, 101, 102
Spectral peaks, 118
Spectral photometer, double beam, 110
Spectral reflectance, 102
Spectral signatures of mineralogy, 109
Spectrometer
— Cary, 117
— IR, 115
Spectroscopy, 27
— absorption, 27
Spectroscopic states of ions, 181
Spin-orbit coupling, 62
Spin multiplicity, 62
Spin-spin interaction, 178
Spurious bands, IR, 115
Strip chart recorder, 142
Structure
— of calcite, 6
— of halite, 174
— of olivine, 65
— of orthopyroxene, 67
Structure factor, 133, 173, 174, 175
Structural defects, 72
— F-centres, 72

Surface factor, 171
Symmetry and parity coupled, 114
Symmetry determination, 167
Symmetry of continuum, 167
Symmetry of discontinuum, 168
Symmetry operation, 169

Target, 132, 133
— chromium, 132
— cobalt, 132
— copper, 132
— iron, 132
Transition dipole moment, 71
Tristimulus, 89
Transmission optics, 110
Transmission spectra, 117

Ungerade, 182
Uniaxial minerals, 28, 36
Universal stage, 45, 46, 47, 48
— Federov, 46
— Emmons, 46

Valence band, 79
— theory, 59
Variation method, 13
— single, 13
— double, 13
Vibration direction, 36
Vibrational absorption, 123

Water bands, 118, 122
Wave front, 25
Wave properties, 24
Wave surface, 25, 26
— negative crystal, 25
— oblate spheroid, 26
— positive crystal, 25
— prolate spheroid, 26
Weissenberg camera, 162
— method, 160, 167
— indexing, 162
— pattern, 166
— photographs, 162, 165

X-ray, 126, 130
— hard, 130
— soft, 130
— generation, 130
X-ray diffraction, 133, 134, 153, 190
— powder method, 134
— of cesium chloride, 175

X-ray diffractometer, 142
    — proportional counter, 142
    — scintillation counter, 142
X-ray diffractometer, 140
X-ray intensities, 171
X-ray fluorescence, 109, 129, 132
X-ray line intensities, 202
X-ray spectroscopy, 179

X-ray studies, 196

Zero layer line, 159
Zero level Weissenberg films, 164
Zone axis, 159
Zonal omission, 169
Zoning, 49